Network Industries and Social Welfare

Network Industries and Social Welfare

The Experiment that Reshuffled European Utilities

Massimo Florio

OXFORD
UNIVERSITY PRESS

OXFORD
UNIVERSITY PRESS

Great Clarendon Street, Oxford, OX2 6DP,
United Kingdom

Oxford University Press is a department of the University of Oxford.
It furthers the University's objective of excellence in research, scholarship,
and education by publishing worldwide. Oxford is a registered trade mark of
Oxford University Press in the UK and in certain other countries

First Edition published in 2013
Impression: 1

British Library Cataloguing in Publication Data
Data available

ISBN 978–0–19–967485–5

As printed and bound by
CPI Group (UK) Ltd, Croydon, CR0 4YY

Preface

Somewhere in Italy a housewife cooks pasta with natural gas imported from a distant place in Russia. The gas travels through a network of pipelines and pumping stations located in several countries. These infrastructures belong to different firms, supervised by different national regulators. Complex contracts and legal arrangements with Gazprom, the Russian monopolist, are required to eventually deliver the gas to the burner tip. The housewife does not know if her bill is based on long-term oil-indexed prices, or on supply and demand equilibria. Most users of electricity do not know who owns their supplier. They are unaware whether those Kilowatts needed for their computer or for their washing machine are offered by the same company that manages the power plants and supplies the household, or otherwise. Consumers have no notion of the existence of spot electricity markets and how they influence their bill. Telephones, broadband connections, and mobile phones allow us voice communication with somebody in a different continent, to read e-mails, and to access the news. However, users cannot say to what extent the bill they pay is the result of the functioning of competitive markets or of the decision, regarding interconnection charges, of a benevolent regulator. Policy-makers, instead, should be aware of all of these issues since they have played a crucial role in reshaping the functioning of network industries.

In Europe, a dramatic change of ownership, structure, regulation, and market opening of essential public services has taken place in less than 20 years. This is not the outcome of spontaneous market mechanisms, but is an entirely top-down policy experiment, mainly born in London, pursued in Brussels, and forced onto more or less reluctant players by laws, directives, regulations, administrative and judicial decisions.

The European reform paradigm revolves around three cornerstones: privatization, unbundling, and regulated liberalization of the network industries. These changes have been implemented to various degrees in the 27 Member States of the EU, and have been widely imitated elsewhere. It seems a good time to ask some plain questions: Are European citizens happier after the reforms than before? Do they pay less? Do they get a better quality of service?

This book tries to answer these questions. The origin of the book can be traced back to a short visit I paid several years ago to the Department of Applied Economics, Cambridge (UK), to present the preliminary results of my research on British privatizations (*The Great Divestiture*, MIT Press, 2004). A small, informal seminar was organized by Professor David Newbery. The discussion that followed was lively and interesting. A comment on that occasion was something that I have pondered over for a while. Somebody remarked that one of the difficulties in any research on privatization is often the lack of a counter-factual history. What would have happened in the UK under public ownership and continued vertical integration of the industries?

In social cost–benefit analysis of investment projects, building the 'business-as-usual' scenario is a crucial issue, and remains so in policy analysis too. There are good, albeit few, examples of this approach in the privatization literature. An alternative path is to turn to a cross-country comparative analysis. In this way, the observation of the consequences of adopting a policy by a country can be assessed against another country that did not implement them, or lagged behind adopting the same reforms. For example, the UK privatized its electricity industry in 1990, while until recently France remained under a vertically integrated public monopoly. After accounting for possible confounding factors, including different technologies and country specific features, can we say that the price performance of the British electricity industry has improved relative to that of France? From this perspective, variability across countries acts as a complement to variability across years (the 'before–after' approach), which per se is often limited by the lack of an appropriate, or evidence-based, counter-factual. Thereafter, I embarked on collaborative research, co-financed by the 6th Framework Programme of the EU. Within the UPP project 'Understanding Privatization Policy: Political Economy and Welfare Effects', the University of Milan team was responsible for studying the effects of policy reforms on consumers. The project needed a combination of cross-country and dynamic analysis of the effects of policy reforms on the prices of household goods in selected network industries of the EU-15 Member States, particularly in energy and telecommunications. This book presents our findings and my own reading of them.

The European Union, before its expansion in 2004 to new members (comprising mostly former transition economies), offers an ideal laboratory to study the impact of the reforms. First, the EU-15 countries are market economies with some institutional variations, but not as extreme as in other regions. The adoption of a more global perspective on reforms would be an interesting, but also a daunting, task. Comparable data across Europe, the Americas, or Asia and Africa are mostly unavailable. Institutions and economic conditions across countries are too different for a meaningful cross-country empirical analysis. Others, however, may try to extend our approach to different regions of the world, as far as data are available.

Second, the Member States of the EU must implement a well defined set of common regulations and directives on network industries; but the timing and framing of the implementation differs across countries and over time, and we can take advantage of this range of variability.

A policy trend that unfolds over the same time—albeit in different ways, in several countries, under a common paradigm, embodied in consistent legislation and similar, but not identical, institutions—is close to an ideal natural experiment, a unique chance for empirical testing of certain reform prescriptions. In doing so one should focus on broad welfare outcomes, rather than on the tiny details of regulation changes.

While my previous research on the UK focused on the welfare effects of privatization on different agents, the research presented in this book focuses specifically on consumer welfare. This means that I mostly disregard other possible important effects of the reforms on tax payers, workers, shareholders, suppliers, and business customers of the industries. This is an important limitation and perhaps the title of the book should refer to consumer welfare only, as suggested by one reviewer. I would argue, however, that a policy reform of network industries that is clearly beneficial (or otherwise) to consumers is unlikely to be too bad (or too good) for a society at large. In fact, the special feature of the network industries I consider is that they provide services of general economic interest, that is essential goods that are crucial to the well-being of virtually any household in a country. Any social cost–benefit test, in such pervasive sectors, should start by evaluating to what extent consumers benefit from the changes. Further research will be needed to establish the effects on other agents, but the 'players' we consider here are by far the most important ones to evaluate the social welfare impact of the reforms. In turn, the latter question can be assessed with a traditional analysis based on price data, and by a less conventional complementary analysis based on consumers' perceptions of fairness of prices and quality of services, through survey data. Both types of analyses have their own limitations, but the combination of objective and subjective evidence gives us a fair idea of what we need to know about the welfare effects of the reforms.

The UPP project team has included a number of colleagues in the Department of Economics, Management, and Quantitative Methods[1] of Università degli Studi di Milano and elsewhere. Carlo Fiorio, who has been involved in the project from the beginning and has closely collaborated with me in its coordination, Emanuele Bacchiocchi, Chiara Del Bo, Pier Alda Ferrari, Marco

[1] Formerly Department of Economics, Business and Statistics.

Gambaro, Giancarlo Manzi, and Silvia Salini. Other team members, who have often been post-docs or PhD students in Milan and Pavia include: Rinaldo Brau (now at University of Cagliari), Lidia Ceriani (now at Bocconi University), Simona Grassi (formerly at Boston University, now at University of Lausanne), Riccardo Puglisi (formerly at the Department of Politics, MIT and now at Pavia University), Raffaele Doronzo (now at the Bank of Italy). Paola Annoni (now at ISPRA), Stefania Lionetti (now at Università della Svizzera Italiana, Lugano), and Ambra Poggi (at the Collegio Carlo Alberto, Turin) have also collaborated. The overall UPP project was managed by the FEEM Fondazione ENI Enrico Mattei (Milan) and included other teams: CERGE-EI at Charles University (Prague), IFO (Munich), OECD (Paris), University of Amsterdam, University Pompeu Fabra (Barcelona), which focused on different topics of the research project.

I am grateful to Bernardo Bortolotti, Carlo Scarpa, and other colleagues of the UPP network for providing several comments and overall support to the Milan project team. Additional grants were obtained by the Italian Ministry of Education (PRIN 2002) and by the University of Milan (First 2007). Helpful insights were offered by a subsequent study on services of general interest for the European Parliament.[2]

Most chapters draw on papers authored by members of the Milan UPP team. In a sense, this is a collective book. The individual contribution is acknowledged in each chapter. As the final drafting of the book was made by me, I am solely responsible, however, for the overall synthesis, policy implications, and any remaining errors, regardless of the many helpful comments provided by the team members, or received over several years by participants in a number of seminars or conferences.[3] I am also very grateful to many colleagues who have joined the Milan European Economy Workshops and contributed to lively discussions.[4] Particularly helpful comments on some chapters have

[2] A joint research venture of DEAS, University of Milan, Ciriec International (Bruxelles), PPMI (Vilnius) and CSIL (Milan).

[3] Barcelona (UPP workshop at Pompeu Fabra, 2006), Bilbao (4th International Conference 'Development in economic theory and policy', 2007), Prague (UPP workshop, 2007), Milan (MEEW 2007 and 2011, DEAS Staff Seminars 2008, 2009, UPP–FEEM Closing Conference), Brussels (European Commission, DG Internal Market and ETSU Conference 2007), Cambridge, UK (Spring Research Seminar 2008), Istanbul (37th EARIE Conference, 2010), Lisbon (ETUC Conference, 2008), Munich (LMU Seminar Management and Microeconomics, 2008 and CESifo & OECD Conference on Regulation, 2010), Pavia (Siep 2008), Santander (Facultad de Ciencias Económicas y Empresariales, 2009, 2011), Rome (Cornell University in Rome, 2010), Boston (APPAM Conference, 2010), Dresden (TUD, 2012), Montevideo (Transnational Institute, 2012).

[4] In addition to other colleagues mentioned elsewhere in this preface, I would particularly mention: Barbara Antonioli (University of Lugano), Guglielmo Barone (Bank of Italy), Janice A. Beecher (Michigan State University), Germa Bel (University of Bacelona), Luc Bernier (Université du Quebec), Giuseppe Bognetti (DEMM), Martin Cave (Warwick Business School), Daniel Diaz Fuentes (University of Cantabria), Sean Dougherty (OECD), Xavier Fageda (University of Barcelona), Ben Fine (SOAS, University of London), Paolo Garella (DEMM), Jean Michel Glachant (University of Paris), David J. Hall (University of Greenwich), Claus Kastberg

been provided by Laura Ammannati (Università degli Studi di Milano), Sébastien Bumbolo (Adriatic LNG), Judith Clifton (University of Cantabria), David Evans (Oxford Brookes University), Mario Genco (CSIL, Centre for Industrial Studies), Tobias Kretschmer (Ludwig Maximilians University, Munich), Robert Millward (University of Manchester), Avner Offer (Oxford University), Steve Thomas (University of Greenwich), and Mildred Warner (Cornell University). I am particularly in debt to them.

I briefly discussed some of the policy implications of this research for Europe at a very unusual 'closed door' symposium at NYU (New York, 10–11 April 2010), organized by the late Tony Judt. Tony had recently published an extended version of his last lecture on the European welfare state in the form of a concise, vibrant book (*Ill Fares the Land*, Allen Lane, Penguin Books, 2010) where he also criticized privatization policies in Europe as I do here. Because of his already failing health condition, the small group of a dozen invited participants was aware that the symposium was probably the last opportunity to discuss with Tony our ideas[5] on the role of government as provider of services. He passed away a few months after the meeting. Tony Judt's masterpiece (*Postwar. A History of Europe since 1945*, Penguin Press, 2005) highlights the risks our societies incur when we downplay the unique potential role of good government and democracy in shaping collective plans and social cohesion and we become overconfident about market mechanisms. The bottom line of this book is a critique of the dominant policy paradigm. Services of general interest, as those supplied by network industries, such as energy, electronic communication, transport, and water are not usual business, where markets and 'light' regulation are socially efficient. These services have an important political dimension, and are often part and parcel of the inclusive economic fabric that shapes citizenship. I deeply regret the lost opportunity to further discuss these issues with Tony.

Nielsen (Copenhagen Economics), Andrea Goldstein (OECD), Nico Grove (LMU), Kyriacos Kyriacou (European Commission), Marc Laperrouza (Ecole Polytechnique de Lausanne), Jonathan Leape (London School of Economics), Eduardo Ley (World Bank), Emanuele Lobina (University of Greenwich), Andrea Mairate (European Commission), Marco Malgarini (ISAE), Antonio Massarutto (University of Udine), David McDonald (Queen's University), Raffaele Miniaci (University of Brescia), Franco Morganti (Network Economy), Dominik Möst (University of Dresden), Antonio Nicita (University of Siena), Gabriel Obermann (University of Wien), Donal Palcic (University of Limerick), Arnold Picot (LMU), Michael Pollitt (Cambridge University), Pippo Ranci (Catholic University of Milan), Eoin Reeves (University of Limerick), Barbara Sak (CIRIEC International), Malcolm Sawyer (University of Leeds), Anna Thomasson (Lund University), George Tridimas (University of Ulster), Jeremy Turk (British Telecom), Catherine Waddams Price (University of Norwich), Johan Willner (Åbo Akademi University), Andrea Zatti (University of Pavia). I apologize for not mentioning here several other participants to the Milan workshops or other conferences, and who also helpfully contributed to discussions related to this research.

[5] In the concluding chapter I mention some aspects of this inter-disciplinary discourse on service provision and social welfare.

I am very grateful to DEMM staff: particularly to Stefania Scuderi for patiently and competently helping me in editing a number of earlier versions of this book and managing the network of co-authors of different chapters. Michela Argenti assisted in UPP administration and support activities, including the organization of several Milan European Economy Workshops in the last years. Alessandra Cavada has patiently elaborated several versions of over 80 tables and figures.

I am also grateful to Adam Swallow and to the OUP staff for their excellent support, and to the anonymous reviewers for very useful comments and advice.

My students of the course on EU Industrial Policy in Milan have contributed to this book through many lively discussions and their own research, and I am grateful to the European Commission LLP Programme for having granted to me, in 2011, a Jean Monnet *ad personam* Chair for graduate teaching on this topic. In spite of my criticism of some aspects of EU policies on network services, I still think that Europe is an interesting laboratory of change, and this book is a tribute to the European idea.

MF
Milan, September 2012

Acknowledgements

The permission by authors and publishers to draw from the following UPP papers is acknowledged:

- Emanuele Bacchiocchi, Massimo Florio, and Marco Gambaro, 'Telecom reforms in the EU: prices and consumers' satisfaction', *Telecommunications Policy*, **35**, 2011, pp. 382–96 (Chapter 4).
- Lidia Ceriani and Massimo Florio, 'Consumer's surplus and the reform of network industries: a primer', *Journal of Economics,* **102**(3), 2011, pp. 111–22 (Chapter 3).
- Pier Alda Ferrari, Carlo V. Fiorio, Massimo Florio, and Silvia Salini, Consumers' Attitudes on Services of General Interest in the EU: Accessibility, Price and Quality 2000–2004', in Marrelli, M., Padovano, F. and Rizzo, I. (eds) (2007) *Servizi Pubblici: Nuove Tendenze nella Regolamentazione, nella Produzione e nel Finanziamento,* Franco Angeli (Chapter 7).
- Carlo V. Fiorio and Massimo Florio, 'Would you say that the price you pay for electricity is fair? Consumers' satisfaction and utility reforms in the EU 15', *Energy Economics,* **32**(2), 2011, pp. 178–87 with permission from Elsevier (Chapter 7).
- Carlo V. Fiorio, Massimo Florio, and Raffaele Doronzo, 'The electricity industry reform paradigm in the European Union: testing the impact on consumers', in P. Arestis and M. Sawyer (eds) (2008), *Critical Essays on the Privatisation Experience,* International Papers in Political Economy series, Palgrave Macmillan, pp. 121–59 reproduced with permission of Palgrave Macmillan (Chapters 2 and 5).
- Massimo Florio, Lidia Ceriani, and Raffaele Doronzo, 'Privatization, unbundling and liberalization of network industries: a discussion of the dominant policy paradigm in the EU', DEMM Working paper 2009–09[1] (Chapters 1 and 2).

[1] Available at http://ideas.repec.org/p/mil/wpdepa/2009-09.html [accessed January 2013].

- Simona Grassi and Riccardo Puglisi, 'Regulation and consumer satisfaction from public services: an individual fixed effects approach', DEAS Working paper 2008–21[2] (Chapter 8).
- Ambra Poggi and Massimo Florio, 'Energy deprivation dynamics and regulatory reforms in Europe: evidence from household panel data', *Energy Policy*, **38**(1), 2010, pp. 253–64 (Chapter 9).

[2] Available at: http://ideas.repec.org/p/mil/wpdepa/2008-21.html [accessed January 2013].

Table of Contents

List of Figures

List of Tables

List of Abbreviations

AEEG	Autorità per l'energia elettrica e il gas—Authority for Electricity and Gas
AC	Alternating Current
ACER	Agency for Cooperation of Energy Regulators
ADSL	Asymmetric Digital Subscriber Line
AT	Austria
AT&T	American Telephone and Telegraph Incorporated
ATM	Asynchronous Transfer Mode
ATRI	Aggregate Telecommunications Regulatory Indicator
BDT	Telecommunication Development Bureau
BE	Belgium
BEREC	Body of European Regulators for Electronic Communications
BG	British Gas
BT	British Telecom
CEGB	Central Electricity Generating Board
CEEP	The European Centre of Employers and Enterprises providing Public Services
CEER	The Council of European Energy Regulators
CNE	Comisión Nacional de Energía
COFOG	Classification of the Functions of Government
CPI	Consumer Price Index
CPS	Carrier Pre-Selection
CCS	Carbon Capture and Storage
CCGT	Combined Cycle Gas Turbine
CCT	Clean Coal Technology
DE	Germany
DECC	Department of Energy and Climate Change
DEFRA	Department for Environment, Food and Rural Affairs
DEMM	Dipartimento di Economia, Management e Metodi Quantitativi—Dept. of Economics, Management and Quantitative Methods
DK	Denmark

DMS	Demand Side Management
DSL	Digital Subscriber Line
DTE	Directie Toezicht Energie (The Netherlands Office of Energy Regulation)
EC	European Commission
ECTR	Indicators of Regulation in Energy, Transport and Communications
EDF	Électricité de France
EBN	Energie Beheer Nederland
EBRD	European Bank for Reconstruction and Development
ECHP	European Community Household Panel
EIA	Energy Information Administration
EIB	European Investment Bank
EMR	Electricity Market Reform
ENI	former Ente Nazionale Idrocarburi
ENEL	Ente Nazionale per l'Energia Elettrica
EPEE	European Partnership for Energy and the Environment
EPS	Electrical Power System
ERDF	European Regional Development Fund
ERGEG	European Regulators' Group for Electricity and Gas
ES	Spain
ESOP	Employee Stock Ownership Plan
ESPON	European Spatial Planning Observation Network
EU	European Union
EURI	European Union Regulatory Institutions Database
EU-SILC	European Union Statistics on Income and Living Conditions
FCC	Federal Communications Commission
FI	Finland
FiT CfD	Feed-in Tariffs with Contracts for Difference
FR	France
FTTH	Fibre to the Home
GDF	Gaz de France
GDF-RT	Gaz de France Reseau Transport
GDP	Gross Domestic Product
GJ	Gigajoule
GMM	Generalized Method of Moments
GR	Greece
GRTN	Gestore della Rete di Trasmissione Nazionale
GSM	Global System for Mobile Communication

GW	Gigawatt
GWh	Gigawatt hour
HH	Henry Hub
KPN	Koninklijke KPN N.V., Royal KPN NV. Dutch landline and mobile telecommunications company
KV	Kilovolt
KWh	Kilowatt hour
ICT	Information and Communication Technology
IE	Ireland
IEA	International Energy Agency
IP	Internet Protocol
ISDN	Integrated Services Digital Network
ISO	Independent System Operators
IT	Information Technology
IT	Italy (in figures)
ITU	International Telecommunications Union
LEC	Levelized Energy Cost
LED	Light Emitting Diode
LISREL	Linear Structural Models with Latent Variables
LLU	Local Loop Unbundling
LNG	Liquefied Natural Gas
LU	Luxembourg
MOM	Market Opening Milestones
Mtoe	Millions tonne of oil equivalent
MUSA	Multicriteria Satisfaction Analysis
MW	Megawatt
NAM	Nederlandse Maatschappij Aardolie
NGN	Next Generation Networks
NGT	National Grid Transco
NL	Netherlands
NMS	New Member States
NRA	National Regulation Authority
OECD	Organization for Economic Co-operation and Development
OLED	Organic Light-Emitting Diode
OLS	Ordinary Least Square
OPEC	Organization of Petroleum Exporting Countries
PCA	Principal Component Analysis

PPI	Private Participation in Infrastructure
PSTN	Public Switched Telephone Network
PT	Portugal
PTO	Public Telecommunication Operator
R&D	Research and Development
REC	Regional Electricity Company
REE	Red Electrica Española
REGREF	Indicators of Regulatory Reforms
SE	Sweden
SGEI	Services of General Economic Interest
SGI	Services of General Interest
TEU	Treaty of European Union
TFEU	Trattato sul Funzionamento dell'Unione Europea
TIGF	Total Infrastructures Gaz de France
TLC	Total Levelized Costs
TPA	Third Party Access
TSO	Transmission System Operator
TV	Television
UK	United Kingdom
UMTS	Universal Mobile Telecommunications System
UPP	Understanding Privatization Policy project
US	United States of America
USD	United States of America Dollar
USO	Telecommunications Universal Service Obligation
V	Volt
VAR	Vector Autoregression
VAT	Value Added Tax
VSDL	Very High Speed DSL
VoIP	Voice over IP
WHO	World Health Organization
Wi-Fi	Wireless Fidelity
WiMAX	Worldwide Interoperability for Microwave Access
WLAN	Wireless local area network
WLR	Wholesale Line Rental
WTA	Willingness to Accept
WTP	Willingness to Pay

Part I

Paradigms, Facts, Predictions

1

Introduction: The policy paradigm

1.1 Introduction[1]

This book takes stock of the widely held view that the privatization and liberalization of network services in the last 20 years have been clearly beneficial to European citizens. Network services such as electricity, natural gas, telecommunications, railways, and water are often crucial ingredients of economic well-being. They offer society the opportunity to coordinate, over time and space, large and complex flows of essential goods. To do so, management of complex systems, capital-intensive infrastructures, and long-term investment planning are needed. The telephone requires switch and voice synchronization between couples (or groups) of users. Postal services require certain delivery points and timetables. Airways work with slots at airports, specified corridors, and flight schedules. Gas pipelines need pressure balanced through pumping stations. The electricity supply requires instant control of voltage, current intensity, and other physical parameters of the flows of electrons over the wires. This ability to effectively manage service delivery is embodied in physical or intangible nets, mutually connecting millions of users who are scattered in different places. Investment decisions over these infrastructures and their critical nodes must consider long time horizons. Construction may take several years, hence the return on investment will be delayed far into the future. In the meantime, the world will change, and the risk profile of the investment will be influenced by considerable uncertainty.

Despite their widely different technological environments, pipelines of natural gas, optical fibre cables, high-voltage lines, aerials transmitting electromagnetic waves, air traffic control systems, and so on share (in variable proportions) the features of natural monopoly. These were identified more than 100 years ago by Farrer (1902): a combination of economies of scale, high

[1] Section 3 of this chapter draws, with many changes and integrations, from Ceriani, Doronzo, and Florio (2009).

capital-intensity, non-storability with fluctuating demand, location specificity and associated rents, and direct connection to users, who often consider the goods as necessities.

Economies of scale are associated with decreasing the average cost of provision, but networks also show increasing returns to scale with respect to social benefits (Shy, 2001). Thus, the scale factor has two dimensions, one on the supply side, the other on the demand side. The former is related to the fixed investment costs, the latter to the connecting capacity. The larger the number of connected users, the more helpful the service is to the users. The interconnection of several points over the space multiplies the opportunities of exchanges. For some services, the social benefit of extending the net increases more than proportionally with the number of points/users connected and this fact often implies a network externality.

Our societies are deeply influenced by this capacity of linking users through networks. Some of the most important innovations of the last few decades are supported by new types of connecting infrastructures, including communication satellites, GPS services, and shared computing facilities. History, however, offers early examples. Roads and water works offered possibly the most ancient demonstration of the social value of extended networks.[2] The political and economic cohesion of the Roman Empire was greatly enhanced by its admirable public aqueducts and long-distance roads. The remains of some of these infrastructures are still visible today. In Italy some water facilities and highways are direct successors of their Roman ancestors (an extreme case of 'legacy infrastructure'). The Internet is a more recent example of the social impact of the extended multi-focal capacity of connection. Its quasi-public goods nature was perhaps a surprise to many accustomed to thinking in terms of traditional bilateral markets, but the low marginal cost of access is typical of any network, as the burden of infrastructure investment cost and operation is shared by many, possibly several millions of users, and the benefits widely distributed.

On the supply side, the natural monopoly feature is at the core of network services, because extending the net leads to decreasing unit costs, at least until there is congestion. Sub-additive cost structures are the typical 'textbook' market failure, because the first comer is able to offer a price (or a tax-financed scheme) lower than any other entrant, given common information on technologies and equal managerial skills.

An industry, under a natural monopoly, seems unlikely to become competitive. This fact was well known to economists and policy-makers of past times, even if their opinions were divided, see for example, Pareto ([1892]2008) on the public and private railways of his times. The arrangement of network

[2] De Luca et al. (2012), discuss this historical evidence.

services for most of the twentieth century was either as a strictly regulated private monopoly, e.g. telephony in the USA and water in France, or as a public monopoly, such as railways in the UK and electricity in France. Why then, in the 1980s and 1990s, at more or less at the same time, did many governments around the world adopt reforms that reversed the previous consensus and tried to inject competition in these sectors?

Europe has been at the forefront of change. Elsewhere, in the USA, Latin America, Asia, and in formerly planned economies, there have been similar reforms, but perhaps nowhere have they been so consistently implemented as in the European Union. In the past two decades, first in the UK, then subsequently in all the other EU Member States, governments have increasingly moved away from the direct provision of public services, from ownership of utilities, and from franchised monopolies. Ministries and independent regulators have shown a greater reliance on market mechanisms, and now consider the network service providers as market players. Former public corporations, and new entrants, are now mostly profit-seekers. There still are many municipally-owned enterprises, public–private ventures, and some important remaining government-owned corporations, but they have been forced to compete with private newcomers and often regard themselves as not very different from their private competitors.

The former public service providers had broadly defined social objectives and missions. This has changed. The government objectives have not entirely disappeared but they have been weakened and have shifted to the market-players. National industrial strategies, consumer protection, environmental and energy concerns, and other polices have been delegated to private or quasi-private firms, through legislation, regulation, taxes, and subsidies.

This process of delegation of policy objectives has been particularly intense in some network industries, including those providing Services of General Economic Interest (SGEI). The latter is a new label, officially adopted by EU legislation, and intended to cancel the memory of what used to be a 'public service', as for the French tradition of administrative law, or the concept of 'public utility' in the Anglo-Saxon tradition. Article 106, second paragraph of the EU Treaty, states that:

> Undertakings entrusted with the operation of services of general economic interest or having the character of a revenue-producing monopoly shall be subject to the rules contained in the Treaties, in particular to the rules on competition, in so far as the application of such rules does not obstruct the performance, in law or in fact, of the particular tasks assigned to them. The development of trade must not be affected to such an extent as would be contrary to the interests of the Union.[3]

[3] TEU, part three, Title VII, Chapter 1, Section 1. Art. 106.

Despite the fact that the timing, speed, and depth of the reform process have varied across industries and countries, a dominant policy paradigm eventually emerged in the 1980s and the 1990s. Helm (2007, p. 9), after reminding us of the definition by Kuhn (1962) of a paradigm in science, observes:

> But paradigm shifts in policy, too: events can conspire to change the historical context to a sufficient degree to make it increasingly hard to reconcile the existing mindset of the policy-makers with the evidence, leading eventually to new objectives and new policy instruments.

He suggests that in future we will see a new landscape.

A paradigm is never a technical issue. It embodies a mixture of observations, reasoning, and beliefs, which need to reach a critical mass in order to defeat a former set of policies and convictions. While the ingredients of the reforms in Europe are observable in different proportions, three core dimensions of the still dominant paradigm can be identified: (i) privatization; (ii) partial unbundling or vertical separation of the net from upstream and downstream activities; (iii) liberalization of entry and regulation intended to encourage (or even force) competition. These components can be combined in different ways in actual policy-making and regulatory arrangements. For example in Europe there has been less unbundling in telecommunications than in electricity, more limited privatization of water services than of airways, and modest liberalization of railways as compared to postal services. Despite this variability across sectors (and countries, and times) of the reforms, their analysis can be conveniently organized along the aforementioned three dimensions, which are regularly monitored by the European Commission, the OECD, and other international institutions.

Newbery (2000), in an influential book that we shall frequently cite, offers a thorough discussion of the reform approach to network utilities. The discussion is supported by several examples, mainly based on the experience of the UK at the end of the 1980s and in the subsequent decade. Newbery states the logical relationships between privatization, unbundling, and competition as ingredients of a new regulatory compact:

a) 'The wave of deregulation that started in 1978 in the US showed that markets were better than regulators at reducing prices and increasing efficiency, and this cast doubts on the social value of regulation'[4] and 'Competition is more effective than regulation at cutting costs to improve productive efficiency, and aligning prices with costs to improve allocative efficiency.'[5]

[4] Newbery, 2000, p. 2. [5] Ibidem, p. 171.

b) 'Restructuring publicly owned network utilities to introduce competition, inevitably raises the question of privatization, and the evidence suggests that while there are good reasons for delaying privatization, it is hard to imagine that effective and efficient competition is sustainable in the public sector.'[6]

c) 'The more drastic reform of structural separation or unbundling separates ownership of the network from that of the services offered over it. This should ensure that all service providers have equal access to the network, which have no reason to favour one over another.'[7] 'Privatization combined with restructuring, preferably involving vertical separation, of public utilities, and liberalizing access to private utilities... offer the twin attraction of enabling competition... and facilitating... price cap regulation for the core network natural monopoly.'[8]

Qualifications and variants of the analysis are then offered. Price-cap regulation is seen as more efficient than the US-style rate-of-return mechanisms to create incentives, but gradually competition should limit the scope of price regulation, except in some bottleneck facilities. Municipal ownership is deemed not necessarily less efficient than private ownership in some services, but its role is of marginal overall significance in the proposed reform strategy. The merits of separation are compared with those of regulated third-party access to the network, and usually the former is preferred to the latter. Solutions tailored for one industry are said to be not always transferable to other industries, for example water and railways, but the overall message of Newbery's book is that competition is desirable and feasible in (most) network utilities.

Later in this chapter, some arguments and contributions supporting the paradigm are discussed, together with some possible objections. This book, however, is not about arguing for or against the principles of what is now a largely shared consensus.[9] It is about the empirical testing of one specific aspect of the reform outcome: benefits for households. In fact, after around 20 years of reforms, and, according to new EU legislation, with more to come in the future, it seems a good time to ask the question: How far are these policy reforms keeping their promise of delivering greater welfare to citizens?

[6] Ibidem, p. 186. [7] Ibidem, p. 188. [8] Ibidem, p. 421.

[9] The rest of the chapter is not going to review the extensive and booming literature on network economics. Some reference texts, specialized journals, and selected papers are cited in the Further Reading Section, at the end of the chapter. See also the Glossary.

The main purpose of this introduction is to briefly remind us of some of the rationales for the paradigm, its assumptions, prescriptions, predictions, and open issues.

1.2 The EU perspective

The perspective of our discussion is a European one, even if some of the issues, and some of the earlier empirical literature that we shall cite, have a more general scope. In Europe, the critical mass for the policy shift was achieved in the 1980s, after the social and political upheavals of the 1970s and the severe oil shocks that destabilized public finances. However there were also other factors, which are discussed below.

Historically, in the EU, the reform of network services has been promoted by different parallel policy trends. Four drivers can be identified. One origin of the story was the overarching goal by the European Commission in the 1980s to create a unique *internal* market for those services that were previously excluded by the Treaty of Rome (1957). The exclusion clause was subscribed to by the six founders of the European Community: Belgium, France, Germany, Italy, Luxembourg, and the Netherlands. The Treaty excluded from some common market rules services such as telecommunications, transport, energy, and water. These services were initially exempt because they were considered of limited interest for trans-boundary competition, and because they were highly sensitive to the political priorities of each Member State. The exemption was abandoned in the new EU Lisbon Treaty (2009), and SGEI are now treated as potentially competitive industries. State aid is forbidden, except under special circumstances, which need to be notified to and approved by the EC. More recently, the EC has also advocated a widespread liberalization of many other Services of General Interest (SGIs) although this proposal has met with considerable resistance in many Member States.

In a sense, these liberalization initiatives are logical developments for the EC, as it tries to extend to all services the successful internal market developments relating to manufactured goods and some commodities that have been at the core of its mission since its inception with the Treaty of Rome. Thus, the reform paradigm has been strongly supported by the common market dimension and the Europeanization of national policies.[10]

[10] The local dimensions of SGIs (i.e. market and non-market activities) and SGEIs are complex. For a synopsis of the discussion see DEAS et al. (2010); CEEP (2010), Obermann et al. (2005), and Bognetti and Obermann (2008).

The second intellectual and political force driving reform has been the UK example under the Thatcher and Major governments (1979–97), and deregulation in the USA under the Reagan administration (1981–89). The UK example was a policy driver that was largely independent from the EU side of the story, as the dialogue between London and Brussels was particularly difficult during the Thatcher years. The British approach to network services in the mid-1990s was perhaps more closely related to policy trends on the other side of the Atlantic, rather than on the other side of the Channel. During the same years there was the divestiture of ownership of British Telecom by the UK government, and in the USA AT&T was forced to divest itself of Baby Bells, its regional telecom operations. Access to natural gas pipelines was liberalized in the USA over the same years as British Gas was privatized. There were however differences between the US and the UK approaches, particularly concerning the role of market regulation. While some of the UK proponents of the reform (e.g. Littlechild, 2000) had a broad vision of dismantling the nationalized industries by large scale divestitures, breaking up monopolistic incumbents, introducing incentive regulation, and opening the markets to competition of new entrants, the actual implementation in the UK was pragmatic and varied widely across sectors and circumstances.[11] And the EU legislation on network industries is perhaps now more comprehensive, intrusive, and compelling than current federal regulations in the USA.

The other two important factors shaping the political climate in Europe were the collapse of the Soviet Union and of other planned economies, and the advent of global financial markets. Both factors helped to convince European policy-makers that state ownership and industry planning of some sectors had to be abandoned in favour of market mechanisms. Policy-makers also realized that there was some political profit to be extracted by divestitures of public corporations and by the new dialogue, if not alliance, with the international financial investors. We shall come back (in Chapter 10) to the relative weights of the aforementioned four drivers in reshaping the network services in Europe. We turn now to sketching the paradigm and its economic rationales, without further discussion of the possibly important political dimensions of the story.[12]

[11] In the USA the States adopted very different reform approaches. For example, until now, several States had not fully liberalized their electricity markets. Network ownership unbundling is in general not mandatory in the USA, as it now tends to be in the EU. Rate-of-return regulation is more pervasive in the USA than in the EU.

[12] Avner Offer (in private correspondence) suggests that a political economy perspective would be appropriate to understand the historical rationales of privatization. This issue is briefly discussed in Chapter 10.

1.3 The paradigm's three pillars

1.3.1 *The first pillar: Privatization*

As mentioned, governments in the last century dealt with market failures in network industries either through public ownership or through heavy regulation of privately owned monopolies. In fact, public ownership can be seen as a regulatory mechanism in itself, when it is difficult to specify a concession contract (or a similar arrangement) with a private entity. Over the past two decades, policy-makers became increasingly convinced of the prevalence of government or regulatory failures, and tended to see market failures as relatively less costly. A steady flow of empirical and theoretical literature aimed at showing that public ownership of vertically integrated firms had not succeeded in decreasing prices in those industries considered to be natural monopolies, and had sometimes increased prices in potentially competitive sectors (Peltzman et al., 1989). As a consequence, the idea that privatization of public monopolies is a necessary condition for the liberalization of markets gained more support.

The theoretical arguments as to why private ownership and market-oriented reforms might lead to greater efficiency are discussed in several strands of research. Pollitt (2002) classifies the main relevant theories as: (i) property rights, (ii) bureaucracy, (iii) influence, (iv) economic regulation, and (v) commitment. On balance, these theories predict that reform will lead to improved economic efficiency. Florio (2004) critically reviews some of the literature.

There are indeed examples of vertically-integrated public enterprises (sometimes sheltered from competition law provisions) which were found to abuse their dominant position in an upstream market to foreclose entry of new competitors in liberalized downstream markets. Among many papers in this vein, Lott (1990) and Sappington and Sidak (2003) suggest that, under a variety of settings, a public enterprise will have even stronger incentives than its private homologue to behave anti-competitively by raising rivals' costs and deterring entry.

This literature has probably had an influence on the intellectual climate in some international organizations, for example see Williamson (1989, 2000) on privatization and liberalization as ingredients of the 'Washington Consensus'. Economists at the OECD suggested that for most network industries, like telecommunications and energy, privatization is a necessary (but not sufficient) condition to obtain a competitive market (Hoj et al., 1995; Gonenc et al., 2001).

The EU has never officially endorsed the mandatory privatization of utilities. The Treaty of Rome required neutrality of the European legislation on issues of ownership of the firms.[13] However, the EC wanted to promote competition, and often the opinion of EU policy advisors was that government-ownership was the enemy of competition. The EC has gradually become suspicious of government control of the incumbent providers of network services, and of any SGEI. One way to weaken the position of government-owned firms was to invoke the fact that the EU Treaty pronounces a general prohibition of state aid to firms.[14] The objective of state aid control by the EC and the European Court of Justice is to ensure that government interventions do not distort competition and intra-community trade. The state aid discipline is in principle compatible with public ownership, when equity finance is provided under certain conditions. The state aid prohibition, however, has often significantly reduced the possibility of access to government equity funds for some public firms, particularly in the transport sector.

This line of attack was probably not very effective, however, in limiting the role of governments in the ownership of energy and telecom providers, because these were usually profit-makers, if not profit-maximizers, under public ownership. In fact, the EU activism in passing legislation intended to constrain the market power of the incumbents was de facto favouring their privatization, which was in any case under the responsibility of the Member States. If the value of the existing assets and future investment of the incumbent is exposed to the challenge of competition, governments would be inclined to shift the risk to private finance. The change in the investment risk profile of network utilities under market opening and independent regulation is hence seen as a rationale for their privatization.

EU directives on transparency may have also played a role. Some principles were already established 30 years ago, through the Commission Directive 1980/723 and were reinforced by Directive 2006/111 on the financial relations between Member States and public undertakings.[15]

More decisively, some of the EU fiscal rules implied budgetary restrictions (more stringent than on the US federal budget) and suggested that the more indebted countries, such as Italy, should use the privatization proceeds from some of the public corporations in the network services to help finance public

[13] Part Six, art. 295 (ex art. 222): 'This Treaty shall in no way prejudice the rules in Member States governing the system of property ownership.'

[14] Part Three, Title V, Chapter 1, Section 3, Art. 92, 'Any aid granted by a Member State or through State resources in any form whatsoever which distorts or threatens to distort competition by favoring certain undertakings or the production of certain goods shall, in so far as it affects trade between Member States, be incompatible with the common market.'

[15] Commission Directive 2006/111/EC, article 3: 'The transparency referred to in article (1) shall apply in particular to the following aspects of financial relations between public authorities and public undertakings: (a) the setting-off of operating losses; (b) the provision of capital; (c) non-refundable grants, or loans on privileged terms; (d) the granting of financial advantages by forgoing profits or the recovery of sums due; (e) the forgoing of a normal return on public fund used, (f) compensation for financial burdens imposed by the public authorities.'

investment, see for example the case of Telecom Italia (Florio, 2006). The recent rescue plans by the EC, IMF, and European Central Bank for Greece and Portugal also involve extensive privatizations aiming at reducing public debt.

1.3.2 *The second pillar: Unbundling*

A second cornerstone of the market opening consensus as advocated, with different tones and meanings, by the EC, by the OECD, and often by the World Bank (see e.g. Yi-chong, 2005, or Estache et al., 2006) is unbundling. This is the separation of non-competitive segments (typically the network core facilities), from those that are potentially competitive (such as production, supply, and maintenance). The alternative to ownership unbundling is regulated liberalization of access, which requires accounting, functional, or legal separation in some form, for example creating a division of the incumbent, with different management and governance mechanisms, see Vickers (1995).

The contractual autonomy of the network owner was then severely constrained by the reforms. In practice the owners of bottleneck infrastructure are exempted from an obligation to contract only if they can show that they have no carrying capacity, or have some other special circumstance. The main reason for unbundling is often presented as a means to prevent network operators from granting privileged access conditions to their own production operations, at the expense of competitors. The stronger the separation (e.g. ownership separation rather than mere accounting separation), the lower the risk of anti-competitive conduct. Gonenc et al. (2001) summarize in four elements the key factors influencing the choice between vertical separation and integration: (i) the presence of economies of scope between the competitive and non-competitive activities; (ii) the availability of information regarding costs both in the non-competitive and in the competitive activity by the regulated firm; (iii) the availability of information regarding demand elasticities, capacity, or quality; (iv) the institutional and legal context.

According to this view, vertical separation should be supported by governments if there are weak or no economies of scope, the incumbent can conceal information, institutions are ineffective at controlling the incentives for the incumbent to delay or obfuscate compliance to regulation, and if legislation is unable to prevent long-term contracts. All or most of these features, as we shall see, are often assumed to be quite common in practice in many countries and industries.

The position of the European Commission in this regard has been clearly in favour of vertical separation between the competitive and natural monopoly segments of network industries.[16] For example in the energy sector, the aim of the first electricity and gas directives was to unbundle these industries to avoid discrimination, cross-subsidization, and distortion of competition. The electricity and gas companies were therefore obliged to keep separate accounts for their generation/production, transmission, and distribution activities. The Member States were initially free to choose between different alternatives: from the most radical that prescribed ownership separation of the monopoly activities from the competitive ones, to a milder legal separation, reached through the creation of different companies under a common holding company, to the weakest version of accounting separation. The aim of separating monopoly and competitive segments of the energy industry was reaffirmed in the second electricity and gas directives, whose provisions were considerably strengthened in comparison to the previous ones. The European Council has proposed ownership unbundling in electricity as the only option for the future (EU, 2007), and, starting from 3 March 2012, Member States must unbundle transmission systems and transmission system operators (EU, 2009a, 2009b). More recently, there has also been European or national legislation pushing for unbundling in the telecom sector, in railways, in the water industry, and in other local public services, again with wide variations around the common general orientation and with regulatory solutions that may differ across industries and circumstances.

1.3.3 *The third pillar: Market opening*

Before the 1980s the network industries were almost everywhere considered natural monopolies because of their large fixed costs. Competition was often considered wasteful. Economic historians (Millward, 2011) have shown, in fact, that private ownership and forms of oligopolistic competition in some network services in Europe, before post-war nationalizations, were often inefficient. Markets literally failed in some industries. The oscillation between high rents under monopolistic or oligopolistic arrangements, and providers' bankruptcy when there were disruptive predatory pricing wars or financial troubles, were politically unbearable. Some of the nationalizations were emergency measures after the collapse of the private service providers in the 1930s, and many earlier municipalizations of local services were motivated by the unreliability and high costs of concessions. In other cases, social cohesion, national security, and defence issues played an important role in shaping the views of policy-makers.

[16] This approach has been repeatedly stated by different policy documents (EU, 1996b, EU, 1998, EU, 2003a, EU, 2003b).

In the last 20 years, however, the perception that technological change could weaken the traditional problems of natural monopoly has often been cited as supporting evidence in favour of market opening. While the monopolistic segments always require adequate sector-specific regulation, in other segments such regulation was supposed to become unnecessary once workable competition was in place. The resulting higher competitive pressure should generate higher productivity leading to lower prices, and ultimately higher economic growth. For example, the European Commission, in a report evaluating the performance of network industries, identifies the main channels through which market opening could lead to greater efficiency:

> Increased allocative efficiency should occur as prices and marginal costs move towards each other, thereby correcting distortions in the structure of production and adjusting the level of output to a social optimum. Increased productive efficiency would arise as firms move closer to their production possibility frontier by using their inputs more efficiently (i.e. producing more with the same inputs, or the same with fewer inputs). Reduction of x-inefficiency would raise the productive performance by reducing managerial slack and improving the organization of work. Finally dynamic efficiency would be enhanced as competition from existing and prospective competitors creates incentives for investment in product and process innovation. (EC, 2007, p. 11)

All these expected efficiency gains should generate manifold structural effects, see for example the position of OECD economists (Gonenc et al., 2001, pp. 23–26 and Nicoletti, 2001, p. 11) and the European Commission (EC, 2007, pp. 12–26) who suggest that the competitive pressure should exert downward pressures on prices and enhance productivity, particularly if cross-border trade is promoted.

Based on the above reasoning, the purpose of liberalizing the network industries in the EU has been pursued over the last 20 years via a steady flow of directives that will be reviewed in the next chapter. The process is still ongoing, as we shall see later on in this book (Part Two).

1.4 Asking the questions

Testing the paradigm on all the above mentioned dimensions, effects, and political underpinnings is largely beyond the scope of this book. Rather, the focus will be on what can be predicted regarding the effect of the reforms on household bills. Prices are the crucial marginal welfare signal to users in developed economies, where quantity rationing is limited, and quality usually (but not always) exceeds a certain standard (this may be different in less developed economies).[17] We structure our discussion below around three questions, one for each of the paradigm's pillars.

1.4.1 *Does ownership matter for prices and social welfare?*

One can argue that privatization is socially desirable per se, because of a political philosophical argument. Private ownership could be seen as a way to enhance individual freedom, which in turn can be regarded as an aspect of social welfare.[18] The core of the discussion here is the constitutional status of property rights, that is their role in the 'social contract'. Ownership is essentially a legal concept: it has emerged over several centuries, and has been defined in many ways according to the overall system of rights and their enforcement. Mainstream economics in the neo-classical tradition have often been less than at ease when dealing with property rights, either of the private firm or of the public providers. See Milgrom and Roberts (1992, chapter 9), for an introductory discussion, and Barzel (1989, 2002) from the perspective of law and economics.

Private property rights, as arguments of individual freedoms and social welfare, are in fact intrinsically subject to the norms, limitations, and enforcement mechanisms that pertain to the sphere of political philosophy. These broad issues are outside the narrow definition of economic welfare as adopted in this book.

We shall then consistently use simple classifications of public versus private ownership, mainly based on OECD regulatory indicators, which will be introduced in the next chapter. This simplification is needed to turn the essentially qualitative information on ownership into a testable variable. The interpretation should, however, be treated carefully.

A different argument for the socially beneficial welfare impact of privatization rests on the view that government ownership distorts agents' incentives to achieve efficiency. Most of the empirical privatization literature tests changes in performance variables after, or because of, privatization, see for example Megginson and Netter (2001) and Megginson (2005).

As mentioned in Section 1.2.1, the core of the incentive argument for the divestiture of public ownership is often presented as follows. Managers of government-owned enterprises are not incentivized to minimize costs, because they do not respond to profit-maximizing shareholders. A different argument is that prices of government owned firms are not cost reflective,

[17] This statement, however, does not imply that the observed prices of the services actually coincide with their marginal *social* values, that is with their shadow prices, in the meaning of applied welfare economics (Drèze and Stern, 1990; Del Bo and Florio, 2012).

[18] This argument can be found, for example, in the writings of members of the IEA in Britain or of the Mont Pelerin Society.

because they are manipulated for political reasons, for example through cross-subsidies in favour of certain social groups. Public providers would then create both productive and allocative inefficiency. Thus, *ceteris paribus*, turning a public entity into private ownership will eliminate that intrinsic source of inefficiency, hence it is desirable on pure economic grounds.

It seems helpful to consider these two different incentive issues separately. Productive efficiency does not exclude price distortions per se.

That private firms use cost-reflective prices (product-by-product or customer-by-customer) is perhaps a myth. Any firm facing a downward sloping demand curve rationally offers discounts to large buyers, or to customers who have certain preferences and information. Price discrimination of this type implies that the mark-up across customer types may vary widely, according to the shape of the cost and demand functions. A multi-product firm, in a dynamic context, can rationally decide to aggressively cross-subsidize some products in its portfolio for strategic reasons, such as to avoid entry of potential competitors, and so on.

The public firm constrained to a balanced budget (the Boiteux type), would be rational in using socially weighted Ramsey pricing, that is a price schedule that is parametric in compensated demand elasticities and in the welfare weights of the users of the service (Bös, 1994). Thus, both non-competitive public and private firms may *efficiently* use non-cost-reflective pricing strategies, to a certain extent, with different implications for social welfare. While selling a product below or above its marginal cost is 'locally socially inefficient', in a second-best dynamic economy under general equilibrium, when prices are not parametric to the firm, the extent, origins, and social implications of any price distortions in one market are complex. Hence, it seems difficult to argue that privatization per se must increase allocative efficiency in the network industries. This motivated (price and non-price) regulation of privatized firms in the first place.

Hence, the efficiency argument for privatization probably rests or falls on incentives for productive efficiency and new technology adoption, more than for allocative efficiency. The key assumption is that, under private ownership, managers respond to profit-seeking shareholders, while they respond to policy-makers under public ownership. Policy-makers may be benevolent or not. If they are benevolent, and well informed, that is they are welfare maximizing principals under symmetric and complete information, there is no need to assume that they must induce the management to avoid cost-minimization. This was the message of Sappington and Stiglitz's (1987) 'Fundamental Privatization Theorem'. If the government is not benevolent, and information is not symmetric and complete, there are two alternatives. Under complete information, if the public firm generates rents, and the policy-maker wants to appropriate those rents, he will still require optimal effort by the management. If the public firms generate losses, but these losses are costly to the policy-maker, for example because he has to raise taxes to cover them, and this harms his popularity in the constituency, again, policy-makers will ask managers to be efficient. If information is asymmetric, it will be up to the manager to appropriate some rents, but this will happen under regulated private ownership as well (e.g. see Laffont and Tirole, 2000).

To predict an *intrinsic* productive inefficiency of a public firm, other assumptions are needed than the alleged greed of the policy-makers. One hypothesis is that there is not just a government with a private agenda, but also a specific interest by the government to extract rents from the firm through some cost-increasing mechanisms, as in Laffont (2005). These mechanisms may include, for example, a coalition with workers' unions. This coalition increases political rent to the politicians, who will be supported by workers at election. The latter would get their rent in the form of wages above the value of the marginal product, or other more informal arrangements. Other examples of cost-increasing coalitions are those between politicians and suppliers of government-owned firms. The suppliers may bribe the politicians. Politicians can also appoint loyal but bad managers in exchange for a generic control on the firm. Asymmetric information and contract incompleteness can further complicate the picture.

Thus, according to these views, both costs and prices can be distorted by 'public ownership'. The merit of the theoretical case for privatization on these grounds needs a benchmark, and this cannot be 'private ownership' in abstract terms. Large organizations are always a complex nexus of contracts, interests, and behaviours. Coalitions between stake-holders are a core governance mechanism in any organization, private or public. Particularly in large, regulated firms, enjoying some market power, these coalitions can share rents and induce inefficiency. For example, executive directors can collude with some shareholders and distribute, officially or unofficially, large and unjustified compensations, which will be treated as costs, hence reducing dividends to other shareholders. They can even be the hidden shareholders of a supplier of the firm they manage. There is certainly no lack of examples of abuse and corruption in private firms, and of concerns for the social and economic impact of bad corporate governance.

A further economic argument for privatization has to do with credibility of commitments and time consistency in a regulated environment. A government that owns a provider of services is expected not to be credible when it fixes the rules of the game and is a player at the same time. Thus a government that opens a market to competition and allows entry of competitors, but still owns the incumbent, is unable to commit to squeezing the profits of the entrants, because it has a vested interest in so doing, if not today then perhaps tomorrow. Thus privatization here is instrumental in competition-enhancing reforms. The argument is in fact more related to regulation than to ownership per se, because it rests on the intuition that there is a fundamental asymmetry in a market where the government is also a provider of the service and appoints the regulator.

Markets with a mixed provision of services are widespread, from education to health, from insurance to transport (De Fraja and Delbono,1989; De Fraja, 2009). It is not self-evident why independent regulators should more likely be captured by public than by private incumbents. The concern arises if the regulator is not benevolent. But why should he prefer to be captured by the government-owned firm or by policy-makers rather than by the privatized incumbent or by the private entrants in a mixed oligopoly? Are the former ready to offer greater side-payments than the latter? Again, this is an issue that is difficult to assess in abstract terms, and any general prediction seems to rest on shaky foundations.

Thus, the impact of inefficiencies on the costs and prices of public or private ownership is not clear cut, and largely depends upon the quality of governance mechanisms (Holmberg et al., 2009) in both environments, and the actual role of competition.[19]

Given the lack of strong priors, the impact of public/private ownership on the prices of network services and consumer welfare should be tested empirically, after carefully controlling for other factors.

1.4.2 *Is vertical disintegration necessary to achieve competitive prices?*

As mentioned in Section 1.2.2, the main argument for vertical disintegration under the new paradigm for utility reform is a pro-competitive one. To establish a fair game, every entrant should have access to the network or to bottleneck infrastructures. This idea is intuitively appealing. But is ownership unbundling a necessary or a sufficient condition for competition? The question does not seem to have a simple or unique answer in the regulatory economics literature, see Baldwin et al. (2012).

If the incumbent's operations show substantial economies of scale and scope, separation of the network may not be *sufficient* to allow socially beneficial entry, see for example Nemoto and Goto (2004) on electricity in Japan or Fraquelli et al. (2004) on energy and water supply in Italy.[20]

One assumption of the paradigm, as mentioned above, is that the natural monopoly features of the industry can be insulated from potentially

[19] For a recent example, see the discussion on privatization of solid waste management and water provision by Bel, Fageda, and Warner (2010), who find that the policy environment and service characteristics explain costs, while private ownership plays no role.

[20] http://ec.europa.eu/internal_market/economic-reports/index_en.htm [accessed 5 January 2013].

competitive segments. Hence, the latter should show constant or decreasing returns to scale. The standard example is power generation in the electricity industry, where small gas-fuelled plants can be as efficient as larger generators. In TV broadcasting, small firms can compete with the major players. There may be counter-examples, however. Nuclear power generation and offshore wind farms seem to be under increasing returns to scale, as the variable costs are much smaller than fixed investment costs. Gas exploration and extraction are best managed by large investors, in order to spread the risks over a number of fields. A large gas field in itself is obviously a natural monopoly, because the extraction and purification variable costs are limited as compared to the fixed costs of the pumping and processing plants. Some hardware and software investments, which are needed to manage mobile telephone traffic, are mostly fixed costs, even if the telephone main network is unbundled. The maintenance costs of rolling stock in the railway service are largely fixed, that is they do not depend upon the occupancy rate of the wagons. Thus there are examples and counter-examples of returns-to-scale of activities upstream and downstream of the main networks. Network externalities of the demand-side further complicate the problem. And so do economies of integration and systemic costs.

Regulators, even after unbundling, have had to occasionally force market entry by asymmetric regulation in order to promote competition in some industries. This is a sort of analogue of protectionism in international trade. Asymmetric regulation imposes higher costs on the incumbent than on entrants. The former has to sacrifice part of his production capacity to the latter, must offer universal services in certain areas or to certain social groups not profitable for the entrants, or is not allowed to offer 'too low' prices on some services. The artificial manipulation of the market organization by the regulator is often needed because the divestiture of the network per se, or regulated third-party-access under functional separation, is far from being a sufficient condition to having entrants ready to compete with the incumbent at lower prices. This is often due to persisting returns-to-scale under certain technologies, as we shall discuss later on in Chapters 4, 5, and 6. As in international trade, ignoring comparative advantages may be costly, and protecting inefficient entrants for too long is a risky policy, as it creates vested interests to defend regulatory rents.

Is ownership unbundling a *necessary* condition for competition? In principle either under public or private ownership of the vertically integrated incumbent the access of entrants can be regulated (as actually happens now in many countries). This Third Party Access (TPA) regulation can be costly or inefficient, or otherwise. On empirical grounds, you can study whether, in a given country and for a given industry, the incumbent uses the ownership of the network to prevent socially beneficial competition. Often (but not always) a private unbundled monopolist will have no interest in opening the network to its competitors. This may happen, and actually happens, with public providers as well, but it is less clear why their management is not instructed by the government in the first place to open the network to third parties. If public corporation incumbents are instructed by governments to behave as private competitors, they will protect their rents, and this will require that other public officials have to fight over their behaviour. Governments actually appoint both regulators and managers of the public providers, so why do they seem to give them different remits?

A distinct, but important issue in the unbundling ingredient of the paradigm is the coordination–transaction costs of a vertically disintegrated network industry. Unbundling involves contracting between separate parties and this activity is not a 'free lunch'. Contracting costs may be large, or small, depending upon sectors and countries. In some industries, for example in gas, the short-run costs of unbundling may be small, but in the long-run they may be substantial if investment in high-pressure long-distance pipelines falls, because of the uncertainty of the demand. Thus one has to balance security of supply concerns against the benefits of removing a bottleneck to essential facilities for competition.

According to Newbery (2000, p. 188), services offered to the network are under a monopsony, as there is just one buyer for such services. Hence, competition in the service provided to final consumers does not generate market efficiency. The incumbent may adopt inefficient in-house production and earn a profit in the regulated business. These situations need further inquiry. If there are potential cost savings in the procurement of services to the network owner, why doesn't the still vertically integrated and privatized incumbent buy the service from third parties, wherever it costs the least? In other words, the question is: why should the incumbents keep in their area of control some upstream and downstream activities? While lock-in situations, perhaps related to path dependency, may be observed in the real world, a systematic protection of inefficient in-house operations would contradict long-term profit maximization of the vertically integrated firm. Under private ownership, a profit maximizing management would buy from external suppliers the services available at a lower cost, divest inefficient operations, and earn a wider margin when reselling the service downstream.

If in-house service production is efficient, because of economies of coordination and persistent economies of scale of the operations, then asymmetric regulation would offer the entrants a form of regulatory rent. Thus, specific circumstances would tip the balance of the relative social benefits of ownership unbundling versus network liberalization and TPA, with and without asymmetric regulation.

Under public ownership of the incumbent or optimal regulation, with a management that has a clear mandate to allow the entry of competitive supply through the network, unbundling per se is costly. If regulation is suboptimal or there is corruption, nothing can be predicted in general about the outcomes of the different restructuring and regulatory options.

Thus, it seems that a number of hypotheses are needed to arrive at the prediction that unbundling will help, indirectly, to deliver lower prices to the consumer. In fact this prediction rests crucially on the third, and most important, ingredient in the recipe: market opening to competition.

1.4.3 *Does competition deliver lower prices to consumers?*

In the context of network industries providing SGEI, there are at least two possible different meanings of market opening. Under *soft liberalization*, governments allow some entry, under a regime of concessions or licences, and in a permanently regulated environment. The extent of competition remains under strong control by the government or the regulator. The former promotes binding legislation, while the latter often takes decisions on the number of competitors, prices, and investment, and—directly or indirectly—on their market shares and returns on equity. Under *hard liberalization*, there is no difference between building a power plant, establishing a company for the resale of water, fixing prices of telecom services, sinking investment in the water cycle, and starting operations, say, in the tourism industry, in the manufacturing of furniture, in the retail sale of computers, or in car production.

Soft (or regulated) *liberalization* can work or not, depending upon how benevolent, informed, competent, and wise the government and the regulator are. Under this regime the government does not give up an active role in the industry. Rather, it exerts its control in a way that is different from direct provision and planning. The crucial change is that government does not establish the legal monopoly in some or in all segments of the industry. Laffont and Tirole (1993) see public procurement as close to regulation in this context. The government 'buys' the service on behalf of the citizens from private providers, but has to allow some information rents to them.

Hard liberalization takes a different route. Here the government actually divests itself of any responsibility for the management of the industry, except for a generic role of market supervision. This role is not intrinsically different from the one that applies, for example, to the food industry for public health reasons, or in the tourism sector to avoid fraud, or in retail sales to avoid market foreclosures. Some proponents of utility reform have gone so far, in the EU and elsewhere, as to think that this form of full liberalization is feasible and desirable. Others have serious reservations. The question, however, cannot be settled by an abstract discussion.

If for good reasons governments consider certain services as 'essential' to society, this has wide consequences for the desirability and functioning of market mechanisms. There are three aspects to consider: exit conditions for both providers and consumers, universal accessibility, and price affordability for the users.

Free *exit* is not a minor feature of a competitive market. The regulatory debate on network services in the EU has focused on *entry*, because in all the Member States an incumbent operator was already in place. Thus the issue was that of challenging that operator. The functioning of competition implies, however, that if for some reason *all* suppliers find it no longer profitable to stay in the market, they should be able to get out, at a reasonable cost. This is difficult in network services for two reasons. First, because of the sunk costs of the owner of the core assets (and possibly long-term financial obligations related to those investments). Second, because if the government considers the service 'essential', somebody should provide it in any case, perhaps receiving subsidies if it is no longer profitable at current prices.

In some network industries there is, in fact, virtually no exit option for either the individual consumer or the core firms. Electricity and some public transport or telecom services, for example, in a sense are vital. The recent sad events of a major earthquake and tsunami in Japan (11 March 2011) show that restoring some network services is a priority that precedes providing even emergency health care, shelter, and food for the survivors, because without transport, energy, and telecoms, it becomes difficult to intervene in an affected area. More generally, for a household or a community cut off from access to these services, because of insufficient income, or because of geography, this amounts to social exclusion and increased hardship. A breakdown in electricity or water supply in a city seems, in this context, to be qualitatively different from the shutdown of, say, restaurants, business consultancies, or footwear factories.

The borderlines between which services have exactly this essential nature varies across time and countries. It is not a technological feature of the industry, or of its outputs, but of political consensus. Once a service is deemed to be un-interruptible by government and citizens, then a full, permanent, 'hard' liberalization is a relatively risky option. Any commitment of the government to not intervene is particularly less credible. As a consequence, what we usually observe in the domain of network services is soft or regulated liberalization, often under mixed oligopoly, an arrangement that can be far from competitive. See for example Hefetz and Warner (2008, 2011) on local public services in the USA, where they focus on market and public interest characteristics of services and discuss the transaction costs issues involved by different arrangements.

The current global crisis can only support the view that governments need to stay in the game if private provision fails.

Accessibility of the service is also a political objective. This usually takes the form of universal service obligations for the incumbent, or for some specific firm, which receives a subsidy from the state to serve certain areas or groups. This is again an indication of the special nature of the industry. Affordability has a political dimension as well. While competition is expected to achieve lower prices on average as compared with the situation before market opening, it is not guaranteed that this will happen for any segment of the market. If governments are committed to ensuring that the household expenditure on a certain package of service should not exceed a given share of their income ('fuel poverty' for example has been defined in this way in the UK and elsewhere, see Chapter 9) this also has consequences for the nature of market equilibria, see for example Clifton and Diaz Fuentes (2010a).

The multinational dimension of some providers adds complexity to the analysis, see Clifton and Diaz Fuentes (2010b). Governments can make, and actually do make, mistakes in their policies on energy, transport, telecommunications, and water. However, the relationship between market design and the existence of public policies related to some network services seems a crucial one.

An important example is energy policy. Helm (2007) argues that by the turn of the century the mounting concerns about security of supply and climate change are going to determine a 'New Energy Paradigm'. Similar concerns apply to the challenges of dealing with the 'Digital Divide' in the context of the information society. The broad issue of the relationship between policy objectives and network services will be discussed in the concluding chapter.

Given the uncertainty surrounding some of the generic assumptions and predictions of the policy reforms of network services, one should turn to empirical analysis and carefully test the hypotheses and the outcomes in terms of prices and welfare changes, see for example Kwoka (2002, 2005, 2006, 2008) on the reform of electricity industry in the USA and Armstrong and Sappington (2006) for a wide review of the literature.

1.5 Structure of the book

After this introduction, which has set the research questions about ownership, vertical integration, and market opening, there are nine chapters:

- Chapter 2 presents some stylized facts on the reform trends of network industries in the EU-15. To track actual policy changes in a testable way, some comparable summary indicators are needed. The OECD reform indicators are reported, and will be consistently used throughout the book. The chapter presents a bird's eye perspective on energy and telecommunications reforms in the EU countries before enlargement. Moreover, to immediately introduce our main focus on reform outcomes, the chapter also shows price trends over a time span of 1990–2007 years in 15 countries and consumer satisfaction for the three industries that we are going to consider in greater detail.

- A simple thought experiment is presented in Chapter 3. What would you expect in terms of consumer surplus when you adopt a reform of a formerly vertically integrated state-owned monopoly? The illustrative model in this chapter is simple. It uses standard microeconomics to show the effect of reforms under very general assumptions. The message is that the conditions for the paradigm to deliver lower consumer prices are perhaps less than obvious, even without explicitly invoking asymmetric information. This simple conceptual model is applied in the subsequent three chapters. Principal-agent issues are also discussed in this framework. This concludes Part One of the book
- Part Two is about empirical testing of correlations between consumer price changes and reform indicators performed by a variety of models for each of the three sectors. Chapter 4 deals with the effect of reforms on residential prices of telephony. This service was the starting point for privatization policies worldwide, following the divestiture of British Telecom in 1984. Chapter 5 describes the effect of reforms on residential electricity prices. Finally, Chapter 6 considers natural gas, a relative latecomer in the reform process, which poses interesting policy issues, including that of security of supply. The core finding of these three chapters is that public ownership in the EU is often associated with lower prices than under private ownership, particularly in energy. Supply and demand shifters, and country features, are usually much more important than vertical disintegration and liberalization as determinants of prices, with some differences across the three services, (for example liberalization seems to have worked better in mobile telephony than elsewhere).
- Part Three adds further dimensions: perceived fairness, quality, and affordability. Chapter 7 complements the previous analysis and explores consumers' perceptions on price fairness and tests the correlation with the reforms. The estimation uses several waves of Eurobarometer data, a large cross-country yearly survey sponsored by the European Commission. This analysis is conducted for all three sectors considered in the above chapter, through a set of different empirical models. Some

counter-intuitive results are found, which are broadly consistent with the objective evidence on price determinants. In countries that have embraced the reforms, after controlling for individual characteristics, consumers do not appear to be happier about the price they pay compared with consumers in countries lagging behind them.

- The subjective evaluation of the perceived quality changes complements the previous empirical analysis of survey data. Consumer welfare is not only a matter of prices and quantity, but also of quality of services. Chapter 8 discusses some evidence on the relationship between reforms and customer satisfaction from this point of view. This is an elusive issue, and the results need to be interpreted with caution. Perception of quality does not appear to be correlated to any of the reform dimensions, but there are interesting variations across industries and countries.
- Distributive issues in network services are politically sensitive. Given the nature of essential services, price changes may raise concerns related to adverse or beneficial distribution effects across social groups and regions. Chapter 9 discusses some evidence that points to such effects. While social affordability issues due to tariff changes following the reforms affect only a relatively small minority of the population in Western Europe, it seems that the effect is not negligible. Moreover, the affordability dimension of the reforms is important in the recently accessed EU countries, and in other regions of the world that have adopted the paradigm.
- The concluding chapter of the book sums up the findings and offers a critical evaluation of the policy design adopted by the European Commission and by other international institutions, from the perspective of consumer welfare. From this angle, the evidence of success of the reform paradigm of the 1980s and 1990s seems to be more limited than expected. Possible reasons and remedies are discussed.

The bottom line of the book, in terms of the policy issues discussed in the last chapter, is that, to a certain extent, network industries are 'political'. Their services are related to important economic welfare issues that are the core of the European social model and of the 'EU 2020' goals. Thus, any rethinking on the ownership, structure, and performance of these industries should consider the reasons why energy and telecommunications—as possibly transport, water, and some other services—still pose special challenges to policy-making. An imperfect competition outcome may be better or not than other forms of regulation, but does not exempt governments from their responsibilities. This reconsideration of a government's role may be the beginning of a new paradigm, one that moves away from market fundamentalism, and focuses on long-term plans and strategies.

1.6 Further reading

According to Economides (1996) 'The literature on networks is so extensive that it is futile to attempt to cover it.' This comment is even more appropriate now, after the intensive research and policy debate of the last two decades. Thus, this section, and the similar ones in the remaining chapters, are based on an unavoidably personal selection of suggested reading, without any ambition to exhaustively scan the field. Having said this, Economides (1996, 2008) could be the entry point to focus on the concept of network. The more recent paper adds to the first a review of some anti-trust issues and case studies, mainly from the USA, while the former is still helpful in clarifying the core concepts and theoretical propositions of network economics. The Glossary at the end of the book offers a short synopsis of some concepts and additional references to introductory texts.

Models in network economics are often framed in incentive theory, see Laffont and Tirole (1993) or Laffont and Martimort (2002), or other versions of modern microeconomic theory of industrial organization and regulation. An accessible and up-to-date overview of the field in a policy perspective is offered by Finger and Künneke (2011). This well designed handbook, after some introductory chapters, reviews reform trends in telecommunications, radio spectrum, Internet, electricity, gas, oil industry, railways, air transport, roads, public transport, ports, water, and the postal service. It also offers country and regional studies, including the EU, USA, Latin America, China, India, Russia, Australia, other developing countries. The reform paradigm is discussed in detail in some excellent books, for example Vickers and Yarrow (1988), Baldwin et al. (2012), Newbery (2000), and Helm (2007). Shy (2001) covers the electronic communication and other sectors, Laffont and Tirole (2001) deal with reform of telecommunications. Several case studies are provided by ten Heuvelhof et al. (2009); on energy policy see Helm (2007), and Glachant and Lévêque (2009). Cameron (2007) offers a wide discussion of regulatory issues in the EU, with an emphasis on legal aspects.

For a survey of literature on market structure, firm strategies, and public policy in network industries see Koksi and Kretschmer (2004). On the role of public enterprises, see Cremer et al. (1989), Sappington and Sidak (2003), De Fraja and Delbono (1989), and Bradburd (1995). De Fraja (1993, 2009) offers a review of the old and new literature on mixed oligopoly, which is often the market arrangement for network industries in many EU countries. See also Martin (1993) and Beiner et al. (2011).

Some official reports by the European Commission are informative, for example 'Evaluation of the performance of network industries providing services of general economic interest', Brussels, several years.[20] Bianchi and Labory (2006) offer an overview of the current approach to industrial policy in Europe.

There are many scholarly field journals that regularly publish analyses on some of the topics briefly mentioned in this chapter, for example: a) *Energy Economics, Energy Policy, Energy Journal, Resource and Energy Economics*; b) *Communication Research, Communications and Strategies, Information Economics and Policy, Information Society, Telecommunication Policy, Telecommunication Systems*; c) *Annals of Public and Cooperative Economics, Competition and Regulation in Network Industries, Fiscal Studies, Industrial and Corporate Change, International Journal of Industrial Organization, Journal of Industrial Economics, Journal of Regulatory Economics, RAND Journal of Economics, Review of Industrial Organization, Review of Network Economics, Utilities Policy*. The bibliography at the end of the book shows other journals of interest.

[20] http://ec.europa.eu/internal_market/economic-reports/index_en.htm [accessed 5 January 2013].

2

Stylized facts

2.1 Introduction[1]

The parallel changes in industry structures, regulation, and performance of the European network industries in the last 20 years have been wide and complex. This chapter reviews some reform trends, particularly of ownership, industry structure, market opening, and changes of prices, and of consumer satisfaction for telecommunications, electricity, and gas services in the EU-15, and some other sectors for comparison.

Regulatory trends are tracked through a summary reform index, based on an OECD international database that runs from 1975 to 2007. Other information on reforms comes from several reports of the EC on the implementation of their directives. Price data are from EUROSTAT, while consumer satisfaction is summarized by Eurobarometer survey data. The presentation of facts is selective, and unavoidably very simplified, but additional insights on reforms, restructuring and prices will be offered by Chapters 4 (Telephony), 5 (Electricity), and 6 (Gas).

The rest of the chapter has the following structure. Section 2.2 presents the OECD reform indicators, and compares them with other sources. Section 2.3 deals with telecommunications; Section 2.4 with electricity; Section 2.5 with natural gas; Section 2.6 is about price trends and their correlation with aggregate reform indicators; Section 2.7 is about consumer satisfaction; and Section 2.8 concludes.

[1] Part of this chapter draws, with several adaptations and integrations, on Ceriani, Doronzo, and Florio (2009) and Fiorio, Florio, and Doronzo (2008).

2.2 Reform indicators

The former REGREF, now ETCR,[2] regulatory dataset has been created by the OECD. It is publicly available and collects information about indicators of privatization, liberalization, and unbundling of SGI across all the OECD countries, which do not cover all the EU-27 and includes some non-EU members. We focus here, however, on EU-15 only, for reasons that have been briefly mentioned in the preface. The OECD includes countries such as Japan and the USA, that for many reasons we consider difficult to compare with the EU Members States, because of scale and wide institutional differences.

The revised OECD indicators of regulation in energy, communication, and transport (ECTR) were released in 2009, for details see Conway and Nicoletti (2006). All ECTR regulatory indicators range from a minimum of 0 (corresponding to full deregulation) to a maximum of 6 (corresponding to the most restrictive conditions for competition). The indicators provide a measure of the degree of public ownership, vertical integration (energy only), entry regulation, and market concentration in the EU-15. One can either use the ECTR scores, which are based on a somewhat arbitrary 'cardinalization' of often categorical variables into the 0–6 scale, or use the sub-indicators as categorical variables or create several aggregate regulation measures starting from intra-sector indicators. In some cases continuous variables, for example on market shares of the incumbent provider, are also available. In Chapters 4 to 9 we shall consistently take advantage of these data.

The reform patterns by countries and industries are discussed in the next sections. As a first step, Figures 2.1 and 2.2 show which countries have progressed most in the adoption of the overall paradigm. This is achieved by using the combined OECD regulation index. This index aggregates indicators on ownership, vertical disintegration, and market opening for each country.[3] By way of comparison, we also add the combined index in airlines, post, and rail—three SGEI which are often cited as other network services on the reform track.

a. Figure 2.1a shows that for telecommunication,[4] looking back to 1990, only Spain and the UK did not show the standard industrial organization of combined public ownership and monopoly prevailing before the reforms. In contrast, in 2007, all the countries seem to have adopted a

[2] ECTR, Indicators of Regulation in Energy, Transport and Communications, available at http://www.oecd.org [accessed 5 January 2013].

[3] Details of the aggregation of sub-indicators are given in Chapters 4, 5, and 6. Typically this is done by giving equal weight to the main themes, for example ownership of the incumbent, unbundling, etc. Each theme is further broken down in specific items (typically one dozen for each sector).

[4] See below Section 2.3 and later Chapter 4.

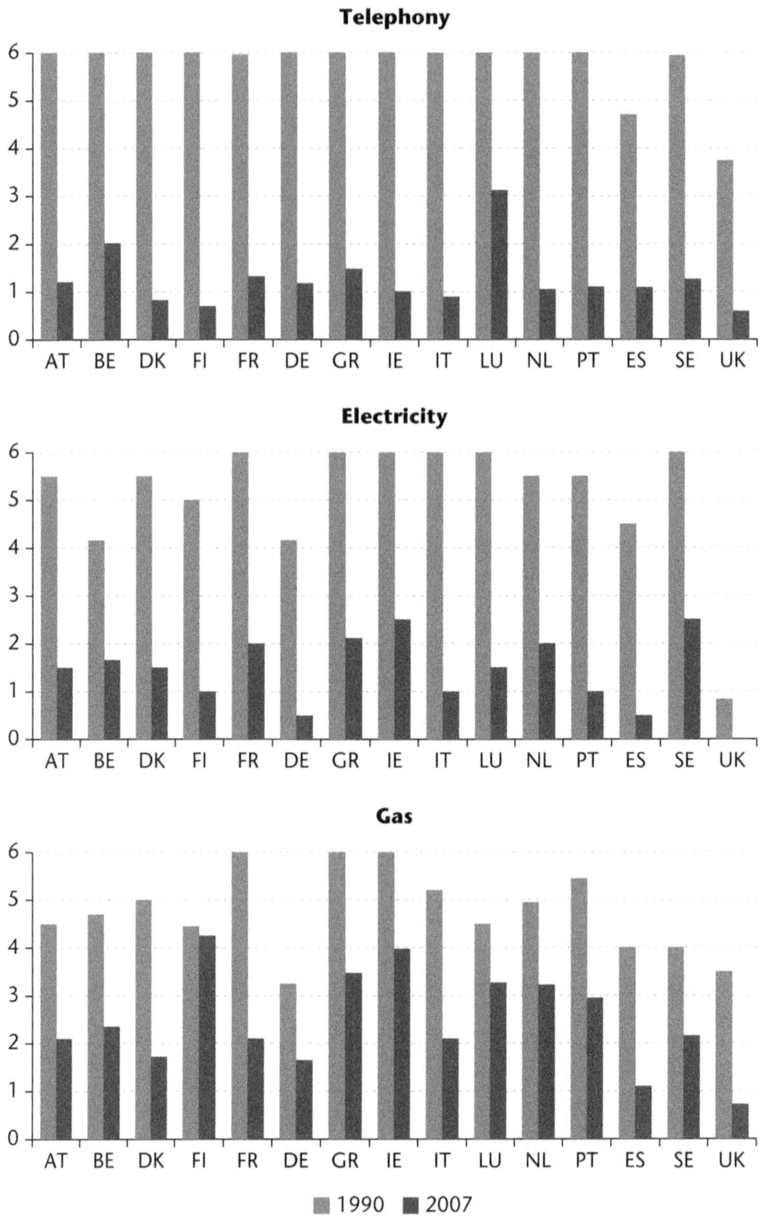

Figure 2.1. ECTR indicators, selected countries, 1990–2007

Source: own elaboration of ECTR scores.
6 = minimum reform progress
0 = maximum reform progress
See list of abbreviations for each country

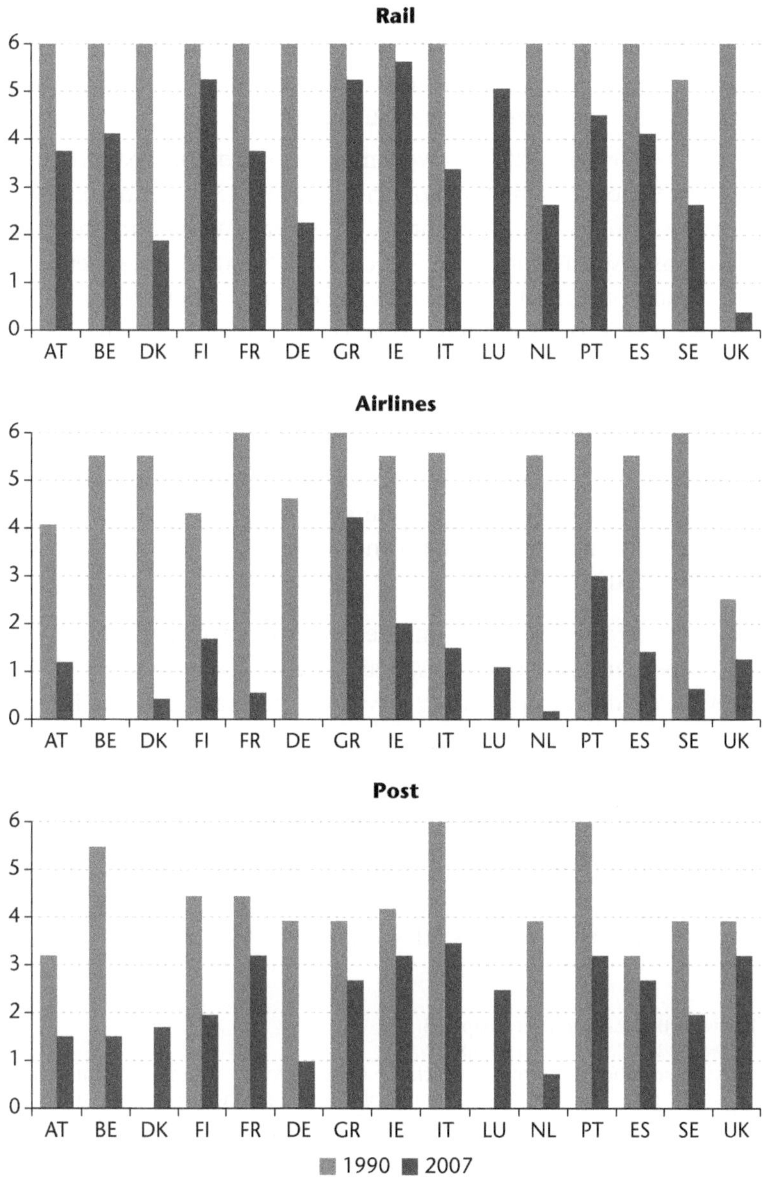

Figure 2.2. ECTR indicators for rail, airlines, and post service, 1990–2007

Source: own elaboration of ECTR scores.

similar reform pattern, with the UK still being most privatized-liberalized, and Luxembourg midway in the reform process. Thus, the convergence of telecommunications toward the reform paradigm across countries and over time has been substantial.

b. Figure 2.1b, for electricity,[5] shows a much more complex pattern. In 1990, the starting points were different across the EU-15. For example Belgium, Germany, and Spain did not have the typical ante-reform organization, which was instead adopted by France, Italy, Sweden, and others. The UK had already privatized the industry, before any other country. By 2007 there is some cross-country convergence, but clearly less than in telecoms, with only the UK having fully implemented the reforms.

c. Gas,[6] as shown in Figure 2.1c, was usually not a government-owned monopoly in 1990, except in France, Greece, and Ireland. In 2007 the variance of industry structures is still high across countries. Thus the three industries can be clearly ranked along the reform path, with telecommunication the most advanced, electricity midway, and gas the least advanced.

The former trajectories can be compared with changes between 1990 and 2007 in transport and postal services, which are also considered by EU legislation as network industries providing services of general economic interest.

a. The earlier broad pattern of rail[7] (Fig. 2.2a) was similar to telecom in 1990, with fully integrated monopoly. In 2007 the UK showed a mostly fully liberalized and privatized industry, with Denmark, Sweden, Germany and Netherland midway or fairly advanced in the reform process.

b. The initial pattern of public ownership and entry barriers was also similar across countries for airlines[8] (Fig. 2.2b), but was less extreme, as only

[5] See below Section 2.4 and later Chapter 5.

[6] See below Section 2.5 and later Chapter .

[7] Railways are probably the sector for which the rationale for the standard reform package was weakest in the first place (Newbery, 2000), because of the strong economies of integration. This fact didn't discourage the EC from considering liberalization. The EC (1996) White Paper 'A stategy for revitalising the Community's railways' was a first step, followed by EC (2002), another White Paper, European Transport Policy for 2010. This paved the way to directives aimed at creating a railways market in Europe (first 'package': 2001/12/EC; 2001/13/EC; 2001/14/EC). A second package was approved in in 2004 and a third one in 2007. The progress in practice, if not in national legislation, has been slow, except in the UK. For example there is no full interoperability, and harmonization of safety standards, etc. despite the creation of a European Railway Agency. Since 2010 international passenger services are open to competition, and domestic passenger traffic will follow. On infrastructure endowment differences across countries, see Chapter 9.

[8] The market opening of airways started in the USA in the late 1980s. Three European directives provided for liberalization in the EU: first in 1987 tariffs were liberalized, and code sharing allowed. Then Regulations 2342/2343/2344/1990/EC opened the market for international passenger and freight transport. A third set of regulations was passed in 1992, and further enlarged the scope of the previous packages. In parallel there was a liberalization of airports, which can be seen as the

France, Portugal, Greece, and Sweden had adopted the full previous paradigm in 1990. In 2007 the change was dramatic for airlines but much less impressive for rail. Again, only the UK was a fully reformed country in 1990.

c. Postal services[9] (Fig. 2.2c) show a very different story. In 1990 only Italy, Portugal, and Belgium were under full public monopoly, while the other countries allowed some form of competition. Liberalization and privatization increased over the years, in all the three sectors, but in 2007 there were still large differences across countries.

The OECD work of tracking reform progress has been fairly careful, but obviously it contains possible problems of measurement and of aggregation. The Milan UPP Team has compared the OECD indicators with other sources, particularly the Market Opening Milestones (MOM) database—prepared by Copenhagen Economics for the European Commission—which is not in the public domain[10] and is used for advisory work for the EC. The MOM database has not been updated, but for the years that overlap with ECTR the correlation is high. There are other sources from other institutions and individual researchers, see Chapter 4, 5, and 6, but unfortunately they either do not cover all the network services, or do not cover the same variables and countries. Hence, ECTR/OECD seems to be the best available source in the public domain for empirical analysis of the reforms.

A different, regularly updated source on the progress of reforms, is the EC reports on the implementation on its directives, which we shall cite below and in the next chapters. These reports are very helpful, but have not been

essential facility defining the infrastructure-based network industry. Council Regulation 1995/93 deals with the allocation of slots. A market approach to the allocation of slots and the creation of independent airport regulators are the current objectives of the EC in this industry. Safety standards will be harmonized by EASA, a European agency, perhaps an embryonic EU regulator. Most important is the European Single Sky project supported by the EU aimed at creating a EU wide traffic control (in fact structured in large functional areas). CO2 emissions are another area of European regulation that is going to have an impact on the airways industry.

[9] The postal service used to be managed along with telecommunications by national government, typically through monopolistic public enterprises. Currently different activities have taken divergent routes: postal financial services have been mostly privatized everywhere and often compete with commercial banks; parcels are often managed by international companies; collection and delivery of mail is still mostly managed by public enterprises. The EC started the liberalization process in 1992 (Green Book on Postal Service). Directive 1997/67/EC liberalized express and parcel and defined the universal service obligation for the incumbent. Directives 2002/39/EC and 2008/06/EC, respectively in 2002 and 2008 decreased the possibility for Member States to establish a monopoly protection for mail, to be abolished not later than 2013. Perhaps surprisingly, some government owned incumbents such as Deutsche Post/DHL and Dutch Post/TNT profited from the change and are now major global players and have even acquired their main private competitors (Finger and Finon, 2011). Interestingly, the substitution, but also complementarity relation, between post and telecommunications question the division of regulatory approaches in this area.

[10] We have been able to access these data, courtesy of the European Commission.

designed for scholarly research, and have differences of coverage and structure across sectors and years.

The OECD also publishes some country reports with more detailed information, for example the Communications Outlooks,[11] and other evidence is provided by the International Energy Agency (IEA), the International Telecommunications Union (ITU), and other sources that we will use later on.

Obviously, any attempt to track the reforms by simple aggregate indicators has limitations, hence treatment and comment on the indicators for empirical analysis must be careful and the interpretation prudent. The discussion below combines the presentation of the indicators that we use, with a more qualitative, albeit very brief, presentation of some facts for each of the three services which are the main object of this book. Other data sources are cited in the Further Reading Section.

2.3 Reform trends in the telecommunications industry

2.3.1 *The UK example and EU directives*

The change of the regulatory environment in which telecommunication industries operate in the EU has been a complex process, often led by the European Commission. Within the EC, a working group on telecommunications was active in the mid-1980s, and was later embodied in DG XIII Information Society. In 1984 a common EU standard was adopted for digital communication (ISDN) and other technical standards. In 1987 the EC Green Paper (EC COM(87)290) advocated a legislative programme that included full liberalization of the equipment sector and progressive opening of services, and harmonized European measures with respect to network access, interoperability, and interconnection. In 1990 the EU Directives 387 and 388 provided for opening the market for private networks and leased lines; though not yet for voice telephony.

Other EC Green Papers, first on satellite communications in 1990 and then on mobile and personal communications in 1994, extended the same regulatory principles in these two areas. In 1996 the Full Competition Directive paved the way for the liberalization of fixed telephony from 1998 (1996/19/EC). Spain, Portugal, Greece, and Luxemburg obtained a delay to implement the Directive and adopted liberalization reforms some years later, while other countries anticipated the Commission pressure and began to open the telecommunication market during the 1990s. Directive 2002/19/EC regulated the access to networks and introduced the obligation to establish an independent

[11] http://www.oecd.org/document/44/0,3746,en_2649_34225_43435308_1_1_1_1,00.html [accessed 5 January 2013].

national regulation authority (NRA). Directives 2002/22/EC and 2002/27/EC further extended consumer protection and abolished some exclusive rights, respectively. These directives were updated by the new Directives 2009/136/EC and 1340 and in 2010 the Body of European Regulators for Electronic Communications (BEREC) was established in order to coordinate NRAs.

Removing the legal barriers to market entry and establishing pro-competitive regulation at European level was a crucial step. Elsewhere this process was promoted by several other international organizations, particularly by the OECD and the World Bank. In the phase prior to 1998, some policy and legal coordination across countries with substantially different starting points in the liberalization process was achieved (Cawley, 2003). The establishment of EU legislation and translation into Member State law was, however, only the beginning of a long and often cumbersome restructuring process.

At European level the Commission has been actively engaged in follow-up work on checking and enforcing implementation both on the regulatory side and on anti-trust decisions. The 1999 Telecommunication Regulatory Review aimed to bring all communication infrastructures and services into a single framework and to improve cooperation between the Commission and national authorities. The directives that were finally adopted in 2002 followed the idea of moving from heavy ex-ante regulation to lighter ex-post anti-trust decisions, and the three core topics were authorization, access, and interconnection. The framework directive deals with aspects common to all areas and the data protection directive extended the coverage of data protection and privacy measures to the Internet.[12]

The basis of the regulatory framework was established at a time when telecommunication was not in mainstream European policy. However, the 2000s saw the emergence of closer economic integration as a major item on the European political agenda and the growing competition in telecommunications became part of a broader effort to sustain the diffusion of the Internet and electronic commerce. The focus of EU legislation was shifted towards content related issues such as security, intellectual property, and privacy. More recent legal developments, which will be discussed in Chapter 4, will be mainly felt in the future.

[12] There are in fact seven directives in the 2002 package: (a) the Framework Directive, that sets the main principles; (b) the Access Directive, that regulates interconnection of entrants to the networks owned by the incumbents; (c) the Authorization Directive, that substitutes the previous system of detailed and individualized licences with a generic provision allowing operation; (d) the Universal Service Directive, stating the requirements of service in terms of access, affordability, and consumer rights; (e) the Privacy and Electronic Communication Directive, providing for the protection of personal data; (f) the Radio Spectrum Decision, setting the principles of the policy in this area; the Commission Competition Directive, related to the role of the EC as for competition issues. The 2009 package updates these directives. In most cases it requires national legislation to comply, see European Commission (2009) and below, Section 4.2.

The UK is the most obvious example of a reform frontrunner, but Finland, Sweden, and Denmark were also early adopters of market opening approach, see Figure 2.2. The UK started its liberalization process ahead of all the European countries. After the decision in 1981 to reform its telecommunication sector (Communication Act), the UK government proceeded cautiously, granting Mercury Communication (part of the Cable&Wireless group, privatized before British Telecom) the right to enter the market, first (1982) only in national calls and some years later (1988) also on international calls, thus creating a duopoly. In 1984 BT was privatized, floating on the stock exchange 52% of the equity previously owned by the state. The remaining equity was floated in 1991 (26%) and 1993 (22%). The government also appointed the sector regulatory authority, Oftel (Telecommunication Act), later transformed into Ofcom with the broadening of its mandate to television. In parallel, from 1983 onward cable TV operators were granted the right to use their television network to provide telecommunication, while BT was barred from offering TV services until 2001. In 1985 two licences to provide mobile services were granted to Racal and to the incumbent BT. Despite the focus on a more appealing market and the strong protection offered by the franchised duopoly, Mercury's market share grew slowly and in 1995 it was just 10% of the total in terms of revenues.

The duopoly policy was abolished in 1991 and since then the UK government has implemented the EU telecommunication directives. Thus, although the UK's telecommunications policy was initially developed independently of other governments and institutions, it was later adapted to meet external obligations. However, with few exceptions, the UK has led, rather than followed, EU liberalization measures. With its pioneering policy, the UK actually opened the telecommunication market 18 years before the majority of European countries did so.

2.3.2 *Changes in the market structure*

With the liberalization of fixed telephony many newcomers entered the voice market, but in each country only a very small number of operators gained a significant position as the incumbents lost their market dominance only very slowly. Telecommunications are complex systems where different elements interact with each other and, on the whole, exhibit significant network economies. Therefore there is not a single act or passage that opens the market, but a set of measures and a continuous policy implemented by governments, regulators, and EU institutions that enable new operators to offer competitive services.

In recent years the area of interconnection and local loop unbundling has gained importance. And eventually the abatement of switching costs for users, as in number portability, has been viewed as a way to improve the competitive outcome.

The control of the client's telephone number is in fact a significant barrier to switching. Number portability enables the subscriber to retain their number when they move from one operator to another. Fixed number portability has continued to play an important role in encouraging competition. By October 2006, 31 million mobile subscribers and 15 million fixed subscribers had ported their number since the introduction of this possibility. According to one EC report, number portability depends heavily on the behaviour of the dominant incumbent (12th Implementation Report, 2006).

Another move was unbundling the local loop. This refers to a series of regulatory measures aimed at providing access to the incumbent's local network, the less duplicable part of telecommunication infrastructure. Since it is usually too costly to build the peripheral part of the network, this is a source of significant market power for the established incumbent. A trade-off can emerge in this context. On the one side the availability of a local connection at a controlled price enables new competitors to offer telecommunication services, particularly broadband access with DSL technology; on the other side, unbundling can be detrimental to competition by retarding the roll out of competing infrastructure (Beranes and Bourreau, 2005).

In Europe, until recent times local loop unbundling failed to give a strong push to competition in the market for fixed voice telephony. Competition in fixed voice was often based on carrier selection (that requires less investment), probably as a result of the low level of profitability in this area. Unbundling, however, is often seen as offering a large potential as a means to allow broadband access to end users by entrants without local networks. Moreover as new technology like VoIP gains ground, the intensity of competition in voice telephony can be expected to increase (De Bijl and Peitz, 2005).

One way to highlight the change of market structure in fixed telephony, which is our preferred telecom case study in this book, is simply to look to the market shares of the incumbent company as reported in several editions of some EC reports, such as the 'Report on the Implementation of the Telecommunications Regulatory Package' 5th edition (1999), or the 'European Electronic Communications Regulations and Markets' 12th edition (2006), and subsequent updates, with the changes in the title reflecting also the change in the scope of EU regulations.

According to this source (that for some indicators and years overlap with the OECD/ECTR source) the share of the incumbent's retail revenues in 1997 was around 100% in all EU-15 Members, except in the UK where it was then around 85% for local calls (with only Sweden and Denmark having very modest market opening, around 5% or less of market shares for the entrants). The picture is similar for national long distance calls (with 77% of market share for the incumbent, 86% in Sweden, and modest market opening in Denmark and Finland). For international calls, in 1997 there was a more visible market opening. In the UK 45% of the revenues were collected by the entrants, 33% in Sweden, and there was a modest market opening also in the Netherlands and Finland. In less than ten years the change was dramatic. In 2005, the share of the entrants in the above mentioned three segments was near to 40–50% in the UK (according to segment), and it was between 20–30% in the remaining countries. In Germany the incumbent lost to the entrants even more than in the UK, particularly for international calls, where 61% of the retail revenues went to the entrants. Interestingly, the shares of the incumbent in mobile telephony, which were generally around 40–60% in 1997, declined less dramatically, or even increased in subsequent years. These trends will be further discussed in Chapter 4.

2.3.3 *Reform trends by country*

Against this complex story, Table 2.1 and Figure 2.3 offer a synopsis of changes by country. As mentioned in the previous section, the OECD provides regulatory indicators about three dimensions: public ownership, entry regulation, and market structure.

At the beginning of the period we consider, in 1975, telecommunications were run by public monopolies in all the EU countries, with the only exception being Spain, where the government owned just the 35% of shares in the public telecommunications operator.[13] Nothing changed until 1982 when, as mentioned, the UK took the first steps towards privatization of British Telecom and the establishment of a duopoly, to be removed in 1991 with wide liberalization of entry.

Finland and Sweden have followed a similar pattern. In 1991, the legal conditions of entry into the trunk telephony market allowed a duopoly in Finland and competition in Sweden but with no third party access.[14] The industry ownership remained with the public sector until 1999, when the industry eventually became privately-owned, competitive, and open to TPA.

Albeit with different timing, all the other countries followed two of the paradigm's steps: privatization and then market opening. Italy privatized Telecom Italia in 1991, followed by Denmark in 1992, and the Netherlands in 1993. In Italy the government gave up the 50% of its PTO shares at the beginning of the reform process, while Denmark gave only the 11% (49% in 1993) and the Netherlands 44%. While by about 2007–08 the telecommunication industries in all the countries were competitive, privatized (with the

[13] PTO hereafter.
[14] TPA hereafter.

Table 2.1. Telephony regulatory changes. EU-15, 1975–2007

Competition	Access	75–81	82–83	84–85	85–90	91	92	93	94	95	96	97	98	99	00	01	02–03	04–07
PUBLIC																		
Monopoly	No TPA	AT BE DK FI FR DE GR IE IT LU NL PT SE UK	AT BE DK FI FR DE GR IE IT LU NL PT SE	AT BE DK FI FR DE GR IE IT LU NL PT SE	AT BE DK FI FR DE GR IE IT LU NL PT SE	AT BE DK FR DE GR IE LU NL PT	AT BE DK FR DE GR IE LU NL PT	AT BE DK FR DE GR IE LU NL PT	AT BE DK FR DE GR IE LU PT	AT DK FR DE IE LU	AT FR LU	AT LU	AT					
	TPA												LU					
Duopoly	No TPA		UK			FI	FI											
	TPA							FI										
Open	No TPA					SE	SE	SE	SE									
	TPA								FI	FI SE	FI SE	FI SE	FI SE	LU	LU	LU	LU	LU
PRIVATE																		
Monopoly	No TPA	ES	ES	ES	ES	IT ES	DK IT ES	DK IT ES	DK IT NL ES	BE GR IT NL PT ES	BE DE GR IE IT NL PT	BE FR DE GR IE IT PT	BE DE GR IE IT PT	GR PT	GR			
	TPA															GR		
Duopoly	No TPA			UK							ES	ES						
	TPA				UK								ES					
Open	No TPA									DK	DK	NL		IE	IE	IE	IE	
	TPA					UK	UK	UK	UK	UK	UK	DK UK	DK FR NL UK	AT BE DK FI FR DE IT NL ES SE UK	AT BE DK FI FR DE IT NL PT ES SE UK	AT BE DK FI FR DE IT NL PT ES SE UK	AT BE DK FI FR DE IT NL PT ES SE UK	AT BE DK FI FR DE GR IE IT NL PT ES SE UK

Source: own elaboration of OEDC data. Shaded area: public or mostly public ownership.

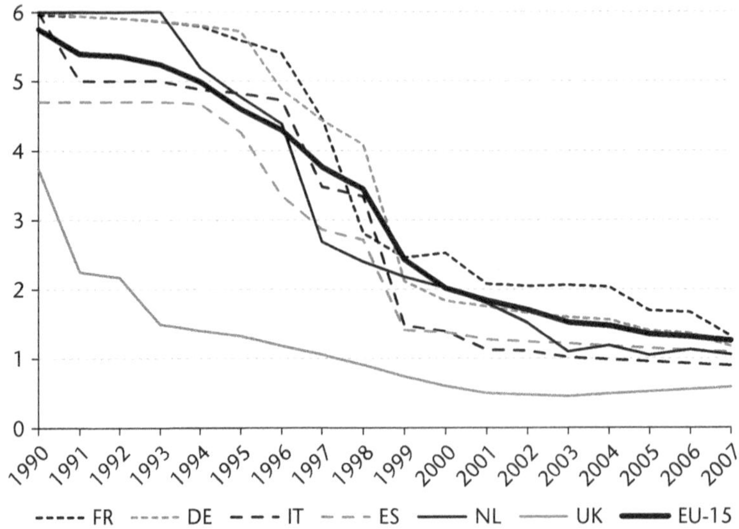

Figure 2.3. Telephony. ECTR indicators, selected countries, 1990–2007
Source: own elaboration of ECTR scores.

exception of Luxembourg), and open to third party access, some differences still remained. For example, the degree of private participation is heterogeneous, and the government still owns 54% of the PTO shares in Belgium and 50% in Sweden, while there is full private ownership in Spain, Italy, the Netherlands, Ireland, Denmark, and the UK. While recorded as private in the OECD indicators, the German and France incumbents, respectively Deutsche Telekom and France Telecom, are still partly owned by the French and German governments, with directly or indirectly a stake of around 30% of the equity. It is difficult, however, to ascertain to what extent this fact influences the behaviour of the management, for example in tariff or wage policies. In general, in this book, we shall not try to adapt or interpret the OECD indicators in dubious cases (as the ones that we have just mentioned), in order to preserve the consistency of the information baseline across the countries and the sectors. The following chapters, however, offer some additional detail on the industries and countries we consider.

The degree of actual (i.e. beyond formal entry regulation) market opening (simply measured as the market share of new entrants) in the trunk telephony market also varies across countries: the average is 36%, the less competitive being Spain with 21% and the more competitive Finland with 60%.

The migration of countries from the northwest to the southeast zone of Table 2.1, or from the top to the bottom of Figure 2.3, is apparent and captures the converging reform trend.

2.4 Reform trends in the electricity industry

2.4.1 *Aiming at a European market*

While it was the UK again that started the reforms in the electricity industry, in most of the other EU countries the momentum for change was mainly provided by EU legislation. The aim of the first EU electricity directive of 1996[15] was to gradually introduce competition in order to create a European market for electricity. Some principles were established, more or less based on the British experience: some form of unbundling of the production, transmission, and supply stages, the introduction of transparent rules for licensing new generation plants, the gradual extension to final customers of the right to buy electricity directly from the producers and the right of access to the network by new entrants, and regulation of tariffs. In 2003 a new directive repealed that of 1996.[16] The aim of this new directive was to speed up the integration process and the development of competition and therefore it was more detailed than the old one regarding several issues, the most relevant of which were the clearer rules about the unbundling of production and transmission stages (especially the separation of the grid from generation and supply); TPA; the compulsory creation of a national energy regulator; the immediate opening of the market to all customers; some provisions on security of supply, universal service obligations, and environmental protection. In particular, the directive required market opening to all non-household customers by 1 July 2004 and to all customers by 1 July 2007.

This Electricity Directive was due to be transposed into national legislation by Member States by July 2004. Its requirements aimed to yield a market-led industry organization. However, several Member States were slow in implementing these measures. At the end of 2005, Greece, Luxembourg, Portugal, and Spain had still not fully notified the Commission of the legal measures taken for the purposes of transposition. In November 2006 the European Commission published a report on the functioning of the internal market in electricity and gas. This report confirmed that cross-border competition was not yet sufficiently developed to provide customers with a real alternative to the nationally established suppliers. Key indicators in this respect were the absence of price convergence across the EU and the low level of cross-border trade. This situation was seen by the EC as a direct consequence of the failure of Member States to implement the second electricity directives on time or with sufficient determination. Furthermore, inadequate use of existing infrastructure and insufficient interconnection between many Member States prevents real competition from developing, despite the political commitment of

[15] First European Electricity Directive no. 92, 1996.
[16] Second European Electricity Directive no. 54, 2003.

the European Council, made in 2002, to achieve an import capacity of at least 10% of internal consumption.

Out of the EU-15 Member States, 11 still have companies that have a strong or dominant market position. In some cases, these companies now appear to have more market power than they did before the first directive was passed (see below). Some features of the industry in selected countries are presented in Chapter 5, along with more recent changes in the EU directives (the Third Energy Package), which reinforce the unbundling process.

2.4.2 *Different national paths*

In spite of the EU's attempts to establish a unified approach to electricity reforms, history and national institutions played their role in shaping patterns of change. In this section we briefly mention the reform milestones in the UK, France, Germany, three Scandinavian countries (Sweden, Denmark, Finland), and two Mediterranean countries, Italy and Spain. As mentioned above, the UK was the first European country to implement a comprehensive reform of the electricity sector at the end of the 1980s. The liberalization process started in 1989 with the Electricity Act and was completed in 1999. The reform process included the breakup of the Central Electricity Generating Board (CEGB) into two generating companies (PowerGen[17] and National Power) and the National Grid Company. So the National Grid, PowerGen, and National Power were privatized in the early 1990s together with a network of RECs (regional electricity companies).

The privatization process ended in 1995 in England and Wales, but Scotland and Northern Ireland followed a different path. As regards the market opening at its beginning, the right to choose the electricity supplier was granted only for the major users; however the threshold level was progressively reduced, reaching complete liberalization in 1998. The Electricity Act also established the introduction of a wholesale market and the complete separation of transmission activity from generation. In the network services (transmission and distribution) the price was regulated by a price cap, under the control of OFFER, a regulation agency, now OFGEM.[18] It is interesting to contrast the British experience with that of France and Germany.

In contrast to the UK, France was one of the latecomers in implementing the EU directives. The experience of liberalization began only in 2000, when the government approved law no. 108, which modified the pre-existing structure created by the law of 1946. The reform introduced a regime of regulated TPA, a wholesale market, and a progressive opening of the sector to competition with

[17] Now known as E.ON UK after acquisition by E.ON.
[18] Green (2007).

the possibility for the consumer of choosing the retailer. The threshold level was established at 100 GWh in 2003 and was gradually reduced in subsequent years. The elimination of any threshold was foreseen for 2007.[19]

Germany began to implement the EU directives in 1998 with the Energy Act, and completed the reform in 2005 with the Energy Industry Act. The reform introduced full market opening: each consumer had the right to choose his own supplier without any limitations or volume threshold. The supplier could also be a producer, a regional distribution company, or a municipalized company. An electricity exchange was also set up, but only in the August of 2005 was a regulatory body, the Federal Net Agency, installed with a mandate to control the industry prices.[20]

The three small Scandinavian EU countries are notable for a combination of liberalization and continued public ownership of the industry. In 1996 there was a radical reform in Sweden following the experience of Norway.[21] After the enacting of the Law for Electricity Supply, a full liberalization of generation and a regime of authorization were implemented. The transmission system remained in public hands, in a monopoly regime managed by a non-profit state company, and distribution and supply were mainly provided by around 280 municipalized firms. The retail market was fully opened, and since 1996 consumers have been able to choose their own supplier without any threshold level. The separation between transmission and distribution operators has been effective since 1992, but the industry only became completely unbundled in compliance with the second EU electricity directive.[22] Finland decided to reform the sector in 1995 with the Electricity Market Act. From then on the market was progressively liberalized and since 1997 the consumer has been able to freely choose their own supplier. The reform also established the division between the transmission and generation firms, although the distribution companies are not yet fully unbundled.[23] The reform experience in Denmark is very similar to that of the other Nordic countries and began in 1996 with the Danish Supply Act, but was completed in 2001 when the law established the complete rights for any type of consumer to choose their supplier. The country's electricity sector is characterized by the presence of public local companies, and despite some privatization the overall ownership structure still remains mostly public. The separation between generation and

[19] Commission of European Community (c).
[20] Commission of European Community (c).
[21] Norway was the first Nordic country to liberalize its electricity market in 1991, following the British model in most respects, but without privatization. The state still plays a leading role through local distributors and in some cases also public producers.
[22] Thomas (2006) and Commission of European Community (c).
[23] Thomas (2006) and Commission of European Community (c).

transmission activities was established in 1996, but the full unbundling of the industry, including distribution, was completed in 2001.[24]

We conclude this brief overview by considering two Mediterranean countries: Spain and Italy. The reform of the energy sector began in Spain with the privatization of two public electric utilities. From 1988 to 1999 the privatization of Red Electrica and Endesa was completed. In 1994 the Law 40/1994, known by the acronym LOSEN, was enacted with the aim of liberalizing the sector. This law mandated the legal unbundling of the transmission network and created an independent joint public–private transmission system operator[25] Red Electrica Espanola (REE). REE offers regulated TPA to both the transmission network and the distribution network. On the generation side of the industry, LOSEN was quite permissive in terms of the maximum market share that any given Spanish generating company could control. Therefore the subsequent years were characterized by an acceleration of the trend toward consolidation: from over 35 independent regional generation companies in 1990 only 5 were left by 2002. Law 54/1997 accelerated the process of liberalization. Full market opening, including for domestic customers, and regulated TPA have been implemented since January 2003. An energy regulator was also created: the Comisión Nacional de Energía (CNE). However, important elements of the EU directives were not transposed, and the adoption of the implementing legislation has been constantly delayed over the last few years. Spain has therefore been one of two Member States, with Luxembourg, subject to the general infringement procedure from the European Court of Justice[26] for non-communication of transposition measures for both the gas and electricity directives.

In Italy, the transposition of the first EU directive was realized in 1999. The main changes were: the introduction of competition in generation and supply to final consumers; the vertical separation of the production stages; and the creation of a wholesale market. The reform of 1999 did not provide for an immediate full market opening: in 2000 only customers with a consumption of over 20 GWh were able to join the free market. The threshold was progressively reduced and full liberalization was achieved in 2007. The 1999 decree also imposed the vertical disintegration of ENEL, the incumbent, still under the control of the government through a strong stake in the equity, with the creation of TERNA, a listed company under government equity control. This company owns the transmission system almost entirely and initially was completely owned by ENEL. The management of the grid was instead devolved to GRTN (Gestore della Rete di Trasmissione Nazionale) a public owned society. The GRTN guarantees open access to the network, according to

[24] Thomas (2006) and Commission of European Community (c).
[25] TSO hereafter.
[26] Commission of European Community (c) and (d).

the conditions and tariffs determined by the regulator. A decree of May 2004 approved the reunification of the transmission network's management and ownership. This was done to eliminate coordination problems between GRTN and TERNA. In June 2004, 50% of TERNA's capital was sold to the public for EUR 1.7 billion. To ensure the independence of the new enterprise called TERNA Rete Elettrica Nazionale spa, as of 1 July 2007 non-transmission energy companies have been prohibited from owning more than 20% of the shares of the new combined company (voting rights must not be higher than 5%). The same decree entrusts TERNA with all of GRTN's responsibilities.[27]

In terms of market shares in generation, the variability across countries is very high. According to EUROSTAT, Economy Structural Indicators, in 2005, the percentage of the total electricity generated by the largest generating company ranged from a minimum of around 20% in the UK, to around 90% in France. Germany, Finland, Italy, Denmark, and Spain were all in the range 26–36%, and the remaining countries show production shares of the largest generator in excess of 40% (with Belgium at 85%, Sweden 47%, etc.).

2.4.3 *Reform trends at a glance*

Figure 2.4 and Table 2.2 show an overview of electricity reform trends in the EU-15 Member States, again based on the OECD indicators.

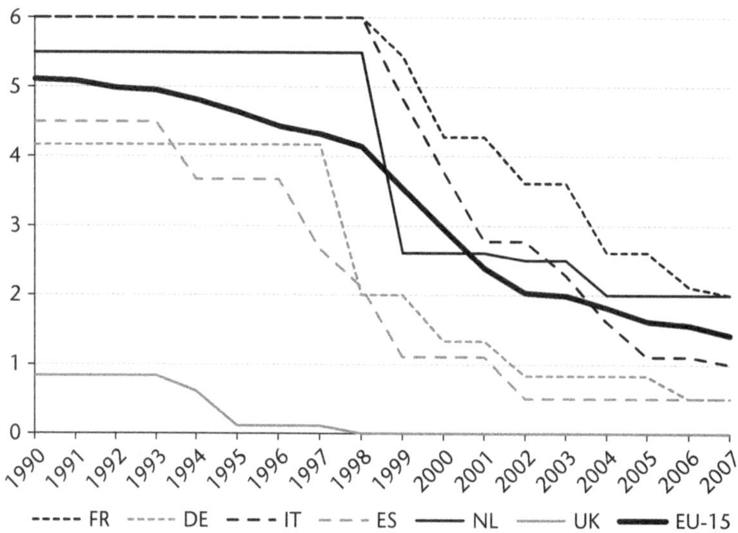

Figure 2.4. Electricity. ECTR indicators, selected countries, 1990–2007
Source: own elaboration of ECTR scores.

[27] Ferrari, Giulietti (2005) and Commission of European Community (c).

Table 2.2. Electricity regulatory changes. EU-15, 1975–2007

Access	Ownership	75–88	89	90	91	92	93	94	95	96	97	98	99	00	01	02	03	04	05	06	07
INTEGRATED																					
No TPA	Public	FR GR IE IT LU NL PT SE UK	FR GR IE IT LU NL SE UK	FR GR IE IT LU NL SE	FR GR IE IT LU NL	FR GR IE IT LU NL	FR GR IT LU NL	FR GR IT LU NL	FR GR IT LU NL	FR GR IT LU NL	FR GR IT LU NL	FR GR IT LU NL	FR LU								
	Mostly public	AT DK FI	AT DK FI PT	AT DK FI PT	AT DK FI PT	AT DK FI PT	AT DK FI PT	AT DK FI	AT DK	AT DK	AT DK	AT DK									
	Mixed	DE ES	DE ES	ES	ES	ES	ES														
	Mostly private	BE	BE	BE	BE	BE	BE	BE	BE	BE	BE	BE	BE								
Negotiated TPA	Public				SE																
	Mixed			DE	DE	DE	DE	DE	DE	DE	DE	DE									
Regulated TPA	Mostly public												AT DK	DK							
ACCOUNTING SEPARATION																					
No TPA	Public						IE	IE	IE	IE	IE	IE	GR IE	GR							
	Mostly public							PT	PT	PT	PT	PT									
	Mixed												PT								
Negotiated TPA	Mixed							ES	ES	ES											
	Private											DE	DE	DE	DE						

Regulated TPA	Public										IT	FR IE LU	FR GR	FR	FR				
	Mostly public						FI	FI				AT IT							
	Mixed												LU	LU	LU	LU			
	Mostly private											BE							
SEPARATE COMPANIES																			
Negotiated TPA	Public			SE	SE	SE	SE												
	Mostly private								ES	ES									
	Private													DE	DE	DE	DE		
Regulated TPA	Public							SE	SE	SE	NL SE	NL SE	IE NL SE	IE NL SE	IE NL SE	FR IE NL SE	FR IE NL SE	IE NL SE	IE NL SE
	Mostly public								FI	FI			AT DK IT	AT DK GR IT	AT DK GR IT	AT DK GR IT	AT DK GR	AT DK FR GR	AT DK FR GR
	Mixed										FI	FI PT	FI PT	FI PT	FI PT	FI PT	FI IT LU PT	FI IT LU PT	FI IT LU PT
	Mostly private	UK	UK	UK	UK	UK					ES	ES	BE ES	BE ES	BE ES	BE ES	BE ES	BE ES	BE ES
	Private						UK	UK	UK	UK	UK	UK	UK	UK	UK	UK	UK	DE UK	DE UK

Source: own elaboration of OECD data. Shaded area: public or mostly public ownership.

According to OECD classification for the electricity industry, we can detect three levels of unbundling ('no unbundling' or 'integrated market', 'accounting separation', and 'separate companies'), three levels of market opening ('no market opening', 'negotiated TPA', and 'regulated TPA'), and five types of industry ownership ('public', 'mostly public', 'mixed', 'mostly private', and 'private').

The starting and the ending points of the reform process are heterogeneous with respect to the ownership structure. Nevertheless, at the beginning of the period nine out of fifteen countries had a government-owned incumbent and another three had a mostly government-owned one, while none was entirely private. At the end of the period only Ireland, the Netherlands, and Sweden still retained government ownership of the incumbent. Switchers include France and Greece which became 'mostly public' from public monopoly; Italy, Luxembourg, and Portugal which ended up with a mixed ownership industry, and the United Kingdom with a fully privatized industry. Austria and Denmark kept a mostly public sector, while Finland switched from mostly public to mixed. Germany moved from mixed to private and Spain from mixed to mostly private.

All the EU-15 countries, with no exception, have started with a vertical integrated incumbent and no TPA, and have ended up with an unbundled industry or regulated TPA.

Unbundling was implemented in two steps (first accounting separation and then separate companies) in Austria, Belgium, Finland, France, Germany, Greece, Ireland, Italy, Luxembourg, Portugal, and Spain. On the other hand, Denmark, the Netherlands, Sweden, and the United Kingdom immediately moved directly to ownership unbundling.

With respect to market opening, only Germany, Spain, and Sweden opted for the intermediate step of negotiated TPA, while the other 12 countries switched to regulated TPA all at once.

Almost nothing changed until the beginning of the 1990s, when the United Kingdom forged ahead to an unbundled mostly private and then fully private electricity industry, with regulated TPA. Germany and Sweden followed. Germany opened the market via negotiated TPA in 1990, but this was a limited move. The industry was then privatized (but there still are many municipally owned companies) and partly unbundled (by accounting separation) in 1998, and the unbundling procedure came to an end in 2002. In 2006 Germany moved from negotiated TPA to regulated TPA.

Sweden followed by taking fewer steps than Germany to reach market opening and full unbundling, but has never fully privatized electricity. In 1991 Sweden opened the market by Negotiated TPA, and the year after completed the unbundling procedure. Regulated TPA replaced Negotiated TPA in 1996.

2.5 Reform trends in the natural gas industry

Differently from telecommunications and, to a certain extent, electricity, the natural gas industry is deeply influenced by geography, or more precisely, by geology. After World War II only manufactured gas production was widespread in Europe, and virtually no natural gas was produced. The country that changed the history of the industry was the Netherlands, where a supergiant field was discovered at Groningen, and production started in the 1960s. Norway and the UK followed, where gas was discovered in association with exploration for oil offshore. In Italy and elsewhere there are minor fields. Hence, the EU needs to import substantial quantities of natural gas from a small number of non-members countries, notably Russia, Algeria, some Middle-East countries, and others. The state, according to a centuries-long legal European tradition (that goes back to Roman law), is the owner of development rights for natural underground resources. Hence, governments usually gave the gas development concessions to public or private enterprises or, in the case of the Netherlands, to a public–private organization, under a franchised vertically integrated monopoly. This pattern has changed over the last 20 years, with widespread privatization and market opening. Most European countries started to reform the gas industry after the EU Directive 98/30/EC in 2000. This directive provided for initial liberalization steps. This was repealed by Directive 2003/55/EC, now replaced by Directive 2009/73/ec. The core provisions of the 2003directive, that was in force in the period analysed in this book, were related to market access, storage facilities, role of the system operator, functional separation, public service obligation and consumer protection, independent regulators, benchmarking and monitoring by the EC, and security of supply. For additional details, see Chapter 6.[28]

2.5.1 *The UK privatizes gas*

The starting point of the reform of the gas industry in Europe was the divestiture of British Gas. In parallel, the opening of the gas market in the UK was carried out well in advance of the requirements of the EU directives. The process began with the Gas Act in 1986 which disposed for the privatization of the vertically integrated public monopolist British Gas and for the creation of a regulatory authority (OFGAS).[29] The reform was virtually completed by

[28] Source: http://europa.eu/legislation_summaries/energy/internal_energy_market/l27077_en.htm [accessed 5 January 2012].

[29] At present, the Regulatory body for the British energy industry (England, Wales, and Scotland) is the Energy Markets Authority, which operates through the Office of Gas and Electricity Markets (*Ofgem*). This was implemented in 2001 from the merger of the gas (*Ofgas*) and electricity (*Offer*) regulatory bodies.

1998 when all domestic customers were given the right to choose their supplier. In 1997 British Gas separated its retail division. It became BG plc and included the distribution and transmission network activities (Transco) while the smaller retail division, which had a small production division, but no network, became Centrica. In 2001 BG plc separated the Transco network division as a distinct company called Lattice, but in 2002 it merged with the National Grid Company to become National Grid Transco (NGT). The regulator therefore required NGT to separate the distribution and transmission sector. The distribution network was split into eight different regional businesses. The privatization of British Gas as a vertically integrated company, however, prevented new entries for many years. The company operated with long-term contracts both upstream and downstream, with specific price-adjustment mechanisms. It was not until the middle of the 1990s that the government required British Gas to progressively reduce its market share in the industrial market, where it had to be replaced largely by oil companies. A spot market was encouraged by the regulator, as the oversupply of gas made the traditional long-term contracts unattractive. The main feature of the retail UK gas market was then the progressive process of horizontal integration with the electricity market: all significant suppliers offer gas and electricity as a dual package. In the industrial market, the combined cycle technology in electricity generation was a key feature of the structural change. Centrica still holds about 60% of the residential gas market, with the rest of the market going to the five major electricity companies: RWE/NPower, EON/Powergen, EDF, Scottish and Southern, and Scottish Power. Gradually, most of the market was arranged through spot bilateral contracts through a unique virtual hub, the National Balancing Point. Physical interconnection with continental Europe (the UK pipeline) has created a complex relationship between the short-term pricing mechanism in the UK, and the typically long-term contracting in the rest of EU. We shall go back to the specific features of the UK reform path in Chapter 6. In fact, this structure has a parallel in the USA, but is very different from the one prevailing in the rest of the world. It is interesting here to compare it with the history of the only other gas-rich EU Member State: the Netherlands, which is the other vantage point to understand the EU reforms.

2.5.2 *The Netherlands nationalizes the network*

Unlike the UK, that until relatively recently was not interconnected to continental Europe, the Netherlands exports about half of its gas production to France, Germany, Belgium, and Italy. This has happened through long-term contracts, and virtually no market opening for three decades. The incumbent was Gasunie, a company owned by the Dutch state (50%) and by Exxon-

Mobile and Shell (25% each). On 1 July 2005, Gasunie was formally split into two companies, a network company that continued to be known as Gasunie and a purchasing and sales company for natural gas, Gasunie Trade and Supply.[30] The Dutch state bought out Shell and Exxon-Mobile's holding in the network company, while the ownership of the purchasing and sales company remained unchanged (see Chapter 6 for additional details). Thus, in fact, the gas network is fully under the government's ownership.

The Dutch government from the beginning took a very prudent approach in the planning of the resource utilization. Part of the gas rent has to be set aside for the future depletion times and alternative energy policies. Under the continued strong control of the government on the industry, actual market opening was delayed. A gas regulator (DTE) was only set up in 1998, 12 years later than the UK. After more than a decade of formal compliance with EU directives, the gas sector is currently still largely controlled by the incumbent Gasunie and its spin-offs. The dominant position of the company is largely unchallenged due to its strong position in terms of production, while the *modus operandi* of the industry has changed gradually in the last decade, particularly because of a larger use of spot markets.

2.5.3 *France, a reluctant reformer*

In contrast to the UK, France was one of the latecomers in implementing the EU gas directives and in adopting the reform paradigm. The main texts governing activities related to gas are the 2003 and 2004 Energy Laws. These laws lifted the monopoly on import and export, introduced TPA, extended the powers of the French regulator in order to include the electricity and gas sector too, and ensured the transposition of the legal unbundling obligation applicable to the TSOs. Despite the adoption of the two directives and the legal separation implemented, the European Commission has repeatedly evaluated that the unbundling of gas industry in France remains insufficient to avoid uncompetitive behaviours. The market opening was, however, for 37% of users in 2003, 73% in 2007, and a switching rate in the industrial segment incomparable with that of the UK: virtually zero for households, and with a minor role for spot markets (even if increasing in the last few years). Gaz de France (GDF), now merged with Suez, was fully public owned until July 2005 when around 20% of the shares were sold by initial public offer. There are two TSOs in the country: Gaz de France Reseau Transport (GDF-RT) and Total Infrastructures Gaz de France (TIGF). They have been unbundled in legal terms only. In the retail segment there is limited competition: Gaz de France

[30] Now called Gas Terra.

has not separated its retail activities and dominates the market for industrial and household consumers, with some entry from other big European players more recently, see Chapter 6 for additional information.

2.5.4 *Germany: The strategic importer*

Germany offers a very different case history compared with the UK or the Netherlands. The country, with its huge demand for energy to sustain its leading manufacturing industry, arranged its gas industry under a group of five upstream private monopolies, in the wholesale trade, and a large number (around 700) of local, often municipally owned, retail organizations. The government's enthusiasm for liberalization was modest. The legal arrangements before the reforms were based on the Gesetz zur Forderung der Energie Wirtschaft (1935) and the Energie Sicherunggesetz (1974). For over 50 years competition was considered wasteful, prices were regulated, and long-term concessions created regional/local monopolies, under the control of the Anti-trust Authority (Bundeskartellamt). The Big Five, until recent years (when spot prices were largely below the long-term oil-indexed ones because of oversupply of gas) earned large rents, while their customers where happy to enjoy relative price stability. The country transposed the second European gas directive by the Energy Act of July 2005. In compliance with the provisions of the directives, the Act established a regulatory authority (the Federal Net Agency) with competences in the gas and electricity sector. This happened 20 years after the UK. Despite the official start of liberalization in 1998, only one gas retailer has a market share above 5% and the recorded switching rate of industrial customers and households is very limited (7% and about 2% respectively).[31] For a country critically dependent on imports, particularly from Russia, the core concern has traditionally been that of security of supply and stability of the system. In the UK, the regulation of the industry was instead deeply influenced by two features: domestic gas production and the close relationship of gas pricing with the electricity spot markets. In Germany, until the 2008 crisis that created a significant gap between oil-indexed gas prices and spot prices, the attraction of the British reform path was limited.

2.5.5 *A Mediterranean perspective*

Two Mediterranean countries have adopted a completely different approach to the development of the gas industry. In Italy, after 40 years of the vertically

[31] In December 2008 the German regulator forced several gas suppliers to reimburse consumers, following an allegation of unfair pricing.

integrated public monopoly on gas by ENI, the legislation had to change in compliance with the EU directives. The first step towards the liberalization of the gas sector was the approval of Law 481 of November 1995 establishing the Italian regulatory Authority for Electricity and Gas (AEEG). At the same time, the partial privatization of ENI was performed. The transposition of the two European gas directives was made by several other legislative measures, taken on different occasions. The most relevant was the decree No. 164/2000 that gave a strong impulse to competition, liberalizing the activities of importation, in compliance with EU Directive 30/98CE, exportation, transportation and dispatching, distribution, and trade of natural gas. The incumbent player is still ENI, partially owned by the government, and not unbundled (until 2012), as it concurrently controls most of the imports, the domestic, and some trans-boundary pipeline networks, through a controlled company (there are ongoing plans, from 2011, to divest the latter), and a share of the distribution. Since 2002 any individual company can control no more than 75% of production and import of gas. This share has to be reduced by 2% per year, until around 60% in 2010. Moreover no individual company can sell more than 50% of total supply to end consumers. The TSO should have complied with third party access from 2000. In practice the compliance is limited.

While there are hundreds of mostly municipally owned retail companies, interestingly the main competitor of ENI is the incumbent electricity company ENEL, also under majority control by the government, with the third player partly owned by EDF, the French state-owned incumbent, and A2A, a big partly municipally-owned company based in Lombardy. Thus, the Italian gas industry can be described as an oligopoly among mostly public enterprises, with a very limited use of spot market, and recently an increasing role of imports by maritime transport of liquefied natural gas,[32] following the Spanish example.

The Italian case contrasts significantly with the ownership and structural features of the Spanish gas industry, which before liberalization was dominated by one integrated private company, Gas Natural. The country imports its gas mainly via pipelines, largely from Algeria, but has decided to develop a diversification strategy based on LNG, transported by ships, and regasified through special plants. This has helped sustain the consumption of gas that has doubled in the last decade. LNG is usually traded through oil-indexed long term contracts. Hence, even if Spain has developed in recent years a wholesale gas market, this is still illiquid. The regulatory approach was relatively tough in limiting the dominance of Gas Natural in the last decade. The government forced some competition by allowing entrants to acquire gas from the

[32] LNG hereafter.

incumbent, which was forced to sell 25% of its long-term contracted gas to six new entrants to promote competition. After this process, the market share of the incumbent was reduced from virtual monopoly to less than 50%. Despite the fact that the gas retail market has been fully open since January 2003 the overall setting is far from being considered highly competitive. The switching rate since market opening is only about 5% and Gas Natural still strongly dominates the market for households with a share over 70%. Full market opening, including for domestic customers and regulated TPA and also for gas storages, has been effective since January 2003. Ownership unbundling, at least partially, of gas transmission system operators was implemented, as well as legal and accounting unbundling of distribution system operators. An energy regulator has existed since 1998 (CNE). In spite of this much more pro-active approach, as compared with France and Italy, the industry is still basically a private oligopoly.

We shall discuss again these six core EU countries for the gas industry in Chapter 6. We turn now to tracking the changes, as in previous sections, by the OECD regulatory indicators.

2.5.6 *An overview of gas reforms*

A brief history of gas sector reforms by country can be summarized by tracking four sets of indicators: ownership ('public' or 'private'), the market share of the largest firm operating in the production/import sector ('greater than 90%', 'between 50% and 90% or local monopolies', and 'no dominant player'), the degree of vertical integration, or unbundling ('no separation', 'accounting separation', and 'ownership separation'), and the level of market opening ('no TPA', 'negotiated TPA', and 'regulated TPA'). Figure 2.5 and Table 2.3 offer these trends at a glance.

In comparison with the other two sectors previously discussed, it is apparent that natural gas is still behind in the paradigm implementation schedule. Even so, a common trend among countries is clearly detectable. The starting point of the story is consistent for the EU-15 along the market opening and integration dimensions: in all countries (except Denmark, which is the outlier throughout the period of study) we see a vertical integrated industry with no TPA. Moreover, with the exception only of Germany, the industry was originally ruled by a monopolist, since the largest firm operating in the production/import sector had a market share of more than 90%. As in the case of electricity, there is wide heterogeneity with respect to ownership: eight countries had a government-owned incumbent, seven a private one. Governments still have some control of the players in Austria, the Netherlands, Italy, France, and elsewhere. If we seek to find a common path among the different countries, we can identify the following. At first, there was full or partial privatization for the

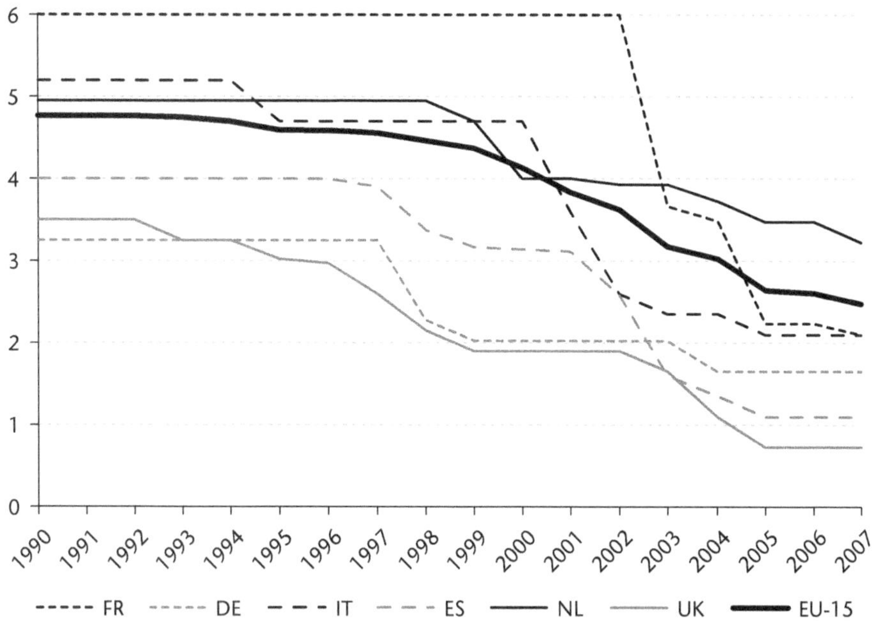

Figure 2.5. Gas. ECTR indicators, selected countries, 1990–2007
Source: own elaboration of ECTR scores.

countries starting with a public incumbent. Second came market opening. In some cases, as in Belgium, Ireland, Italy, Austria, the Netherlands, the United Kingdom, and Denmark, this was achieved in two steps, from no TPA to negotiated and then regulated TPA; for all the other countries it was done in just one step. In 2007, all the countries completed the process with regulated TPA, except Germany which stopped the reform at negotiated TPA.

Unbundling was the third step everywhere (except in Finland, Luxembourg, Ireland, and the Netherlands where the industry remains vertically integrated).

All this should have led to a decrease of the market share of the incumbent. Belgium, France, Italy, and the Netherlands started with an industry dominated by a monopolist, and moved to an industry where the market share of the incumbent is between 50% and 90% or where there are local monopolies. The UK, which already started with a more competitive industry, ended up with no dominant player. Denmark moved directly from monopoly to a competitive market.

Table 2.3. Gas regulatory changes. EU-15, 1975–2007

Access	Competition	Owner-ship	75–85	86–87	88–92	93–94	95–97	98	99	00	01	02	03	04	05	06	07
NO UNBUNDLING																	
No TPA	Monopoly	Public	BE DK FR GR IE IT PT UK	BE DK FR GR IE IT PT	BE DK FR GR IE IT PT	BE DK FR GR IE IT PT	DK FR GR PT	DK FR GR PT	DK FR GR PT	DK FR GR	FR	FR					
		Private	AT FI LU NL ES SE	AT FI LU NL ES SE UK	AT FI LU NL ES SE	AT FI LU NL ES SE	AT BE IT FI LU NL ES SE	AT BE IT FI LU NL SE	AT BE IT FI LU	FI IT LU PT	FI PT	FI PT	FI	FI			
	Some Competition	Private	DE	DE	DE	DE	DE										
Negotiated $_{(N)}$ or Regulated $_{(R)}$ TPA	Monopoly	Public					$IE_{(N)}$	$IE_{(N)}$	$IE_{(N)}$	$IE_{(N)}$	$IE_{(N)}$						
		Private			$UK_{(N)}$	$UK_{(R)}$			$NL_{(N)}$ $SE_{(R)}$	$BE_{(N)}$	$IT_{(N)}$	$IT_{(R)}$			$FI_{(R)}$	$FI_{(R)}$	$FI_{(R)}$
	Some competition	Private							DE	DE	DE	DE	DE				
LEGAL SEPARATION																	
No TPA	Monopoly	Private							ES	ES	ES	ES		PT			
Negotiated $_{(N)}$ or Regulated $_{(R)}$ TPA	Monopoly	Public											$DK_{(N)}$				
		Private					$UK_{(R)}$	$UK_{(R)}$	$UK_{(R)}$	$UK_{(R)}$	$UK_{(R)}$	$BE_{(R)}$ $ES_{(R)}$ $UK_{(R)}$	$BE_{(R)}$ $IT_{(R)}$ $NL_{(R)}$ $ES_{(R)}$ $UK_{(R)}$				

ACCOUNTING OR LEGAL SEPARATION

Negotiated $_{(N)}$ or Regulated $_{(R)}$ TPA	Monopoly	Private	AT$_{(N)}$	AT$_{(N)}$	AT$_{(R)}$	AT$_{(R)}$				
ACCOUNTING SEPARATION										
No TPA	Monopoly	Public		GR	GR	GR	GR	GR		
		Private					PT	PT	PT	
Negotiated $_{(N)}$ or Regulated $_{(R)}$ TPA	Monopoly	Public		DK$_{(N)}$ IE$_{(R)}$	DK$_{(N)}$	IE$_{(R)}$	IE$_{(R)}$	IE$_{(R)}$	IE$_{(R)}$ GR$_{(R)}$	IE$_{(R)}$ GR$_{(R)}$
		Private	NL$_{(N)}$ SE$_{(R)}$	BE$_{(N)}$ LU$_{(R)}$ NL$_{(N)}$ SE$_{(R)}$	LU$_{(R)}$ NL$_{(N)}$ SE$_{(R)}$	LU$_{(R)}$ SE$_{(R)}$	AT$_{(R)}$ BE$_{(R)}$ DK$_{(N)}$ IT$_{(R)}$ LU$_{(R)}$ NL$_{(N)}$ ES$_{(R)}$ SE$_{(R)}$ UK$_{(R)}$	BE$_{(R)}$ DK$_{(R)}$ IT$_{(R)}$ LU$_{(R)}$ NL$_{(R)}$ UK$_{(R)}$	BE$_{(R)}$ DK$_{(R)}$ IT$_{(R)}$ LU$_{(R)}$ NL$_{(R)}$ UK$_{(R)}$	BE$_{(R)}$ DK$_{(R)}$ IT$_{(R)}$ LU$_{(R)}$ NL$_{(R)}$ PT$_{(R)}$ UK$_{(R)}$
	Some competition	Private					DE$_{(N)}$	AT$_{(R)}$ FR$_{(R)}$ DE$_{(N)}$ ES$_{(R)}$ SE$_{(R)}$	AT$_{(R)}$ FR$_{(R)}$ DE$_{(N)}$ ES$_{(R)}$ SE$_{(R)}$	AT$_{(R)}$ FR$_{(R)}$ DE$_{(N)}$ ES$_{(R)}$ SE$_{(R)}$
OWNERSHIP SEPARATION										
Negotiated $_{(N)}$ or Regulated $_{(R)}$ TPA	Some competition	Public				FR$_{(R)}$	FR$_{(R)}$			

Source: own elaboration of OEDC data. Shaded area: public or mostly public ownership.

2.6 Price trends and correlation with reform indicators

While EU institutions and national governments changed the regulatory environment and the organization of the industries, service prices for households changed as well. Is there some evidence that the reforms had an impact on consumer prices?[33] Clearly, prices of telephony, electricity, and gas are influenced by many factors, and have their own dynamics. But is there any symptom of correlation, if not of causality, going from the reforms to prices? We introduce this question by some simple examples.

Figures 2.6a, 2.6b, and 2.6c show for selected countries nominal, net-of-tax household price trends in Euros, between years 1998 and 2007, from EUROSTAT. For fixed telephony the price we consider here is for a 10 minute local call, for electricity the price for a household with a yearly consumption of 3500 KWh, for gas the price for a household with a yearly consumption of 83.7 GJ.[34]

As for telephony, the price of a local call in the UK has decreased from around 0.50 Euros for ten minutes of connection in 1998 to around 0.40 in 2007. In contrast, the price was lower in 1998 and also slightly falling in other major EU countries such as France, Germany, Italy, and Spain. In all these countries it is now lower than in the UK. The next chapter shows the details of price trends for other telephony services, and uses additional data sources.

As for electricity, the price of the KWh in the last 17 years has oscillated in the UK between a minimum of slightly more than 0.08 and a maximum of 0.12 Euros. In France the range has been between 0.09 and 0.10, in Germany between 0.12 and 0.14, and in Italy between 0.15 and 0.20.

The price of gas in the two producing EU countries, the UK and the Netherlands, has oscillated between respectively 5.5 and 11 Euros in the UK, and 5.9 and 12 Euros in the Netherlands. In Germany, the range has been 6 and 14 Euros, in Italy between 7 and 12 Euros.

Inspection of Figures 2.7a–c suggests that there is no clear evidence of simple correlation between price levels and reform indicators in 2007. The aggregate ECTR indicator (see Section 2.2 and Figures 2.1a–c) goes from six ('unreformed' country) to zero ('fully reformed'), hence a negative correlation between prices and reform would imply that most countries lie on a northeast–southwest zone, while in fact half of the countries lie in the northwest–southeast zone (high relative prices and intensive reforms).

[33] Given our research focus, we do not report changes of prices for the business sector, which were also possibly influenced by the reforms. It is likely that some businesses were able to get lower prices from the liberalized market in several countries, particularly major companies with some bargaining power. Establishing how much of these cost savings have then been transmitted to households through price mechanisms is largely beyond this study.

[34] Consumption thresholds are determined by statistical conventions.

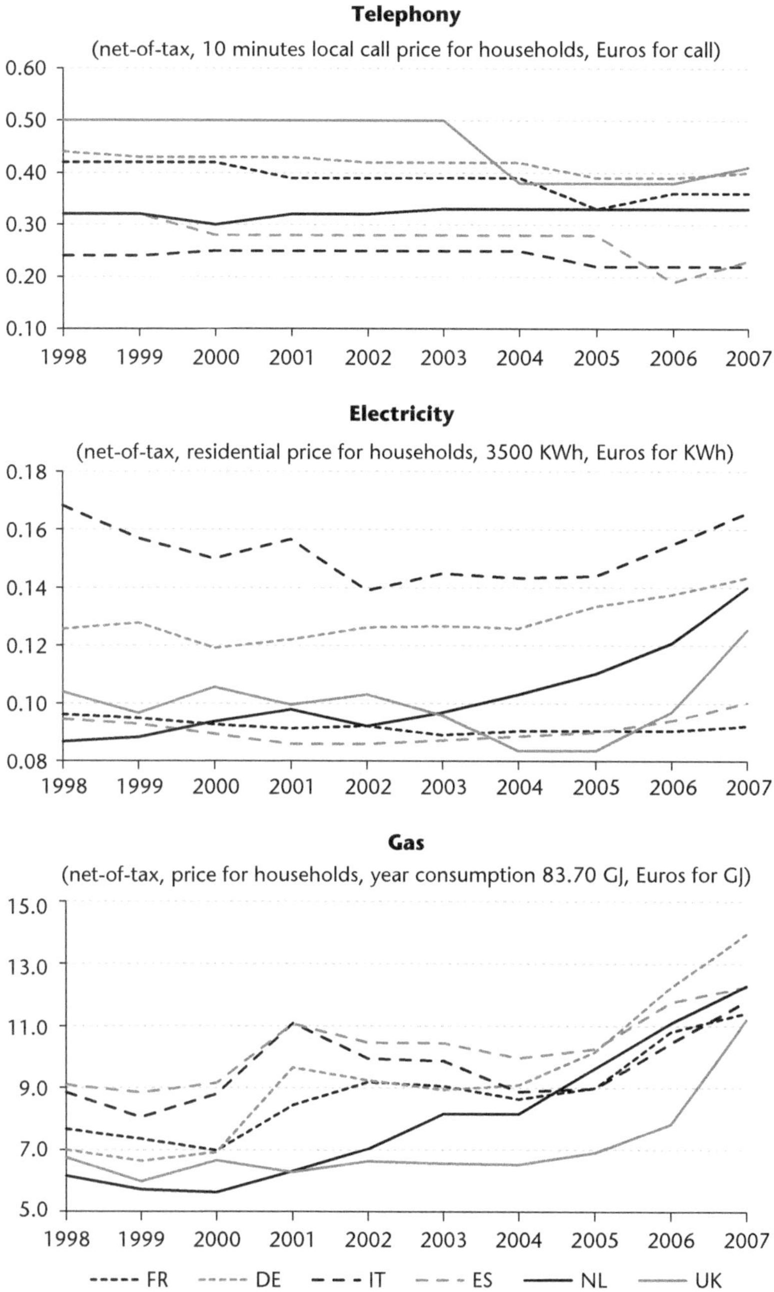

Figure 2.6. Price trends, 1998–2007

Source: Eurostat.

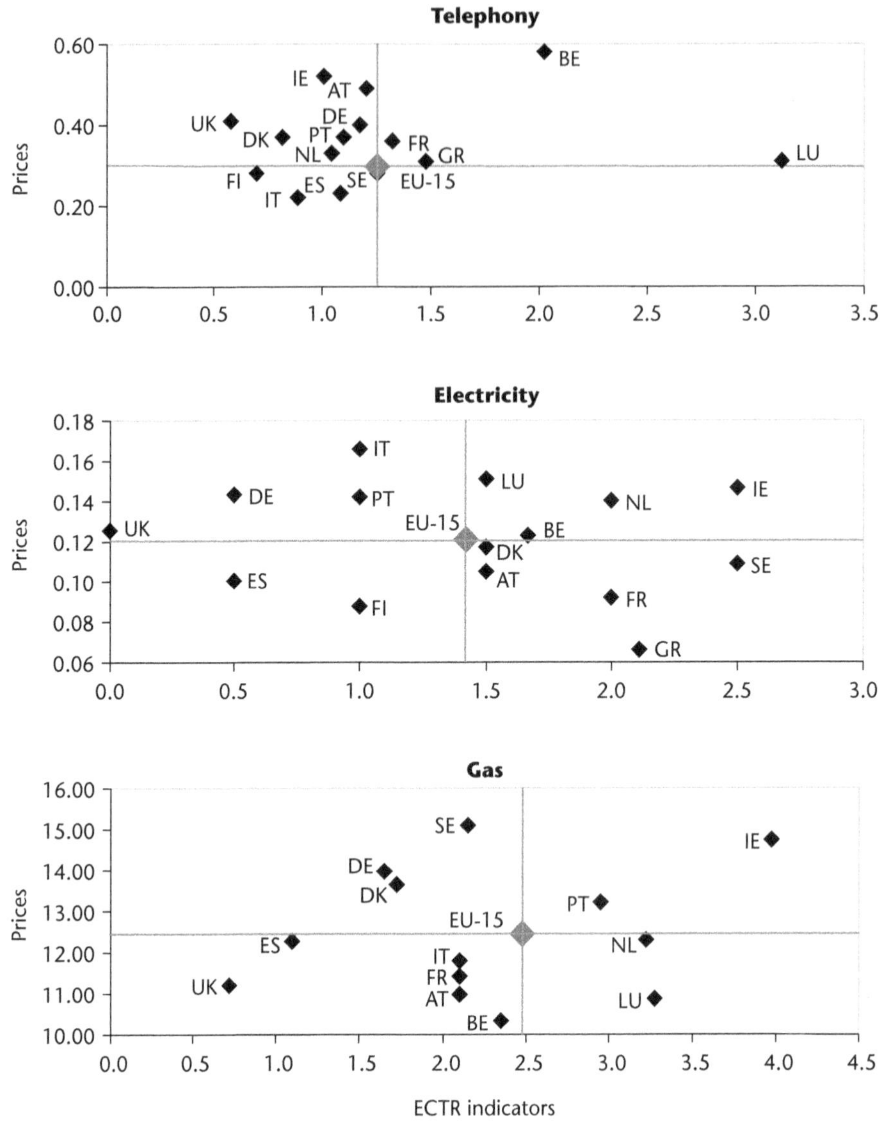

Figure 2.7. Correlation between price and reform indicators in the EU-15, 2007

Source: own elaboration of prices and ECTR score. All prices are net-of-tax. Telephony prices refers to 10 minute local calls, Euros per call; electricity prices refer to residential, price for household's consumer, 3500KWh/Euro; gas prices are household's at 83.70 GJ/Euro. The ECTR indicator is the aggregate reform score. For Greece and Finland gas prices are not available. Average prices and reform indicators are shown respectively on the horizontal and vertical lines.

2.7 Consumer satisfaction

Finally, it is interesting to see to what extent consumers are happy with the prices they pay, a theme that we shall discuss in greater detail in Chapter 7.

Figure 2.8 shows the proportion of respondents to the Eurobarometer survey that state that they are dissatisfied or fairly dissatisfied with the service prices they pay. While between 2000 and 2006 this proportion decreases considerably, there are interesting variations across sectors and consumers and countries where dissatisfaction for telephony services is higher include Italy, Finland, France, Portugal, and Spain. Dissatisfaction with electricity prices is higher than the EU-15 average in Germany, Finland, France, Portugal, and Sweden. As for the sector of natural gas, dissatisfaction is higher in Spain, France, and Germany. How these findings correlate with the regulatory and industry natures of the countries will be discussed in Part Three of the book, but again simple visual inspection, and comparison with Figures 2.7a–c, show high dissatisfaction in both reformed and unreformed countries. Perceptions of quality will be discussed in Chapter 8.

2.8 Summing up

In conclusion, across the EU-15 countries, both in the energy and telecom services, (as in other network industries mentioned in Section 2.2), there is wide variability of reform patterns, industry organization, and performance, in a double perspective. The starting points were different, and after two decades convergence towards the new paradigm still varies across industries and countries. Interestingly among the services that show the greatest and the lowest cross-country convergence, these are respectively telecommunication and post, which were often under the same departmental control in the 1970s. It is this variability along a common policy trend that we will exploit in the next chapters.

Energy and telecommunications providers, unlike other network industries, were or still are almost always and everywhere profitable under public ownership. This implies that, 'on average' (i.e. at an aggregate corporate or departmental level), the government-owned entities that provided electricity, gas, and telephony in the EU before privatization more than covered their costs, hence their budgets were usually independent from transfers by the central or local government. In principle, their investments could have been financed by capital markets, either through the issuing of bonds or through raising

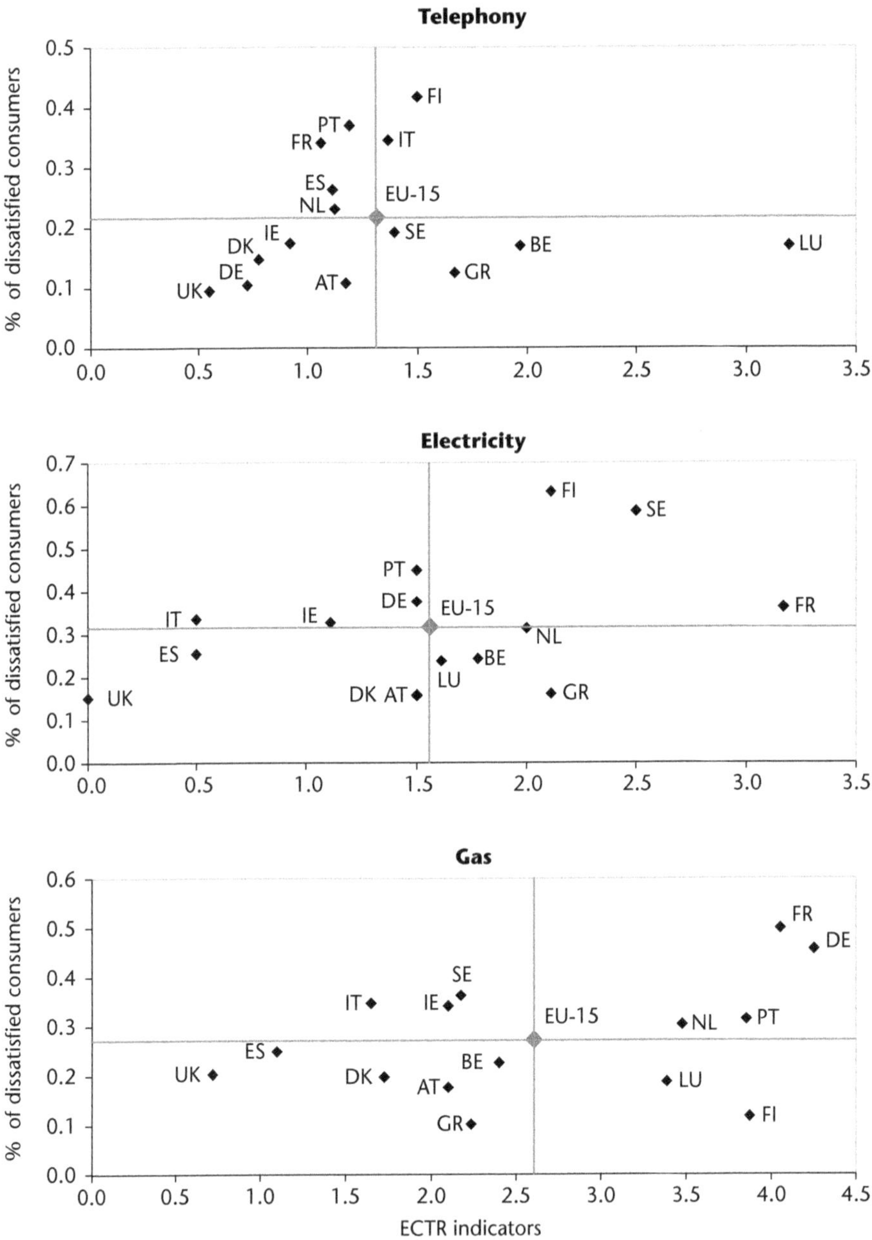

Figure 2.8. Correlation between the proportion of dissatisfied consumers and reform indicators, EU-15, 2007

Source: Eurobarometer surveys and ECTR sector indicators.

matching equity capital, while still remaining under the control of the government. This makes the need to study the impact of the reforms for taxpayers less compelling.[35]

This either usually was, or still is not, the case, however, for local transport, rail, and in some countries post and water services, even with some differences between countries. The performance of postal and water services has broadly improved in more recent years. Some public transport activities are now able to cover their costs because of tariff increases and restructuring. In the 1980s and 1990s, however, there was often a clear divide. Some services needed public subsidies, others offered a 'dividend' to the public budget. Hence, the privatization and liberalization of energy and telecoms cannot be fundamentally seen as a way to solve industry-specific financial problems, even if the privatization proceeds have been utilized by governments to solve their cash needs elsewhere. Thus, it seems reasonable to study the impact of privatization and market opening in those industries where prices were less distorted in the first place. This raises the question of what determines the prices of network services. We have seen in this chapter some trends, and simple correlations with reform indicators, but we need to explore these relationships in more depth.

2.9 Further reading

The literature on regulatory reform in the EU is very extensive. Any suggestion for further reading needs to be very selective, and focus on some specific issues. Finger and Finon (2011) offer a good introductory synthesis of liberalization of network industries in the EU, and mention the core directives. Clifton et al. (2007) and Roland (2008) survey privatization in the EU and elsewhere. Helm (2007), Glachant and Lévêque (2009), Cameron (2007), Kiesling (2009), and Mazer (2007) offer analyses of reform of energy markets. More general issues of network economics, deregulation, and competition, including of telecommunications, are discussed by Shy (2001), ten Heuvelhof et al. (2009), Cameron (2007), Megginson (2005), Fichert et al. (2007), Künneke et al. (2019), Gottinger (2003), Gomez-Ibanez (2006), Viscusi et al. (2005), Spulber and Yoo (2009), Dewenter and Haucap (2007), Henri and

[35] Taxpayers may have been damaged by undervaluation of assets at divestiture (as extensively discussed by Florio, 2004, for the UK), but may have benefited from increasing tax revenue from corporate profits. As the two effects would go in opposite directions, this component of social welfare change can be neglected for the purpose of this study, but would be interesting for further research.

Matheu (2001), Robinson (2007), and MacAvoy (2007). Martin et al. (2005) discuss reform of several network industries in the EU-15.

On the emergence of EU agencies and networks of regulators, see Coen and Thatcher (2008) and Lavrijssen and Hancher (2009), the European position in COM (2008) 'European agencies: the way forward', and the discussion about ACER, the agency for cooperation of energy regulators. It is not clear yet how far these new institutions will speed up convergence of regulatory arrangements in Europe.

Some of the cross-country regulatory databases include: EURI (European Union Regulatory Institutions Database) on telecommunications, 1997–2003, covering the EU-15, used by Edwards and Waverman (2006); Megginson (2005) Appendix 1 on privatizations 1961–2002; on this topic see also FEEM's Privatization Barometer,[36] the PPI (The Private Participation in Infrastructure) database, sponsored by the World Bank, covers 150 countries and many sectors; see also Estache and Goicoechea (2005). On telecom regulation in developing countries see Wallesten et al. (2004), including 178 variables; some databases are based on occasional surveys of national regulators, for example the Electricity Regulations Database.[37]

The OECD and other regulatory databases do not usually cover some of the important sectors that can also be considered network industries. Water reforms are overviewed by Menard and Peroo (2011). They observe that full liberalization in the UK is still an exception, but that private participation is a dominant trend in several EU countries (and elsewhere). Gassner et al. (2009) offer an extensive international survey (over 1200 utilities in more than 70 countries). The sector is characterized by clear economies of scale and more recently of scope (well insulated cables can now run alongside water pipes), a proportion of fixed costs that far exceeds that of other network industries, high economies of density, decentralization of management because of the high transportation cost of water over long distance, health and other externalities in the urban context, and very limited substitution. Massarutto et al. (2007) also discuss the need for quality standards. Menard and Peeroo (2011) compare three reform models in drinkable water service: England and Wales, France, and Germany. They point to widespread 'opportunism' as a source of slow pace of reform and mixed results after reform. For an overview of other sectors see Finger and Künneke (2011), particularly chapters on radio spectrum management (ch. 8) and the Internet (ch. 9) as a complement to the focus in this book on telephony; on the oil industry (ch. 12) for a comparison with

[36] http://www.privatizationbarometer.net/ [accessed 5 January 2012].

[37] http://econ.worldbank.org/WBSITE/EXTERNAL/EXTDEC/EXTRESEARCH/0,,contentMDK:20699165~pagePK:64214825~piPK:64214943~theSitePK:469382,00.html [accessed 5 January 2013].

natural gas; and the chapters on liberalization of different transport services, not covered in this book. See also the footnotes of Section 2.2 on the rail, airways, and postal service. Other transport services affected by reforms include ports, roads, and local public transport, also covered by Finger and Künneke (2011).

3

Welfare effects: A simple thought experiment

3.1 Introduction[1]

After having briefly reviewed some of the reforms of network industries in Europe, and before turning to a more detailed empirical analysis, this chapter offers a simple discussion of prices and consumer surplus changes under different regulatory regimes of the network industries. Using consumer surplus is a shortcut for the measurement of household welfare, but the loss of generality compared with other metrics is usually limited, and there is the advantage of a widely used concept in policy evaluation.

While there is a very wide literature on the regulatory reforms of network industries, a direct comparison of the welfare effects is often difficult to infer. What would be needed is an illustrative framework which allows direct comparison among different states of the world. The story starts with a relatively cost-inefficient, vertically integrated public monopoly and the virtual welfare loss when compared with the first-best is measured. Then unregulated privatization is introduced and the trade-off between cost and allocative inefficiencies is compared. Third, a price-cap regulation is imposed, and we show how stringent the cap should be to recover the allocative inefficiency of private monopoly.

Vertical disintegration is a further step that increases per se transaction costs. Then, a modest liberalization step in the form of a licence given to a second player is considered, and the consumer surplus under the duopoly is found. Finally, we allow for full liberalization, with associated sale costs and competition benefits. In spite of an extremely simple microeconomic framework, the indifference conditions between each reform step and the status quo are non-linear in some key parameters and there is no clear welfare

[1] This chapter (except Sections 3.9, 3.10, and 3.11) draws heavily on Ceriani and Florio (2011).

dominance of one unique reform combination over the others. The results are discussed in the concluding section.

Before going into detail, we start with a graphical representation of our story.

3.2 First sketch of the story

To give some idea of what we have in mind, even if we need to simplify a great deal, let us observe price changes in a simple thought experiment.

As mentioned, we consider, first, a vertically integrated public monopoly; second, a privatized, but still vertically integrated monopoly, without any (binding) price regulation (except for the constraint of no price discrimination across users); third, a price-cap regulation of the privatized incumbent monopoly; fourth, unbundling as a preliminary step to liberalization; fifth, a duopoly market, without price regulation; sixth, full market entry, as an extension of the previous case. One may interpret the different steps either as variations across countries, or across industries, or over time.

Let us assume that the network component is under decreasing average costs and the marginal cost is zero. In contrast, the operation component has constant average costs equal to marginal costs. In a given initial year of our story, the network is a sunk cost: capital has been invested in past times and financed by the taxpayer, who will continue to finance some maintenance costs. Thus, the public monopolist needs only to cover its variable costs. The government wants to maximize the consumer surplus, but also has a private agenda. The public enterprise, in this simple example, implements tariffs equal to the marginal cost of the operations (other assumptions are possible, but we will not explore them here)[2]. While, by construction, there is no allocative inefficiency, the public provider is, however, possibly affected by x-inefficiencies because of rents distributed to managers, workers and politicians. Thus, costs are not minimized, and unit cost c is greater than c^*, a minimum long-term marginal cost (we are not discussing here the complications that arise in the long run because of technology shifts). The cost under public monopoly is given by $c_1 = c^*(1 + \alpha)$, where α stands for inefficiency (this can be zero with a benevolent government and symmetric information). Hence, the price for the network service is simply $p_1 = c^*(1 + \alpha)$, see Figure 3.1a.

Consumer unhappiness (perhaps recorded by surveys such as Eurobarometer, (see Chapters 2, 7, and 8) leads the (elected) government to privatize the industry. The private owner has an incentive to minimize costs. At this early

[2] This assumption implies that there is productive inefficiency, but not allocative inefficiency under public ownership. More complex frames can consider both.

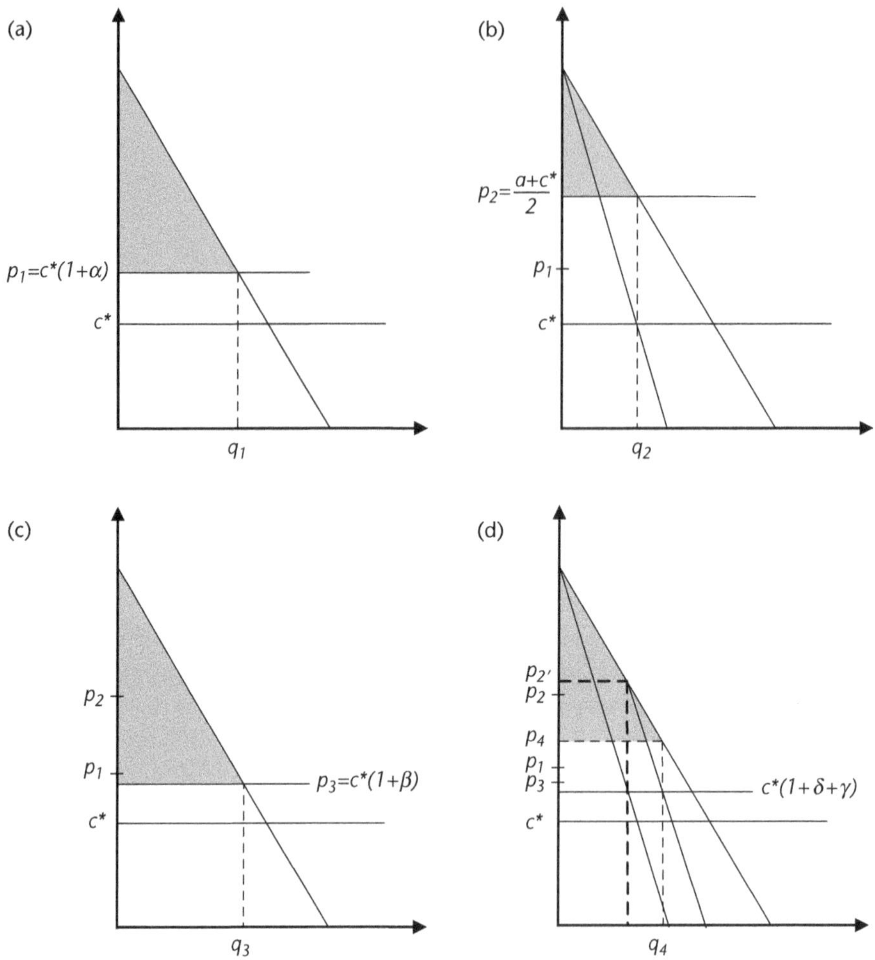

Figure 3.1. Price equilibria under different regulatory regimes

(a) Public monopoly constrained to pricing (p_1) at marginal cost, with some *x*-inefficiency relative to minimum cost c^*

(b) Unregulated private monopoly: price (p_2) increases in this example relative to p_1, but the ordering can be reversed if α is sufficiently high

(c) Incentive price regulation can offer price (p_3) lower than under either public or private vertically integrated monopoly, in this example, but only if the regulator is strong enough and β is low

(d) Market opening to a Chamberlin duopolist after unbundling of the network and lifting the price-cap increases operating costs but forces the incumbent to lower its optimal price p_2' to p_4: in this example $p_2 > p_4 > p_1 > p_3$, but other orderings are possible.

stage of the reform the government is unable or unwilling to effectively regulate prices (e.g. to maximize privatization revenues to the Treasury, or because of lack of experience of the regulator). Thus the industry is now under a vertically integrated private monopoly. The new equilibrium price (see Figure 3.1b) is now p_2. The consumer surplus (and perhaps the perceived fairness of the price) will change accordingly.

Popular dissatisfaction with the first step of the reform pushes the government to appoint a (tougher) regulator with the power to impose a price-cap to the private, vertically integrated incumbent. The regulator imperfectly knows the firm-specific technological constraints and the managerial effort of the regulated firm, and he picks up a price constraint sufficient to offer the private investor an 'adequate' profitability (participation constraint) with some slack. This can be represented by a simple mark-up $(1 + \beta)$ over the first best marginal cost, which is compatible with participation. The new price is $p_3 = c^*(1 + \beta)$. Obviously, $p_1 > p_3$ only if $\alpha > \beta$ Figure 3.1c, where $p_2 > p_1 > p_3$, shows one possible outcome.

The regulator then imposes on the incumbent the separation of the network (examples are electricity in the UK since 1990, Sweden since 1996, or Finland and Portugal since 2000, and fixed telephony in the UK since 2006, etc. see the previous chapter). This new reform is best seen as preparation for granting access to entrants. The simplest case is the establishment of a duopoly under private ownership and the separation of the network. The regulator lifts the price-cap for the operators. There are now some coordination and transaction costs (allocation of time and capacity slots), under unbundling, c^* is no longer achievable, but only $c^*(1 + \gamma)$. Moreover, there is now an additional cost to the entrant because, for instance, he has to advertise his entry. Let us say that δ is this extra cost. We assume that there is some market inertia. Under the Chamberlin (1948) 'small group' competition model, the entrant targets just the demand not satisfied by the incumbent at the original price. Thus the entrant fixes its optimal quantity and price supply of the service based on the residual segment of the demand curve. The optimal price for the entrant is clearly less than for the incumbent's segment (technology is common), and the incumbent must follow the entrant and fix the price at the new lower level. Figure 3.1d shows the new situation, where $p_4 > p_3$.

If γ (loss of economies of integration, related to incomplete contracts and asymmetric information) and δ (duplication of some sales and administration costs) are sufficiently high and α or β are sufficiently small, then the duopoly equilibrium price, p_4, can be higher than each of the other three market structures.[3]

[3] The intuition is clear. EDF, the public monopolist in France, offered electricity prices 'near' to the first-best long run marginal costs (nuclear), c_N^*. To allow entry for a competitor from Germany

The government in our virtual story learns this lesson, and takes further steps to lift any market entry regulation. For example, since July 2007 all EU households have been able to choose a different electricity supplier and switch their contract. To study the impact of full market liberalization on prices, again, we want to be as simple as possible. After having unbundled the network, the regulator is now ready to offer a licence to any new entrant. Let n be the number of electricity or gas providers. There still is no price regulation, and the network service costs are covered by the taxpayer. For the generic *n-th* entrant, profit maximization under Cournot (1897) competition implies $p_5 = c_n/[1-(\sigma_n/|\varepsilon|)]$, where c_n is the marginal cost for the *n-th* firm, p_5 is the price when market entry is free, ε is the price demand elasticity of the industry itself around the equilibrium, and $\sigma_n = q_n/q$, that is the share of the demand covered by the entrant (e.g. Varian, 1987, pp. 452–53). The smaller the entrant's market share, the more elastic the firm's demand, whose elasticity is $(|\varepsilon|/\sigma_n)$. As the market share tends to zero, the condition for profit maximization is the one associated with pure competition. The firm's cost, c_n, however, is again not equal to first-best, because of the inefficiency of unbundling (ϕ now replaces $\delta + \gamma$ of the duopoly example).[4] Having five gas suppliers competing in the same town to capture a share of the demand, while using the same network, accessing the same technology, and in fact probably importing the same natural gas, has two partially offsetting effects. While competition gradually reduces market power, and hence prices, each firm has to face somewhat increased sales and administration costs compared with our benchmark c*. Assume that $c_n = c^*[1+\phi(n)]$, where $\phi(n)$ is a function linearly[5] increasing with the number of competitors. For example, a regional gas supply oligopoly with five competitors, and demand elasticity in the equilibrium point $|0.8|$, and $\phi(5) = 0.05$ (meaning that each entrant has a 5% extra-cost to secure itself a 20% market share), implies $p_5 = c^*(1.05/0.75) = 1.40c^*$. This outcome implies that the vertically integrated public monopoly must have been affected by $\alpha > 40\%$, that is significant *x*-inefficiency, to deliver higher prices than the unregulated oligopoly. Alternatively, under a price regulated monopoly the extra profits permitted should be based on $\beta > 40\%$, that is a very lenient or

(possibly gas powered, $c_N^* < c_G^*$), both the incumbent and the entrant have to incur interconnection network-related transaction costs (currently there is an auction system), plus some output related costs such as advertising. In fact, Percebois (2008) shows that the consumer surplus change can be negative in France if the equilibrium tariff in certain hours is set at a higher marginal cost in Germany, while the producer surplus of EDF will increase. The opposite happens in Germany. If we disregard producer surpluses (which here are duopoly profits), the two combined changes in consumer surpluses in France and Germany do not need to cancel out. Thus, partial market integration may create some losers among the consumers, see Section 3.9.7.

[4] It is worth noting that if the network owner is private or if it is a budget-constrained public firm that applies the rule 'price = average cost' the inefficiency of network unbundling would be even worse, because zero access price is optimal.

[5] Non-linear monotonic functions can be considered without qualitatively altering the results.

uninformed regulator, to get higher consumer prices. With ten competitors $\alpha = \beta = 0.20$.

In fact, the outcome of market liberalization in the network industries in Europe seems to be oligopoly, with relatively few competitors in (until now) poorly interconnected regional markets.[6]

Thus, even in our very simple story of the reform sequence, the prediction of prices under different regulatory regimes depends on a set of (possibly country/industry specific) parameters. Welfare dominance of one regime over the other is not warranted. Even 'full' market opening can deliver higher prices than a regulated monopoly, if entrants are few, elasticity of demand low, and unbundling and related costs high. Our simple thought experiment shows in a precise sense why the evaluation of the reform outcomes of network industries is essentially an empirical matter: it depends upon a set of parameters that easily lead to non-linear outcomes along the reform 'line'.

3.3 An illustrative model[7]

What follows is a slightly more detailed presentation of the story focusing on consumer surplus (i.e. the shaded areas in Figure 3.1). As mentioned in the previous section, the starting point of our exercise is a market where there is a public monopolist. We assume that the public monopolist faces a linear market demand curve:

$$p^D = a - bq \tag{3.1}$$

where $a, b \in \mathfrak{R}^+$are parameters, p $\in$ are prices and $q \in \mathfrak{R}^+$quantities. Without loss of generality, we set $b = 1$. The public monopolist's costs can be expressed by the following separable cost function:

$$c^{Pu} = vq + f^* \tag{3.2}$$

where $v \in \mathfrak{R}^+$, vq stands for *variable* component of the cost function and $f^* \in \mathfrak{R}^+$ for the *fixed* component (we again assume that the network already exists, that is we assume that the investment is a sunk cost, while f^* represents fixed operation and maintenance costs, *including depreciation and cost of capital).

[6] Taking again the example of the electricity market, the combined market share of the two major competitors in 2006 was 68% in Germany, 75% in Spain, and over 80% in Belgium and France, with only the UK having established a more competitive market, which was, however, not interconnected with continental Europe.

[7] This section requires some basic microeconomics and can be skipped.

We will assume that the *variable* costs are linked to the service's provision and administration while the fixed costs belong to the natural monopoly part of the industry (generation, network maintenance).

Therefore, the natural monopoly's component of the industry is characterized by increasing returns to scale (decreasing average costs $\bar{c}^N = f^*/q$) and marginal costs equal zero. On the other hand, the part of the network industry represented by the operator has constant average costs equal to the marginal cost ($\bar{c}^{Op} = c'^{Op} = v$).

Moreover, we assume that the public monopolist's costs are higher than a first-best economy benchmark c^* (that we assume can be inferred by observation elsewhere or by engineering data) because of some x-inefficiencies, in such a way that the public monopolist cost function lies above the efficient one:

$$c^{Pu} = v^*(1 + \alpha)q + f^* \tag{3.3}$$

where $v^* \in \mathfrak{R}^+$is the efficient level of variable costs and $\alpha \in \mathfrak{R}^+$ is a parameter which captures both unintended inefficiency because of a lack of incentives to minimize costs by the managers, and rent extraction by policy-makers and other stake-holders who actually control the firm (perhaps the trade unions).[8]

After considering this social cost of public provision, the managers are instructed to maximize consumer surplus, under strict budgetary constraint. Hence, the production takes place where price equals public monopolist's marginal costs ($c'^{Pu} = v$). Hence, by substitution, where market demand[9] equals public monopolist's marginal costs:

$$q^{Pu} = a - v^*(1 + \alpha) \tag{3.4}$$

Note that, by setting the price equal to the marginal cost, the public monopolist faces a loss equal to the fixed costs f^*. Hence, the State has to subsidize the public monopolist in order for the firm to remain economically viable. We assume here that the excess burden of non-lump-sum taxation for the consumer in order to cover the network's fixed cost is not greater than the social inefficiency of any second-best pricing formula, such as Ramsey pricing or average cost pricing. Hence, we neglect this social cost, because it does not change the results.

The welfare loss of the public monopoly inefficiency is defined as the difference between the first-best consumer surplus (CS^*) and the public monopoly consumer surplus CS^{Pu}:

[8] We restrict the case to α being positive.

[9] We use the Marshallian demand function for the calculation of the consumer surplus because, following Willig (1976) we assume that the error implied by an uncompensated demand function is small.

$$CS^* - CS^{Pu}\left[\frac{(a-v^*)^2}{2}\right] - \left[\frac{(a-v^*(1+\alpha))^2}{2}\right] = v^*\alpha\left[\frac{v^*(2+\alpha)-2a}{2}\right] \quad (3.5)$$

which is obviously null for $\alpha = 0$.

It is interesting to note that the x-inefficiency acts as an indirect tax that creates an excess burden to the consumers, even when the rents by other stake-holders are distributed and increase their welfare.

For the following, as in Bradburd (1995) and Willner (1996), it is convenient to reformulate all the indifference conditions in terms of demand elasticity, rather than using the intercept of the linearized demand function (a). Let η be the absolute value of the demand elasticity:

$$\eta = \left|\frac{\partial q}{\partial p}\right|\frac{p}{q} = \frac{v^*(1+\alpha)}{a - v^*(1+\alpha)} \quad (3.6)$$

Note that we can rewrite the intercept a as:

$$a = \frac{v^*(1+\alpha)(1+\eta)}{\eta} \quad (3.7)$$

3.4 Privatization

As a first reform step, the government privatizes the public monopolist, without, however, imposing a price-cap regulation. The only regulatory rule is that price discrimination is not possible and quality is constant. Since the private owner has an incentive to minimize cost, we assume now that the management is instructed to wipe away any inefficiency and to maximize profits. Let us assume that the privatized monopolist is now able to achieve first-best costs $c^* = v^*q + f^*$. At the optimal choice of output we have marginal revenues equals marginal costs. Given the linear demand described above, the marginal revenue function is:

$$MR(p^{UPr}) = a - 2bq \quad (3.8)$$

The optimal output is:

$$q^{UPr} = \frac{a - v^*}{2b} \quad (3.9)$$

The unregulated private monopolist charges the maximum price it can get at this level of output:

$$p^{UPr} = \frac{a + v^*}{2} \quad (3.10)$$

The consumer surplus is now the area between the demand function and the price charged by the firm:

$$CS^{UPr} = \frac{(a - v^*)^2}{8b} \tag{3.11}$$

Therefore, the variation in consumer surplus obtained by privatizing the industry is:

$$CS^{Pu} - CS^{UPr} = \frac{4(a - v^*(1 + \alpha))^2 - (a - v^*)}{8b} \tag{3.12}$$

which is positive for values of α smaller than $\frac{a-v^*}{2v^*}$. In terms of demand elasticity η, using equation (3.7), the condition for privatization to increase the consumer surplus becomes:[10]

$$\alpha \geq \frac{1}{\eta - 1} \tag{3.13}$$

Hence, if α is small enough, an inefficient public monopolist will deliver a higher consumer surplus than an efficient private unregulated monopolist. Note that we could reformulate this result (and also the following results in this chapter) in terms of the cost reduction produced by privatization, instead of the terms of the cost rise being linked to the public monopolist, as in Willner (1996). Let μ be this cost reduction so that $v^* = v(1 - \mu)$. Therefore, the condition for privatization to be welfare improving becomes

$$\mu \geq \frac{1}{\eta} \tag{3.14}$$

Therefore, assuming, for example, the price elasticity $\eta = 2$, the public monopolist's inefficiency parameter α must be greater than 1, or, using the alternative formulation, μ must be greater than 0.5 for a welfare improving privatization without price-cap regulation. In other words, the privatization of very inefficient public monopolies will be beneficial to the consumer even without strong price regulation. A moderate or modest inefficiency of the public enterprise, however, does not justify this step per se, because the allocative inefficiency will be greater than the effect of the cost decrease. Therefore, in order to rank the two different markets based on consumers' surplus evaluation, we need to be able to compare the *x*-inefficiency of the public monopolist with the allocative inefficiency of the private one. Note, moreover, that price elasticities for public utilities could be very low, between zero and one: if

[10] This implies $\eta > 1$ which is expected for a profit-maximizer monopolist, who will never supply the market in the lower elasticity section of the demand.

this is true, therefore, privatization as we modelled it in this first reform step would perhaps never be beneficial, whichever public monopolist level of inefficiency.

3.5 Price regulation

As a second step, the government understands that greater benefits for the consumer can be achieved by regulating the maximum price that the monopolist can charge. A regulator is appointed. For simplicity, we model this price-cap as the incentive compatible (not necessarily optimal) mark-up ($\beta \in \Re^{+}$) above the first-best market's price:

$$\hat{p} = p^{*}(1+\beta) = v^{*}(1+\beta) \tag{3.15}$$

The regulated private monopolist sets production by satisfying the demand at the price-cap:

$$q^{RPr} = a - v^{*}(1+\beta) \tag{3.16}$$

In this way, and as long as the condition $\eta \leq \frac{1+\alpha}{2\beta-\alpha}$ is met, it gains an extra profit of $\frac{1}{2}v^{*}\beta[a - v^{*} - 2v^{*}\beta]$ on top of the profit it would get from rationing the quantity to the optimal level (marginal revenues equals marginal costs).

Simple algebraic manipulation shows that the consumer surplus for a price regulated private monopolist is:

$$CS^{RPr} = \frac{\left(a - v^{*}(1+\beta)\right)^{2}}{2} \tag{3.17}$$

Therefore, privatization with regulation increases consumer surplus if and only if[11] $\alpha > \beta$.

Hence, consumer surplus definitely increases in comparison with the unregulated private monopoly. But, again, nothing can be stated, a priori, on the welfare dominance between public and regulated private monopoly. The reform has a positive impact on consumers' welfare only if β (the maximum tolerated mark-up) is smaller than the public monopolist's inefficiency α. On the other hand, the regulatory reform will reduce consumers' surplus.

[11] The second solution of the quadratic expression ($\alpha > \beta$ when $\eta > \frac{2(1+\alpha)}{\beta-\alpha}$) is ruled out because we assume that the private firms are better off by setting the price equal to the price-cap, and therefore that $\eta \leq \frac{1+\alpha}{2\beta-\alpha}$ is true.

3.6 Unbundling

Let us suppose now that the third step of the reform is implemented: the regulator imposes the unbundling of the network operator (see for example Willner (2008)). For instance, it forces the separation of network from operation. This new reform is clearly in preparation for the market opening that will be discussed in the next sections. Hence, we now face two different firms: the first acting as network owner has increasing returns to scale (decreasing average costs $\bar{c}^N = f^*/q$) and its marginal costs equal zero. While the operator has a constant average cost function and constant marginal costs. We can reasonably assume (see for a discussion, Pollitt (2008)) that unbundling the network from operations increases costs since we are losing economies of integration: for example, we are duplicating administration costs or other managerial costs.

Hence, total costs are higher than the first-best cost. We assume that the cost function for the network owner does not change, while the cost function for the operator is now:

$$c^{Op,Un} = v^*(1+\gamma)q \tag{3.18}$$

Where $\gamma \in \Re^+$ is the parameter describing the inefficiency introduced by unbundling the market.

In other words, the unbundling costs are entirely shifted to the operator. Note that consumer preferences are not influenced: the market's demand function is the same as before. The government keeps subsidizing the network owner (as it was before in favour of the public and private monopolist). The operator will maximize his profits under the price-cap constraint. As in the case of regulated private monopolist, with the additional assumption that the unbundling inefficiency γ is smaller than the allowed mark-up β, the supplier satisfies the demand. The resulting equilibrium, including the indifference condition about consumer surplus, is the same as the one of the regulated private monopolist, whereas profits are reduced by $v^*\gamma(a - v^*(1+\beta))$, because of the higher marginal costs.

Clearly, because unbundling per se is here just a cost, the reform is justified only as a preparation to market opening, a step that we consider in the next section.

3.7 Duopoly

In the next reform step the regulator offers a single new licence to an entrant operator, and lifts the price-cap. The incumbent must offer the entrant the right to access the unbundled infrastructure (or there is a separate owner for it). This move establishes a legal duopoly, for example in the consumer's retail market. Under Chamberlin's small groups' model, the new entrant faces the residual market demand left by the incumbent and higher costs (for example, it bears costs for an advertising campaign) than under private monopoly without price regulation, see Figure 3.1d.

This mechanism can be seen as an alternative to price-cap regulation, because the incumbent is forced to adopt the equilibrium price in the entrant's market, and to provide access to the network. On the other hand, as compared with first-best or unregulated private monopoly, there is a cost increase: first, because of the cost of vertical disintegration as in the previous model and—second—because of separate administration and sales costs for the two competitors. We represent this combined incremental cost as:

$$c^D = v^*(1+\delta)q \tag{3.19}$$

where $\delta \in \Re^+$ is the parameter modelling the duopoly inefficiency, $\delta > \gamma$.

The competition arising from the duopoly entails a lower equilibrium price (the incumbent has to follow the price set by the entrant since the good is homogeneous) and higher production than the private monopoly's. The consumer surplus (CS^D) is now the area under the industry demand curve and over the equilibrium price p^D that is established by the optimal production plan of the entrant.

$$CS^D = \frac{1}{2}\left[\frac{3a - 2v^*(1+\delta) - v^*(1+\gamma)}{4}\right]^2 \tag{3.20}$$

With respect to the starting point (integrated public monopolist), consumer surplus can be either higher or lower: the result depends on the relative size of the public monopolist' inefficiency α, the unbundled market inefficiency γ, the duopoly inefficiency δ, and the price elasticity of the demand η. In particular, in order to have consumer surplus improvement under duopoly with respect to the case of public monopolist:

$$\alpha > \frac{\eta(2\delta + \gamma) + 1}{3\eta - 1} \tag{3.21}$$

Note that, even assuming that there are no inefficiencies correlated to unbundling and duopoly ($\delta = \gamma = 0$), if the elasticity ranges between zero and 1/3, consumers are better off under the public monopolist for any level of positive α, and if, for example, $\eta = 1$, the public monopolist must be the 50% more inefficient than a vertically integrated private one.

3.8 Market entry

In the last step of the reform, the consumer retail market is opened to competition. Following the Cournot model with many firms, let us suppose now that the regulator offers *j* licences to new entrants, *j = 1, 2, . . . , n*. For the generic firm, we can write the profit maximization condition (marginal revenues equal marginal costs) as:

$$MR = c' \tag{3.22}$$

$$p(q) + \frac{\partial p}{\partial q} q_j = c \tag{3.23}$$

where *p(q)* is the inverse demand function and *qj, cj* respectively the *j - th* firm quantity and marginal cost.

We can rewrite the second condition as:

$$p(q)\left[1 + \frac{\partial p}{\partial q}\frac{q}{p(q)}\frac{q_j}{q}\right] = c_j \tag{3.24}$$

Recall that $\left|\frac{\partial q}{\partial p}\right|\frac{p(q)}{q}$ is the definition of elasticity of the industry demand curve η, and let $s_j = \frac{q_j}{q}$ be the *j - th* firm's share of total market output. The above condition becomes after simple manipulation:

$$p(q) = \frac{c_j}{\left[1 - \frac{1}{\eta/s_j}\right]} \tag{3.25}$$

We can think of η/s_j as the elasticity of the demand curve facing the single firm: the smaller the market share of the firm, the more elastic the demand curve it faces. Note that if $s_j = 1$, the firm is a monopolist, and the condition boils down to the unregulated monopolist's. If, instead, the firm's market share tends to zero for *n* approaching infinity, the conditions reduce to the first-best solution: price equal marginal costs under a competitive equilibrium. To keep things simple, in our model firms are all similar to each other, and have the same level of marginal costs. Nevertheless, marginal costs can be higher than in our first-best benchmark, because they are an increasing function of the number of firms in the market, because of the loss of economies of scale and integration. The product is still homogenous and each firm needs to advertise its brand, has sales and administration costs, and so on. Note that we index the total cost function by the actual number of firms operating in the industry $c_j = v*(1 + \zeta_j)q$, $\zeta \in \Re^+$ in such a way that: $\zeta_1 < \zeta_2 < \ldots < \zeta_n$. Thus, while—as mentioned—with ζ_1 we are back to the unregulated private monopolist, with a duopoly ζ_2 will be the same as δ in the previous model and so on, and we move gradually towards oligopoly and full competition (we still exclude price discrimination, and we also exclude collusion). The output when j firms operate in the market is:

$$q^{ME} = \frac{j\lfloor a-(1+\zeta_j)v^*\rfloor}{j+1} \tag{3.26}$$

and the market clearing price is:

$$p^{ME} = \frac{a+j(1+\zeta_j)v^*}{j+1} \tag{3.27}$$

Therefore, the indifference condition between public monopoly and the case of oligopoly (j firms operating in the market), becomes

$$\alpha = \frac{1+\eta\zeta_j j}{\eta j-1} \tag{3.28}$$

Note that, under full competition, as the number of the competitors increases, price tends to marginal cost. However, under our assumption (see footnote 6), there is an increasing residual inefficiency due to unbundling. Therefore, the reform outcome welfare dominates vertically integrated public monopoly only if $\alpha > \zeta j$. Inversely, this simple condition shows also the maximum number of licences that the regulator can offer if the reform should be welfare improving as compared with vertically integrated public monopoly. It is not, however, unambiguous whether liberalization is the better of the other arrangements. Our results are summarized in Table 3.1.

We shall use some ideas from this framework when discussing our case studies of reform in the EU.

3.9 Discussion: Objectives, incentives, and regulation

This section discusses the relationship of the previous illustrative model to some themes in the extensive literature on regulatory economics. This literature is often framed in incentive theory and builds on earlier ideas about modern public economics. The discussion below, albeit very selective, further clarifies the motivation, scope, and limitations of the subsequent empirical analysis. When needed, we use the concepts of Section 3.2 (particularly about such parameters as $c^*,\alpha,\beta,\gamma,\delta$, etc.). The discussion is structured around the following five themes: a) objectives of principal(s) and agent(s); b) incentives under public ownership; c) incentives under private ownership; d) vertical separation versus access regulation; e) markets as incentive mechanisms; f) price convergence under market integration.

Table 3.1. Market equilibria, consumer surplus under different regulatory regimes and consumer surplus indifference conditions between different regulatory regimes

	Quantity	Price	Consumer surplus	Indifference condition*
First-best	$q^C = a - \upsilon*$	$p^C = \upsilon*$	$CS^C = \frac{(a-\upsilon*)^2}{2}$	*n.a.*
Public monopoly	$q^{Pu} = a - \upsilon*(1+\alpha)$	$q^{Pu} = \upsilon*(1+\alpha)$	$CS^{Pu} = \frac{(a-\upsilon*(1+\alpha))^2}{2}$	*n.a.*
Unregulated private monopoly	$q^{UPr} = \frac{a-\upsilon*}{2}$	$p^{UPr} = \frac{a+\upsilon*}{2}$	$CS^{UPr} = \frac{(a-\upsilon*)^2}{8}$	$\alpha = \frac{1}{\eta-1}$
Regulated private monopoly	$q^{RPr} = a - \upsilon*(1+\beta)$	$p^{RPr} = \hat{p} = \upsilon*(1+\beta)$	$CS^{RPr} = \frac{(a-\upsilon*(1+\beta))^2}{2}$	$\alpha=\beta$
Unbundled market	$q^{Un} = a - \upsilon*(1+\beta)$	$p^{Un} = \hat{p} = \upsilon*(1+\beta)$	$CS^{Un} = \frac{(a-\upsilon*(1+\beta))^2}{2}$	$\alpha=\beta$
Duopoly	$q^D = \frac{3a-\upsilon*(3+\gamma+2\delta)}{4}$	$p^D = \frac{a-\upsilon*(1+\gamma)+2\upsilon*(1+\delta)}{4}$	$CS^D = \frac{(3a-\upsilon*(3+\gamma+2\delta))^2}{32}$	$\alpha = \frac{\eta(2\delta+\gamma)+1}{3\eta-1}$
Market entry	$q^{ME} = \frac{j[a-\upsilon*(1+\zeta_j)]}{j+1}$	$p^{ME} = \frac{a+j\upsilon*(1+\zeta_j)}{j+1}$	$CS^{ME} = \frac{j^2(a-\upsilon*(1+\zeta_j))^2}{2(1+j)^2}$	$\alpha = \frac{1+\eta\zeta_j j}{\eta j-1}$

* the indifference condition is with reference to public monopoly.

3.9.1 *Objectives of principal(s) and agent(s)*

Regulatory economics often deals with a divergence of objectives between one (or more than one) principal, and one (or more) agent, see Laffont and Tirole (1993). Usual assumptions are profit maximization for the privatized firm, and social welfare maximization for the public firm or the regulator. Then these simplified assumptions are gradually relaxed. Private firms are seen as comprising different types of shareholders, managers, and other stake-holders, each with their own objectives. Public firms, in turn, are seen as managed by individuals responding to ministries or regulators, who in turn are appointed by policy-makers, while these are elected by voters. Variations in this setting pave the way to principal-agent models where the divergence of objectives is supplemented with information asymmetry. One traditional way to represent social welfare in this literature (see e.g. Vickers and Yarrow (1988)) is simply:

$$W = (V(q) - p(q)q)) + w\pi(q) - \lambda T \tag{3.29}$$

where W is the social welfare function, $V(q)$ is *gross* consumer surplus, $p(q)q$ is consumer expenditure, given service prices and quantities, $0 \leq w \leq 1$, is a welfare weight, π is firm profit, λ is the shadow price of public funds, and T is any transfer to (or from) the firm. As $\pi(q) = (p-c)q$, then: $W = V(q)-c(q)q$

when $w = 1$, and there are no transfers. Here the social welfare is expressed simply as the sum of consumer and producer surpluses (firms' revenues and consumers' expenditures cancel out). In general, however, the profit should be discounted by a welfare weight that accounts for the distributional preferences of the benevolent government or regulator (Caillaud et al., 1988). The value of w would be a function of income distribution in the country, and the equity preference of the government. On the rationale for welfare weights, see for example Brent (2007), Evans and Kula (2011), and the discussion in Chapter 9.

There are therefore two possible justifications for our focus on consumer surplus only in this book: either that we assume $w = 0$, or that further analysis is needed on producer surplus. While the latter view is sensible, the former is not entirely ruled out. If *normal* profitability is included in c, the wedge $(p(q)-c(q))$ is socially inefficient. Extra-profits or rents enter with a *negative* sign in some models of regulation, for example Laffont and Tirole (2000, p. 43): 'In words, social welfare is equal to the gross surplus S engendered by the project, minus the total monetary and nonmonetary cost of the project... evaluated at one plus the shadow cost of public funds, minus the shadow cost of the public funds times the firm's rent. The key feature of this social welfare function is that leaving a rent to the firm is socially costly.' See for details Laffont and Tirole (1986, 1993). In turn, this fact is related to the shadow pricing of public funds arising in second-best economies, see Dahlby (2008). Thus, distributional and allocative efficiency consideration would often limit below unity the weight of producer surplus in welfare analysis. If the privatized utility is partly owned by foreign shareholders and we are interested in domestic social welfare, an additional discounting factor is to be considered. Hence, focusing on the consumer side of the social welfare effects of reforms is perhaps less of a limitation than one would think. Under constant $c^*(q)$, as in our illustrative model, the social welfare maximization problem is greatly simplified, but in a more general setting an explicit consideration of the cost function and the welfare weight of profits is needed. This is particularly important for some profit-making public enterprises, such as those that we shall consider in the following chapters. In principle, distributional issues and the shadow price of public funds would make it necessary to fully include producer surplus into W under public ownership. Thus, the empirical evaluation based on consumer surplus only would be conservative.

When moving to general equilibrium, the above discussion should be supplemented by demand and supply complementarity and substitution effects, which are prominent in public economics literature, but often neglected in regulatory economics, which is typically a partial equilibrium analysis. Finally, as Vickers and Yarrow (1988, p. 93) noticed: 'Although prices provide some information about *marginal* utility, they do not say much about the whole area under demand curve (let alone income effects, etc)'.

While this is true in general, the welfare ranking of states of the world and of prices for linear approximations of the exogenous demand function in one specific industry is often the same, hence a qualitative analysis can focus on consumer prices as a shortcut for changes of consumer surplus. Galal et al. (1994) and Jones et al. (1990), show a cost–benefit methodology for using price and demand elasticity information in the context of privatization and regulation.

3.9.2 *Incentives under public ownership*

The α parameter in the benchmark model of the previous sections can be seen as a summary statistic of different incentive issues under public ownership. The traditional literature of optimal tariffs of public utilities was more concerned with allocative issues (see Beato and Mas Colell, (1984) or Atkinson and Stiglitz (1980)) than with internal efficiency of the firm. More recently, the core problems in privatization literature are usually related to the fact that under public ownership principals are not profit maximizers (see the above discussion), and there is neither the market for corporate control, nor the threat of bankruptcy. Hence the incentives to cost minimization are less than under private ownership. It is important, however, to distinguish determinants of the α-inefficiency. First, the principals are the voters, and the agents are the policy-makers, who have a private agenda. Second, managers of the public enterprise are self-interested agents of the government principals. Even ignoring the difficulties that arise in translating voters' preferences in a social welfare function, certainly a principal-agent issue is relevant if the elected policy-makers want to maximize their utility, including the probability of being re-elected, and try to appropriate rents while are in office. For example, in the Laffont (2005) theory of regulation for developing countries, the government welfare function could be a linear combination of W and private utility of the policy-makers U. Hence, an augmented objective function could be written as

$$W' = \kappa U(\pi(q) + (1 - \kappa)(V(q) - p(q)q) + w\pi(q) - \lambda T)) \tag{3.30}$$

where κ is the weight of private utility U of policy-makers in the government objective function W'. An example is when there is a (hidden) dividend for policy-makers proportional to profits of the utilities. Cost-minimization is still implicit, but social efficiency is altered by the distributional factors. A similar reasoning applies to the 'downstream' principal-agent problem of policy-makers and managers of public enterprises. One way to model this is to suppose that the latter have private information on $c(q)$ in the spirit of Baron and Myerson (1982) and of many subsequent papers in regulatory economics under unknown costs. Another way to discuss the issue is to notice that costs depend upon efforts of the managers to reduce them, and these efforts are unknown to the policy-maker. The problem would be further exacerbated if the private utility of the policy-makers were $U(c)$, that is a function of (some) costs of the public enterprise (e.g. employment or wages of workers or of suppliers). In the Laffont (2005) style, hence, the resulting combined welfare function will be:

$$W' = \kappa U(c) + (1 - \kappa)((V(q) - p(q)q)) + w(p + (q) - (\phi - e)(q) - \lambda T)) \quad (3.31)$$

where $c(q) = (\phi \text{-} e)$, ϕ is the exogenous technology or state of nature, and e is the effort of the manager. It is important to distinguish the two determinants of the α-inefficiency, as their origins and possible remedies are different. If the political process is socially inefficient, the cure is upstream. Besley (2006) discusses principal-agent issues focusing on the relationship between voters and elected policy-makers. Good quality of government, measured in some way (see Rothstein, 2008) will contain κ possibly close to zero (benevolent government), but it will not solve the principal-agent problem between policy-makers and managers of public enterprises. The latter can be seen as a variant of the theory of bureaucracy à la Niskanen, for example with the public enterprise managers wanting to maximize budget allocation to their firm. However, in practice, the asymmetric information issues between policy-makers and public sector managers cannot be greater than those between regulators and managers of privatized firms, and between the latter and shareholders. In fact, public ownership can be seen as a way to solve information and monitoring issues, as emphasized by some of the literature on incomplete contracts, see also Bös (1999) and Schmitz (2000) and Schmidt (1996). Del Bo and Florio (2011) discuss some cost–benefit propositions on public enterprises under asymmetric information and two types of corruption: state capture that distorts policies and administrative corruption that distorts their implementation. They conclude that in general the former is less disruptive of welfare calculations than the former. By comparison, we turn then to incentives under private ownership.

3.9.3 *Incentives under private ownership and regulation*

In our illustrative model the β-inefficiency arises when the firm is privatized as a monopoly, and the regulator is unable or unwilling to squeeze prices as close as possible to c^*. The core of the problem in most of the literature is the trade-off between the internal inefficiency of the public firm and the allocative inefficiency of the privatized one. The latter is an immediate consequence of profit maximization under monopoly. Baumol and Willig (1986) have argued against the standard way to assume natural monopoly, and have emphasized the contestable nature of some markets, so that a potential threat of entry is sufficient to limit allocative inefficiencies. Others have emphasized, as mentioned, the monitoring efficiency of banks, creditors, financial markets, and the threat of take-over and bankruptcy. Whatever the merits of these highly debated views, market power is a usual feature of network industries. A simple measure of allocative inefficiency for privatized network services in the benchmark model is $(p^{un}- c^*)/p^{un}$. Clearly, a binding price-cap would imply that β $(p^{un}- c^*)/p^{un}$, but is $\beta = 0$ desirable?

The question is implicit in the discussion by Vickers and Yarrow (1988, p. 35), who present a benchmark model of a public versus a private monopolist, both providing the same homogenous good, and in both cases the managerial incentives to reduce costs are less than optimal. Their model compares the utility function of the manager under private and public ownership in the form (we slightly adapt notation and other details):

$$U^{priv} = (p^{un} - c(e))q - ae \tag{3.32}$$

$$U^{pub} = V(q) - c(e)q - be \tag{3.33}$$

The e variable here represents again managerial effort in the two situations, and a and b are respectively the marginal cost of effort to the private manager relative to the benefit to him of a marginal increase of profit; and b is the same to the public manager when $W = V(q)-c(e)q$ replaces profits. Optimality conditions in both cases would require $a = 1$, $b = 1$, that is the marginal cost of effort to the manager equals his marginal reward. If incentives are imperfect, but monitoring ability by the principal (policy-makers and shareholders) are equal under public and private ownership, then Vickers and Yarrow show that for $a = b$, public ownership will deliver higher service quantity, lower price, higher managerial effort, lower cost, and hence greater social welfare. An isoelastic specification of both prices and costs, takes respectively the generic forms

$$p(q) = \zeta q^{-\varepsilon} \tag{3.34}$$

$$c(e) = \psi e^{-\omega} \tag{3.35}$$

where ζ and ψ are constants, ε is the inverse elasticity of demand, and ω is the elasticity of unit costs to managerial efforts. It follows from derivation of first order conditions of managerial utility maximization that a sufficient condition for social welfare be higher under public ownership is: $a/b > (1-\varepsilon)^{1/\varepsilon}$. A sufficient condition for welfare dominance of private ownership is $a/b < (1-\varepsilon)^{(1-\omega)/\omega}$.

Vickers and Yarrow (1988, p. 38) concluded then that regulation:

> would introduce a new policy instrument into the model, involving a further trade-off between internal and allocative efficiency, and can generally be expected to lead to results that are more favorable to private ownership than are those derived above. Nevertheless, the possibility of regulation is unlikely to affect the central implication of the analysis, which is that privatization is likely to improve social welfare only if it improves significantly keener managerial incentives than does the control system for public enterprise.

Under this perspective, in our illustrative model, the simple indifference condition between the α and β inefficiencies is an easy way to represent the fact that the regulator is nobody else than another public bureaucracy. Hence, many of the abovementioned considerations about distortion of welfare objectives because of private agendas of the principal, and asymmetric information issues, can be repeated.

One side of this problem was noticed by early US authors such as Stigler (1971), who insisted on the risk of regulatory capture, while the other side of the problem has been framed in regulation under asymmetric information. Both strands of literature concur in that they identify precise reasons for what we have labelled the β-inefficiency of regulation. It is however important to acknowledge the difference between the two origins of the problem. Regulatory capture, or non-benevolent regulation, is conceptually different from information constraints to optimal regulation. The former can be increased by the latter, but information rents rise also when the regulator is benevolent. In other words, if a benevolent regulator tries to squeeze the rents of the regulated firms, either under US-style rate-of return regulation (by a very strict definition of a 'fair' return to capital), or under British-style price-cap (by a high x in a *RPI-x* formula), the incentives to cost-minimization are reduced under asymmetric information. In Vickers and Yarrow's 1988 (pp. 89–90) words: 'If a regulator agency had as much knowledge as industry conditions as the firm being regulated, it could simply direct the firm to implement its chosen plan... Indeed, it would then be better simply to appoint the regulatory agency to run the enterprise....'

Clearly, rate-of-return regulation will increase the problem (see the discussion in Laffont and Tirole, 1993, on the power of incentive schemes), but any regulatory approach would face both problems.

3.9.4 *Vertical separation versus access regulation*

The illustrative model of Sections 3.2 and 3.3 points to another source of inefficiency: γ-inefficiency related to unbundling. The most radical form of unbundling is ownership separation; the milder policy is the regulation of access or interconnection rates. We shall further discuss these alternatives in our case studies, but here we need only to restate that neither policy is a free-lunch. Suppose there is an upstream potentially competitive sector, and a downstream one under natural monopoly. A well known problem is the risk of double marginalization, if the upstream segment is less than competitive: in other words the downstream incumbent will apply a mark-up over prices of the upstream services, which in turn include a mark-up. Alternatively, under certain conditions, the unregulated incumbent will ask a prohibitively high price for access to upstream competitors. Thus, the determination of the access rate by a regulator is needed, and again there will be the issue of asymmetric information, which will create some rents. Both double marginalization and information rents of this type disappear under vertical integration. The problem of fixing the 'right' charge for access by a regulator under the milder regime is not easy, and has been widely debated in regulatory economics literature. In some cases it has been described as a 'nightmare' (Newbery, 2000), particularly when the regulator is required to know the reaction functions of the entrants and the incumbent, and when the latter is also price-regulated in its end-product portfolio. According to some early views the interconnection charge should be lower than the marginal cost of the network service. For example, Vickers and Yarrow (1988, p. 74) suggest that the optimal interconnection charge under a duopoly should be (we slightly adapt their result):

$$i^* = f - ((1+\xi)(f+g))/(1-\xi)\varepsilon \tag{3.36}$$

where g is the is the unit cost of producing the upstream input, while f is the unit cost of the downstream input. Thus, $c = (f + g)$ is again the marginal cost of the 'full' service delivery, ξ is a parameter representing the nature of reactions of the competitors, and ε is again elasticity of demand. Hence, for certain plausible combinations of ξ and ε, the charge for the interconnection is below marginal costs for the incumbent, who runs its business at a cost higher than the entrant, whose cost is only $g + i^*$. More recent literature shows a range of different approaches, see for example Laffont and Tirole (2000) for telecommunications. While in a second-best context it is not surprising that one inefficiency must be created to counteract another inefficiency, the concern here is with incentives. If the incumbent is a private firm, its return on investing in the network is squeezed by regulation, and this can lead to under-investment. In Laffont and Tirole's (2000, p. 137) words:

> Access pricing rules are necessarily imperfect... Any access pricing rule in an instrument of regulation of the bottleneck owner's rate of return. As such it governs the owner's incentives to build and maintain the bottleneck.

Vertical separation, either under public or private ownership, cuts the regulatory dilemmas in a different way, but creates other problems. In principle, the carrying capacity of the network can be auctioned and demand–supply equilibria will determine the access price, with the right incentives to the owner of the network. If this separate owner still enjoys scale economies, and there are network externalities, a market failure is likely, as marginal cost pricing will create a loss, and other pricing policies will amount to forms of discrimination, albeit of different type than under vertical integration. Moreover, while ideally one would need a continuous auctioning mechanism for carrying capacity, this is very unlikely to be sustainable for the whole industry, which needs some price stability for medium to long term planning. Our benchmark model has eliminated this problem by assuming that all investment is sunk, and the government covers the remaining costs. This is just a convenient ad hoc assumption, and a limitation of the model.

In any case, both regulated access and ownership unbundling are socially costly, and must be justified by the advantages of competition.

3.9.5 *Markets as incentive mechanisms*

Introducing competition in an industry that would tend to monopoly (or to some form of collusive oligopoly) is a very old issue in economics. It was perhaps its earliest motivation as a new science from the time of Adam Smith, when statutory monopoly in foreign trade and in many domestic industries was the normal state of affairs. Thus, we do not need here to insist on the powerful incentive mechanism provided by competition. We need to justify, however, the conclusion of our illustrative model that if competition is not strong enough, having 'some' market entry can be welfare inferior to public monopoly, or to regulated private monopoly. The benefit side of our story is mainly on service providers' internal efficiency, the cost side on allocative inefficiency when competition is less than perfect. This is the same trade-off mentioned by Vickers and Yarrow (1988) and many others, with some additional qualifications. In the illustrative model the δ-inefficiency is compounded with the γ-inefficiency into a simple variable $\phi(n)$ that can be seen as the social cost of a public intervention that forces entry in industries that tend to be uncompetitive. This variable increases with the number of entrants, while the allocative inefficiency decreases. Thus it is crucial to appraise the relative importance of the two drivers.

As mentioned, the γ-inefficiency is related to the policy of unbundling, but all imperfect markets generate social costs, not just network industries. We suggest, however, that imperfect competition in network industries has some special features. Some of these issues have already been discussed in Chapter 1 and do not need to be further discussed here. There are two points, however, that may need a clarification: information asymmetry at the consumer level and network externalities. The standard textbook model of perfect competition simply assumes product homogeneity and no information search cost for the consumer. Paradoxically, while for most consumption goods homogeneity is a very strong assumption, this is not the case for many network services. It is indeed true that the quality of water, of a telephone call, or of electricity can vary, and we shall discuss this issue in Chapter 8. The variability of quality is, however, certainly less in the services provided by the utilities than in clothing, cars, or tourism services. The normal state of affairs in the latter industries is rather monopolistic competition, with some explicit product differentiation, R&D, and advertising to win customers loyalty. The reward of a search cost to the consumer in this context is the probability of discovering a better combination of price and quality against the alternative of picking up an offer at random. There are well known equilibrium conditions where the search process should stop.

Competition for water or phone calls cannot be of the same type as for furniture or restaurants. Given the relative homogeneity of the good provided by utilities, one would argue that for the consumer it is mostly a matter of searching the lower price. If this mechanism works, we should observe price convergence in network industries, and a powerful squeeze over time of profit margins. Indeed, this is the core rationale of the reforms. Profit-maximizing firms, however, know this and—if they are prevented from colluding by effective regulation—then try to win brand loyalty by obfuscating the essential homogeneity of their supply. This translates into advertising and other marketing expenditure, including the offer of multi-product packages. These expenditures are likely increased with the number of entrants, without actually offering the consumer any substantial price/quality difference. In a precise sense these are wasteful expenditures, which should be seen as a social cost of market opening, when consumers are not well informed. We discuss this point again in Chapters 7 and 8.

This problem is compounded with network externalities, which are trivially solved by vertically integrated monopoly, and—less trivially—are created by market fragmentation. When, for example, a telecommunication provider controls a certain share of the market in terms of the number of switched lines, he also controls the destinations and origins of calls for customers of other providers, and the value of his network will be enhanced if he can internalize this opportunity. If there are ten providers, it is apparent that the social cost of the network externality is greater than if there are just nine; see the glossary for some definitions and propositions on inefficiency of competition in networks. These concerns are indeed much more limited in other industries (while not unknown particularly in the distribution and banking services, for example).

3.9.6 *Dynamics*

The discussion so far has been in terms of comparative static. If we add a time dimension to our illustrative model, the story needs some important qualifications. First, α and β inefficiencies will be subject to dynamic forces. One driver is learning by policy-makers and regulators. Newly established arrangements are probably more affected by information asymmetries than more mature contexts. This argument is, however, counterbalanced by other aspects of the above discussion. States of world, including demand and technology, change over time; hence there is some obsolescence of knowledge of the principals, while the agents will notice the changes earlier. Second, if a component of the α and β inefficiencies is more a matter of distorted objectives, as corruption of policy-makers or capture of regulators, time will not necessarily cure the disease. Learning is a mechanism that works on both sides of the environment, for example the regulated firm learns how to cheat or to make side payments. In general, one important aspect of adding the time dimension to the modelling of network services is that technology tends to be more endogenous, as it will be related to capital investment and R&D expenditures. We have also noticed that with increased competition the demand curve for the individual provider tends to be flatter, but in a progressive economy the demand for the industry will also respond to change of income over time. Water is more an essential good than broadband connection in lagging economies than in developed ones. Going back to the discussion in terms of demand elasticity in our illustrative model, this fact has some obvious consequences for the welfare comparisons of the different regimes. Thus, there will be a long-term tendency of decreasing of c^* and of η, hence real prices should fall, at least if prices of inputs do not increase enough, and if the balance of the reform effects is benign to the consumer. Disentangling the effects of the reforms from the other determinants of prices and consumer welfare needs to explicitly consider the dynamics of different factors.

3.9.7 *Price convergence under market integration*

While the illustrative model implicitly assumes that the market is unique, in the real world network service markets are often not well integrated across countries. There are two issues related to this feature: Is market integration beneficial in welfare terms? Will prices converge with market integration?

The first issue has already been briefly mentioned, see footnote 3. We further discuss this point here, and then turn to price convergence.

Auriol and Biancini (2012) study the welfare implications of market integration in a two-country model when the benevolent regulators are national and in each country there is a regulated monopolist. Their theoretical analysis is related to electricity, but is probably more general when there are non-linear transportation/transmission costs over the network. It shows that the creation of an integrated market is welfare-enhancing only if the production cost differences between the countries is large enough (and compute the threshold under certain assumptions). The intuition is clear. If in the two countries the cost differences are large, then cross-border market opening will benefit both the exporting country (producer surplus increases) and the importing one (consumer surplus increases). The shut-down of the inefficient firm in one country is possible, but mitigated by transportation costs. If, however, the differences of cost are not large enough, fierce competition, while bringing down prices, may also have adverse welfare effects because of wasteful 'business stealing' activities, and it can be shown that welfare may decrease in both countries. Moreover, there will be redistributive effects within each country and under-investment in network facilities, coupled with over-investment in cost-reducing infrastructure. This theoretical analysis points to network infrastructure as a public good, with a related externality, and to the need for a supra-national regulator, in fact a 'federal' social planner managing particularly cross-border interconnection investment.

In this model, as in any market integration model, prices converge across countries. The fact that we do not observe much price convergence in the EU-15, as can be glimpsed going back to Figures 2.6 and 2.7 (in Chapter 2) is an indication that market integration is perhaps not working as expected. Market liberalization within a country that is relatively insulated from another country is a quite different story than liberalization under wide market integration. Cost differences across firms within one country may be much more limited than differences across firms in different countries. This has important consequences. In the Auriol and Bianchini (2012) framework, there may be two alternative interpretations. Cost differences across the EU countries may not be large enough to justify integration. Or they are large but, despite this, the lack of a supra-national social planner (in spite of attempts to create EU-wide regulatory agencies) and political economy constraints (related to the weak democratic legitimacy of European institutions, as compared with national ones) prevents investment in interconnection.

Whatever the exact balance of the two alternative explanations of market integration failure for each industry, the empirical analysis in the rest of the book will extensively take advantage of cross-country price variability and national features of regulation, technology, demand, and of other unidentified country effects. While we cannot directly estimate the parameters of our illustrative model (for example the inefficiency of public enterprises relative to first-best, or the transaction costs of unbundling, etc.), we shall use as far as possible proxies for ownership, vertical integration, market opening, and a set of control variables, having in mind the above discussion.

3.10 Conclusion

Our objective in this chapter was to present a very simple analytical framework to study the welfare effects of different reforms of network industries. The analysis has been carried out in a merely abstract way, but we suggest that it offers some intuitions about possible policy trade-offs. For the discussion of the policy implication of different reform patterns see for example Willner and Parker (2007), Pollitt (2009), and Willner (2011).

The relative size of consumer surplus depends non-linearly upon a list of parameters. We need to know whether and to what extent the public monopoly is inefficient in the initial situation, whether the regulator is strict in setting the price-cap for a private monopolist, and how much (in)efficiency is created by unbundling and market opening. Therefore it seems far from being obvious that a standard package of reforms across countries and industries would be always and everywhere welfare improving.

Reforms of the network industries have both benefits and costs. We have shown this in a very straightforward way, by focusing on process and consumer surplus. Differently from most of the theoretical regulatory literature we have skipped modelling information issues, collusive behaviour, and games between agents, because we want to be simple and general.

In a first-best world, in a one-consumer type economy and in partial equilibrium, a social optimum is achieved when price equals marginal cost, and—in turn—marginal cost is the minimum, given the best technology available and parametric factor prices. Moving away from a cost-inefficient, vertically integrated public monopoly, that, however, sets a tariff at (its own) marginal cost, can be achieved in several ways. If a government is very corrupt or lacks of managerial skills, the welfare gain of an unregulated private monopoly, due to cost decrease, may be higher than the significant allocative inefficiency related to monopoly prices. Under modest initial inefficiency of the public firm, however, the balance between the two types of inefficiencies can be reversed. Price-cap regulation limits monopoly rents but needs detailed information for the regulator to be able to understand the cost structure of the regulated firm, and the regulator can be as corrupt or unskilled as the government that appoints him. In general, price-cap regulation may look better than unregulated private monopoly, but it is not always better than vertically integrated public monopoly, again because of the balance between different types of inefficient outcome.

An alternative to price control is liberalization, which usually is expected to imply some form of unbundling of the network. Limited market opening, for example under a duopoly, decreases allocative inefficiency, but increases some costs related to unbundling of the infrastructure, and duplication of administration and sale operations. Full liberalization can be achieved by adding licences for new entrants. This will strongly decrease monopoly rents, but will often increase management and sale costs. Clearly, and despite the decrease of allocative inefficiency, if the additional costs of unbundling and of loss of economies of scale under market opening are higher than the initial cost-inefficiency of the vertically integrated public firm, the status quo will still be welfare dominant. Moreover, in some cases, unregulated private monopoly or price-cap regulation may be better than inefficient liberalization.

In principle, a small number of parameters and assumptions allow a computation of the consumer surplus associated to each reform pattern in our story. The key parameters in fact are just four: the cost-inefficiency, if any, of the public firm, the price-cap, the cost of unbundling, and the cost of market opening. Given a demand function, and a least-cost technology, it is easy to compute welfare indifference conditions between different reforms, where the benefit side is the change in consumer surplus. Readers are invited to play with some figures they may have in mind and to discover how inefficient the vertically integrated public enterprise must be to justify the reform. The further reading section mentions a small sample of much more complex theoretical ideas, but it seems now helpful to move away from the abstract reasoning and to turn to our empirical case studies in reform.

3.11 Further reading

Obviously the illustrative model we have proposed in this chapter is just a shortcut for a more complex framework. In the literature (e.g. Newbery (2000), Laffont (2005), Rey and Vergè (2008), Rey and Tirole (2007)) there is no straightforward consensus on the effect of regulation and vertical disintegration in particular. According to Newbery (2000) the restructuring of network industries should take advantage of differences in economies of scale of different segments (physical network having sub-additive costs, and upstream and downstream activities operating under constant or decreasing returns to scale). This leads to a paradigm of reform where vertical disintegration of the network is a crucial remedy to market foreclosure. Rey and Tirole (2007) mention divestiture as the extrema ratio measure, with milder forms of vertical separation as an alternative to be assessed case by case. Moreover, they also review potential cases for socially beneficial market foreclosures. Rey and Vergè (2008) show that the welfare effects of vertical restraints are crucially different according to a number of features. They conclude that while the impact on aggregate profits of a 'vertical structure' (and vertical integration can be seen as an extreme case of such arrangement) is positive, the impact on consumer surplus is ambiguous. Willner (2011) and Parker and Willner (2007) present interesting theoretical results which challenge simplistic views of the reforms' welfare outcomes.

Beesley and Littlechild (1983, 1989) present the early conceptual model of the regulatory reform as proposed in the UK. Defeuilley (2009) critically discusses the 'Austrian' foundation of the reform paradigm.

On wasteful competition expenditures when a product is homogeneous, see Klemperer (1987) and on electricity Joskow (2000).

Part II

Testing the Paradigm

4

Testing the paradigm: Telephone services

4.1 Introduction[1]

The telecom sector is not just one among the network industries that have experienced a policy paradigm shift over the last 20 years: it has been its core laboratory worldwide and the one where reform started earliest. If one had to pick a single year as the turning point, 1984 would be the most convenient one, with the parallel divestiture of AT&T in the USA and of British Telecom in the UK. The nature of the divestiture was, however, different across the two sides of the Atlantic. AT&T was a private regulated monopolist that was broken up into seven regional Bell operating companies. British Telecom was a public corporation that, after privatization, came under a regime of regulated duopoly along with another privatized operator, Mercury.

In both cases, institutional changes, away from regulated and integrated monopoly in the USA, and from integrated public monopoly in the UK, were motivated by a willingness to establish competition in the industry.

At the same time, a dramatic process of technological change reshaped the industry. Hence, regulators and law-makers had to frequently adjust their views because of entirely new developments in telecommunications. After more than two decades of policy reforms, what is the dividend for the consumers?

Ideally one would look to detailed case histories, considering long time series and counterfactual histories, see for example Harper (1997), Parker (2001), Galal et al. (1994), Florio (2003), and Florio and Puglisi (2006) on British Telecom. A complementary approach takes advantage of cross-country evidence. Under the latter perspective, as mentioned in the Introduction, the EU offers something near to a natural experiment. On one side, there is one

[1] With Emanuele Bacchiocchi. Sections 4.5 to 4.8 draw, with some adaptations, on Bacchiocchi, Florio, and Gambaro (2011), while the rest of the chapter is new.

actor, the European Commission, that pushes towards a well defined reform paradigm, as embodied in subsequent telecom directives.

On the other side, there are the Member States that, in compliance with the EU legislation, adopt new policies and institutions. During this process, differences emerge across countries in reform design, sequencing, timing, and market outcomes.

One attractive aspect of a cross-country empirical study of telecoms is that, in contrast to other network industries, such as electricity or natural gas, countries are not exogenously constrained in technology adoption by (in part) intrinsic geographic characteristics. This allows us to focus on country differences in ownership, institutions, competition, and other industry features, under common technological environments and shocks.

The structure of the chapter is as follows. Section 4.2 offers a brief overview of technological change and policy reforms in the EU. Section 4.3 is about market structure in selected Member States. Section 4.4 observes price variability across countries. Section 4.5 briefly reviews some earlier empirical literature on testing the impact of telecom reforms on prices. Section 4.6 applies the conceptual framework of the previous chapter to price predictions following reforms. Section 4.7 presents our data sources. Sections 4.8 and 4.9 report the findings. Section 4.10 presents additional evidence about mobile telephony. Section 4.11 concludes.

The main empirical findings suggest that ownership plays virtually no role in determining tariffs, and that competition within the fixed telephony industry is less important than substitution, with mobile telephony, and other factors, as determinants of tariffs.[2] For mobile telephony prices, ownership also has no impact, while larger market shares of entrants have a beneficial effect.

4.2 Technological and regulatory change

4.2.1 *Some features of the technology*

For many decades the telephony service was based on a stable technology environment. The public switched telephone network (PSTN) had a typical structure, as summarized by, for example, Harper (1997), Newbery (2000), and Laffont and Tirole (2000). The final consumer had to rent (or buy) a telephone or, for some business entities, a switchboard. The core of the service was ensuring bilateral voice communication between these apparatuses. This was achieved through a twisted copper pair laid down between the customer

[2] In this chapter we do not consider possible improvements in quality of service, but Chapter 8 adds this dimension.

premises and a first node of the network, a switch. This was, and often still is, the 'local loop', even now when copper is being replaced by optical fibre or co-axial cable. The first switch reached by the wire was then connected to a local exchange, and from there to other customers, identified by a telephone number. Connection to other networks was required for international calls, or within some countries where there were regional monopolies, each with its own network. These telephone networks, and their operators, were usually different from those of the telegraph service, radio, and television, which used air-borne electromagnetic waves.

The relatively quiet life of the industry has been shaken by a series of technological shocks since the 1980s. The first of these was born outside the telephone sector, and derived from the opportunities given by electronic technologies based on the digital coding of signals. Under traditional telephone technology, the user's voice was transformed into an 'analogue' electric signal, and the switches were based on a relatively simple electro-mechanical design that routed the call from one line to another. Advances in electronics allowed the transformation of voice into a digital signal, and the switch into one in a computer. This paved the way for changing the nature of the service, as not only voice but virtually any kind of information could be transformed into a digitalized signal and communicated between users. Content is translated into a sequence of numbers expressed in binary coding, and this potentially allows for a universal communication language understood by any electronic apparatus. The coded information can be managed in various ways: stored, fragmented in packets, re-assembled, compressed, unzipped, and released through a set of appropriate instructions.

This evolution opened the door to a number of related changes. First, the telephone is only one of the possible apparatuses that can manage communication from the user's premise: computers or other electronic machines (faxes were earlier examples) can be alternatives. Hence, the standard telephone offered by the provider had to compete with other devices, all compatible with the digital treatment of the information at the end-point. Second, and more important, the network architecture now offers a range of alternatives. The local loop can still be based on copper, but can also be based on co-axial cable, or on optical fibre, which allows for greater bandwidth communication. The local loop can also be based on an aerial that sends the signal to the switch.

The latter form of communication is the same principle that allows mobile telephony, where the telephone is connected through air waves to a specific 'cell', which is provided by an aerial or a satellite, and from there to other cells, including access to a fixed network. Either optical fibres or microwaves connect the nodes. Hence, there is potentially not just one physical network for each country or geographical area, but a range of different networks. Some of these are 'public', and under certain protocols and arrangements can be accessed by everybody. Other networks are 'private', and their capacity is reserved to certain users.

Cables or waves in principle can convey any kind of digitalized information: messages, financial transactions, television programs, any content of the Internet, combined computing capacity, and so on. Data traffic is now more important than voice in the telecommunication industry of developed economies in terms of revenues, and is still increasing at a fast rate.

The multiplicity of networks, and the fact that in principle all of them can carry information, has blurred the boundaries between activities that were once regarded as different industries. Cable television companies can offer telephone services; fixed telephone companies can distribute television contents; producers of personal computers, laptops, and pads, can sell software for data, image, and voice services. The mobile phone industry has transformed what used to be a voice receiver/transmitter into an all-purpose portable electronic terminal. Thus, there is now a potential convergence of different technologies and industries in a broad electronic communication market.

There are, however, some technical features that make each of the communication services specific still, and these differences should not be overlooked if one is to understand the evolutionary logic of the new industrial panorama. For example: television broadcasting is one-way, voice telephony (including through the Internet) is two-way and needs time synchronization, e-mails are two way but do not need strict simultaneity of communication. Thus, the best design for the networks supporting these services is possibly specific. These differences have a big influence on the quality of the service, and on its costs.

The core of the traditional PSTN (circuit switching) is that it actually reserves a line between two users, who stay connected in real time (even if with a small delay, which is more apparent in long-distance international calls). If the lines are 'busy' the call will 'fail', but when the connection is established, the communication capacity is reserved. When a certain bandwidth capacity is shared under packet switching, there is no longer a reserved connection between users, and specific software is needed to optimize the routing and management of information. This may create congestion, delays, and interruptions of service if there is not the right balancing of carrying capacity and traffic. Internet telephony (VoIP), when poorly managed, provides an example of the low costs, but also of the still not entirely resolved technical problems, of getting real time voice communication without reserving a line connection.

Electronic communication services are also still different from the perspective of the user. For example, mobile and fixed-apparatus services do not need to converge, as obviously a small screen on a smart-phone does not offer the same visual comfort of a large screen at home. The bandwidth needed to transfer a voice call, in terms of bits per second, is a small fraction of what is needed for managing high-definition images in a colour display. Asynchronous transfer modes can circumvent the need to shift from copper to fibre in the local loop for voice telephony to a certain extent, but video-conferencing requires larger bandwidth. The need for security in electronic financial transactions is far higher than in television broadcasting. Thus, while the current panorama of the digital revolution may suggest that in the end there will be just one network of electronic communication, where any kind of information is transferred, it is still unclear whether this convergence is meaningful in economic terms.

This uncertainty has far reaching influences on the organization of the communication industries. Should telephone companies, facing competition from VoIP, transform themselves into producers and vehicles of content? Or are media companies going to become TLC operators? Are the Internet companies becoming the new dominant players in a convergent electronic communication industry? These are difficult questions, and the explosion of technological innovations poses difficult challenges to regulation.

4.2.2 *Regulatory issues*

Before the 1980s the telecommunication industry was almost everywhere considered a natural monopoly because of large fixed costs. Consequently, governments did not allow competition that was considered wasteful. In most countries the industry was either nationalized or, as in the USA, a private monopoly under strict regulation (Melody, 2011). The latter came in the form of cost-plus pricing rules, with wide scope for cross-subsidies. The business user tariffs had a high mark-up to allow for long-run marginal cost pricing or other pricing rules applied to residential users. Tariffs of international and long-distance national calls often (but not everywhere) paid for losses in the local calls' segment of the business. Connection charges were low and uniform, to offer universal access to rural users. According to Laffont and Tirole (2000), these pricing rules distorted incentives for allocative and productive efficiency. The core reforms of the late 1980s and 1990s were divestiture of incumbents, new incentive regulation (usually meaning the use of price-caps instead of rate-of-return targets), and liberalization. At the same time, there was the perception that new technologies, particularly the diffusion of microwave communication, could weaken the traditional case for natural monopoly.

The starting point of liberalization in the EU legislation was modest: the apparatus rented to users by telephone companies. This was strictly standardized and approved by the monopolist, allegedly to ensure technical coordination (and the associated rent accruing to the monopolist, usually in the form of a fixed rental charge).[3] Subsequently, and much more importantly, different providers were licensed to offer the service. Early examples are cable TV operators competing with the privatized British Telecom in the 1990s. Entrants could save on investment costs on proprietary networks, and share the bandwidth of existing facilities with the incumbents. Thus, the liberalization of interconnection was a second crucial step.

This posed a trade-off between investment in the upgrading of existing networks, or the creation of new networks, and the access to those networks by service providers who operate without large sunk costs. Low access price increases competition but disrupts investment in the network. Thus, while the initial focus of the regulators was on capping tariffs to end users, it then shifted to interconnection charges. Number portability and other regulatory measures were also intended to unlock competition where the existence of traffic bottlenecks continues to offer wide opportunities for market foreclosure.

The regulatory paradigm that emerged in the UK was in favour of breaking up the incumbent, privatizing it when a public corporation, or otherwise forcing it to divest part of its capacity in favour of entrants and/or force access for competitors to their networks, establish independent regulators to administer new licence and design price-caps, and enhance competition everywhere. In Newbery's (2000, p. 329) words: 'All the evidence suggests that modern dynamic economies can support competitive telecommunication markets, even for voice over the wire, and greatly benefit from such competition. The main questions are therefore how best to create effective competition.'

In some views, after a transition period, under strong asymmetric regulation against the incumbent in order to protect the entrants, full liberalization can be applied, sector regulators abolished, and their residual powers surrendered to generic competition offices.

Along this line of thinking, international organizations, such as the World Bank, the OECD, and the EC (with some variations between them), started to build something of a consensus telecom reform package, and advised governments everywhere to apply it. The OECD and the EC started to define regulatory indicators and reform milestones and to evaluate governments as frontrunners or laggards in implementing the policy changes (see Commission of the European Community, 2007, and Gonenc et al., 2001). Minority views, often originating from within the industry, for example Harper (1997), who had reservations about full privatization and liberalization, mainly

[3] In the USA this required several legal battles between AT&T and suppliers of new terminals, and decisions of the Federal Communications Commission and of the Courts, particularly since 1968 (Carterfone Decision).

because of technological arguments of sub-additive costs of the network, were paid scant attention.

With the benefit of hindsight, 20 years on, we can see that the reform approach was often based on strong assumptions about the evolution of technology and firms' behaviour. The newly established regulators, the courts, governments, and managers, had to learn from real world evolution, as a number of unexpected technological changes questioned earlier views.

As mentioned, the traditional technology of the industry was based on first, switching, second, transmission capacity, and third, network software. Switching shifted from analogue codes to digital ones well before the reform, in the 1970s and early 1980s, and the change in general was successfully managed by state-owned companies in Europe (Millward, 2005). They were able to self-finance the necessary investment and were often more efficient than their privately owned US counterparts. Mobile telephony was promoted by state-owned incumbents quite early on, well in advance of AT&T and the 'Baby-Bells' on the other side of the Atlantic.

The continuous digital technological progress worldwide has been such that productivity increase was fast almost everywhere. The number of lines per employee, or per unit of invested capital, increased dramatically under public monopoly in continental Europe. For example, Telecom Italia, a listed public corporation with the Italian Treasury as the majority stakeholder, at its divestiture in 1997 was among the best companies in Europe for switched lines per employee, see Florio (2007). For British Telecom, productivity trends were higher in the two decades before divestiture than after it (Florio, 2003). This performance was linked to high growth in demand,[4] investment, and research expenditure. While the integration between incumbent operating companies and their technology suppliers was widely criticized as anti-competitive, the internalization of R&D was in fact high under monopoly (for example AT&T and Bell Laboratories), possibly for a well-known Schumpeterian argument.[5]

While it is not self-evident that divestiture was needed to foster the adoption of the digital technology opportunities, the case for the decline of natural monopoly in some core segments was probably exaggerated.[6]

[4] This also largely explains the long waiting times for connections in post-war Europe.

[5] Sterlacchini (2012) observes that the privatized telecoms are now investing in research much less than they used to. The drop in R&D between 2000 and 2005 was 57% for Deutsche Telekom, 18% for Telefonica, 20% for BT, 12% for Telecom Italia, while France Telecom increased its R&D expenditure by 27%.

[6] Economists tend often to rather extreme oscillations of opinions. For example, nowhere the intellectual battle in favour of (private) monopoly was fought harder than in the USA, when 'In the course of the period reviewed, 1968–1984, AT&T sponsored theoretical and applied research at Bell Labs, employed more than 100 economists and more than 200 small consulting firms, many run by leading academic economists. Most economics departments at leading universities in the USA were represented...' (Melody, 2011, p. 109). In contrast, the Federal Communications Commission employed a half dozen economists. Ironically 'The work by the AT&T economists

The expectation in the mid-1980s was that microwaves would imply much less sunk cost than the traditional twisted pair of copper wires. In fact, cellular telephony boomed, and offered welcome mobility to users. However, it was soon realized that the key issue was the interconnection of wireless and wired transmission, with the former shifting to new systems of data compression, such as asymmetric digital subscriber (ADSL) lines, or to transmission hardware, such as optical fibre. To the surprise of many (including the US regulators), fixed line transmission under the new technologies was much more efficient in carrying advanced services, including data and images, such that the Internet revolution occurred through wired transmission, and only to a limited extent, until quite recently, through wireless technology. This evolution re-established a natural monopoly element in the industry, as any physical network (roads, railways, pipelines, wire, etc) usually exhibits sub-additive costs. The third factor, the evolution of advanced software to optimize signalling, combined with the new role of networks, in a sense reinforced the argument of decreasing average costs in crucial segments of the industry. This is related to the substantial fixed cost of designing, implementing, and maintaining the complex electronic platforms involved in the functioning of the systems and to demand side network effects. Hence, it seems far from certain that technological change has forever wiped away natural monopoly concerns.

4.2.3 *A still evolving panorama*

Technological change in recent years has taken two parallel paths. In wired technology, the existing infrastructure has been improved, either with new software for managing cable transmission of signals, or with new networks (VSDL, Very High Speed DSL, and Fibre to the Home, FTTH), which reduce the cost of carrying data. Next Generation Networks (NGN) is the generic name for innovation in both optic fibre and non-fibre new infrastructure, able to deal with a range of broadband services. Governments have been a factor in these improvements, with substantial subsidies to close the digital divide across regions and types of user in each country, see Picot and Wernick (2007). Belgium, Sweden, Spain, and Portugal are promoting Docsis 3.0 (version 3 of Data Over Cable Service Interface Specification), while Germany, France, and the UK have launched investment plans based on VSDL or

was virtually all special applications or extensions of static neoclassical theory' (ibidem). Several of the well-known contributions by William Baumol on sustainability of natural monopoly, contestable markets, Ramsey Pricing, efficient component pricing rule, and other concepts are closely related to advisory work for AT&T. When the paradigm shifted in the opposite direction, many economists quickly dismissed the early ideas, and enthusiastically embraced the view that natural monopoly is a relatively minor issue in telecommunications.

FTTH. The most important change is probably the diffusion of VoIP, which in turn can now be managed by dedicated operators, who take care of synchronization and other qualitative aspects, or simply bilaterally join two or more users by their personal computers. In some countries, such as France and the Netherlands, around one third of fixed voice traffic is now through managed services over broadband.

In the wireless domain, there has been a rush to take advantage of the available radio spectrum (Melody and Lemstra, 2011), some of it available because of the migration of traditional analogue TV broadcasting to the digital signal. The great advantage of wireless technology is that it is a cost-effective alternative to closing the digital divide in low population density areas, and, second, a simpler interface with mobile services. In some EU Member States—for example in the Czech Republic—one third of the retail broadband lines are now wireless. Mobile 3G (UMTS) operators have improved their networks, and the EU institutions are willing to promote further advances. The latter changes are too technical to be reported here, but they are said to promise an avenue for network competition.

These new evolutions create a number of additional difficult problems to regulators. While the scope and range of services has increased, and the distinction between the telecoms, the media industry, and other activities has become blurred, there has been a multiplication of networks. Providers of electricity, water, gas, railways, highways, and cable television are all potential investors in the telecom networks, as they own a 'hole' in the ground where cables can be laid down. To invest, these utilities need mutual access and some stability of revenues. Moreover, the local loop, that is the final connection of the fixed line with the user, is still the crucial bottleneck, as microwaves cannot yet effectively substitute for the broadband width offered by optical fibres or optimized cables. Thus liberalization, that is allowing the entry of a plurality of players, is far from decreasing the role of regulators, because they should be trying to find solutions to complex two-way access between networks, with continued or new issues of natural monopoly and bottlenecks.

The trade-offs arising in this context are significant. Entry into the local market can happen in different ways, through facility-based entry, resale of services, or unbundling. Each of the solutions has costs and benefits. In some cases promoting competition,[7] paradoxically, can be more expensive than expected. Competition can be artificial, when it is supported by continued heavy-handed asymmetric regulation. In fact, this amounts to a form of

[7] The interplay between country competition authorities, national telecom regulators, the judiciary, including the European Court of Justice, and other players, adds complexity to the picture, and cannot be dealt with here, See Further Reading Section.

indirect subsidy to entrants. Establishing access pricing in this context gives the regulators wide powers. They literally can decide about the long-term profit and loss of companies. In some cases it seems that regulation has gone a full circle, back to implicit rate of return regulation, as network owners have to show their cost allocations, under a number of possible accounting definitions.

The perhaps rather premature announcement of the death of natural monopoly and of heavy public regulatory powers sheds new light on the issue of ownership rights. Public or quasi-public ownership of some parts of the networks is no longer as out of question as it seemed to be in the 1990s. While the EU is neutral on ownership, it asks for separation of the network for operations, and a regulation that in principle offers access to everybody.

The French and German governments have been reluctant to sell their controlling shares in the incumbents. BT has been forced by the regulator to establish a separate manager for the network ('Openreach'), with a special (equality of access) board, and more than 200 binding regulatory undertakings. In Italy, the government has recently given the regulator wide powers over the networks, and has even considered, but not implemented, a form of re-nationalization of the main network.

To deal with the new scenario, the European Commission, after two years of negotiations which started in 2007, introduced a new reform package, approved on November 2009 by the Council of Telecoms Ministers of the 27 Member States. The new reforms will be translated into legislation over the next few years. The new key reforms include: a) on number portability, the right of any European consumer to change a fixed or mobile operator in just one working day; and the contract with the operator should be no longer than 24 months; b) improved information to consumers about the contract they subscribe to; c) wider citizen rights regarding Internet access, strictly constraining the ability of governments to limit access to the Internet for prevention of criminal offences; d) national regulators will be able to set rules on minimum quality standards for network transmission services ('net neutrality'); e) consumer protection against personal data breaches and spam; f) better access to the pan-EU emergency telephone service '112', including for citizens with disabilities; g) greater independence of the national regulators; h) establishment of a new European Telecoms Authority (BEREC, Body of European Regulators for Electronic Communications); i) increased overseeing by the European Commission on the decisions of the national regulators, in order to avoid contradictory rules that can distort competition; l) power of the national regulators to impose functional separation of the network from services (building on the experience of Openreach in the UK); m) acceleration of broadband network connections; n) increasing network competition between legacy infrastructure based on copper-wire, and wireless or fibre Next Generation Access.

These new reforms are ambitious, particularly the last two, and will take several years to be implemented. See Appendix 4.A.2. In the meantime, it seems helpful to review what has been achieved by previous waves of reform.

4.3 Fixed telephony and the European electronic communications market

4.3.1 *Market structure*

The evolution of the industry, summarized in the previous section, blurred the distinction between different services, and an electronic communications market has emerged. In what follows we highlight some aspects of this changing scenario, with a particular (but not exclusive) focus on fixed telephony. The reason for selecting the latter service for empirical analysis is that historically the transition from public monopoly to privatization and liberalization started there. Thus, we have a longer period of observation available, and a richer virtual counterfactual history, as provided by a comparison of industry change across EU countries. In Section 4.10, however, we also provide some empirical analysis for mobile telephony and the latter enters as a control in estimates of fixed telephony prices.

Fixed telephony has faced a strong challenge from change in the industry. In 2008, the EU-27 electronic communication market, including fixed and mobile telephony, voice, data, content transmission, Pay TV, and any broadband services, generated around 351 billion Euros of revenues (see EC, 2010).[8] Revenues of the fixed voice service were around 126 billion (36% of the total), those of mobile 165 billion Euros (including data transmission), Internet services 42 billion, and the remaining 35 billion were the revenues of pay TV services. The traditional fixed voice is thus no longer the primary cash-cow of the industry, as mobile services have taken the lead. Moreover, while voice traffic is declining, data and TV services are increasing. In terms of investment, however, 70% of it is for fixed telephony (with a still minor role for broadband), with more than two thirds of investment made by the incumbent telecoms company, often the former public monopolist. In fact, seven out of ten major global players in the industry are European companies. These include British Telecom, France Telecom, Deutsche Telekom, Telecom Italia, Telefonica (Spain), Telia Sonera (Sweden and Finland), and KPN (the Netherlands).

[8] 15th Progress Report on the single European electronic communication market.

Governments have historically considered communication networks as strategic assets, both for military, economic, and social reasons. This feature, that was true for transport networks, postal, and telegraph services too, was inherited by traditional telephony and is now extended to broadband fibre-networks. As a consequence, the regulatory regime, including subsidies and concessions or other legal arrangements, has strongly influenced the development, on a national basis, of the fixed telecom networks, which remain highly capital intensive infrastructures. In spite of more than 20 years of privatization and liberalization, the fixed telephony service still tends to be dominated by national champions.

In the mobile service, what has emerged from liberalization in Europe is an oligopoly, with ten mobile operators concentrating around 90% of the market, and four main groups owning 'the first and/or the second largest mobile operator in almost all EU Member states. Most of the main mobile operators are subsidiaries of a fixed incumbent. The only large European group which is not a subsidiary of a fixed incumbent has now entered in the fixed market to complement its activities' (EC, 2010, p. 7). This strong linkage between the current industrial panorama, and its past history, shows that one should be careful in seeking to understand the extent and nature of telecommunications competition in Europe. While clearly the Internet revolution and cable TV have created entry opportunities for new players, and mobile services have displaced a substantial share of revenues from traditional telecommunication activities, the core of the infrastructure is still controlled by former telephony incumbents, which cash in on the interconnection rates.

National regulators have tried to bring down fixed interconnection rates, but they have been less successful in controlling mobile termination rates, which remain relatively high (e.g. compared to the USA) and vary greatly between countries. Moreover, given the linkage between mobile services and the fixed network, including the highly capital intensive broadband infrastructure, former incumbents play an important role in all the core businesses of the industry. In terms of broadband lines, around 2009 Telecom Italia had a share of 57% in Italy, Telefonica of 55% in Spain, Deutsche Telekom of 46% in Germany, around the same share as France Telecom in its domestic market. From this perspective the UK is an exception, with 72% of the market share of broadband lines owned by the entrants.

The latter tend to use local loop unbundling (LLU) as their preferred shortcut to challenge the incumbents. LLU is one way to break one of the natural monopoly elements in the industry, that is by attacking it downstream. In the EU, regulators have either forced incumbents to fully divest LLU, or to share the access with third parties, through regulated tariffs and contracts. There are huge differences, however, between Member States, with competition between different platforms as an alternative to unbundling in some countries.

In some cases, subsidies to the incumbents to extend connections in disadvantaged areas (e.g. in the countryside) have been given by governments under the condition of stronger obligations for the incumbent to open its network to competitors.

In spite of this regulatory pressure in traditional fixed voice telephony, and the substitution for mobile and VoIP services, with the market shares of the incumbents stable, or increasing, in recent years, the prices offered by the former monopolists are also stable or increasing.

It may be surprising that service providers challenged by new technologies, and hence by a steady decline of traffic volume, are able to defend prices and market shares. The above mentioned horizontal integration between services, the continued role of expensive infrastructure investment, complex software, and skilled human capital, may converge to explain the outcome. Moreover, while regulators have focused on entry and vertical unbundling of some segments of the industry, many core operators have 'horizontally' bundled the services. Some companies offer customer packages that include fixed and mobile voice, data transmission, and even pay TV. Hence it becomes increasingly difficult for the regulators to understand how costs and prices should be accounted for. According to the Van Dijk Consulting report (2010), 68% of broadband offers were bundled with a range of voice and non-voice services.

The rest of this section focuses on the market shares of the incumbents in the EU for fixed telephony only (but the reader should remember the implications for the broader electronic communications market, given the above mentioned linkages between the traditional and new services).

In terms of market shares of the incumbents (one of the variables considered below in the empirical analysis of telephony prices), the average share of retail revenues in the EU-27, around 2008, was slightly more than 50% for international calls; around 60% of calls to mobiles; and 70% of national calls. Figure 4.1 shows the incumbents' market share in fixed telephony by country in the EU-15, those of main interest in this book. Nowhere is the share less than 60%, with again the UK and Sweden showing slightly less dominance of the incumbent.

This picture in a sense understates the market concentration, as in fact three quarters of users have direct access to the network through infrastructure owned by the incumbent. For example, in the UK, around 78% of subscribers are still with BT, a figure which is the same or very close to the percentage of subscribers with Telecom Italia, or with Telecom France. Hence, after two decades of liberalization, the asymmetry in market shares is still strong between the incumbent and the entrants.

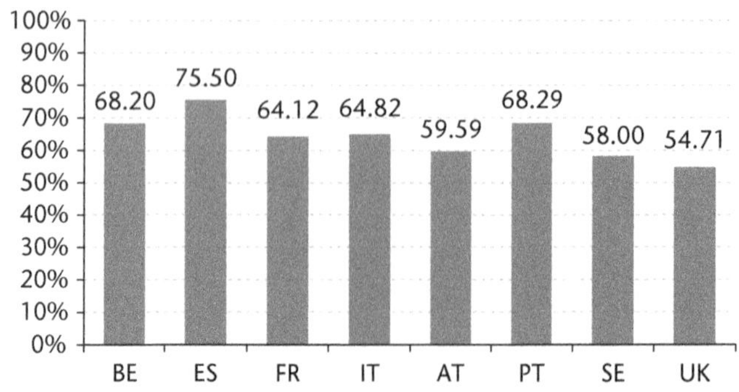

Figure 4.1. Incumbents' market share in fixed telephony market (all types of calls by retail revenues, %). Selected EU countries

Source: own elaboration of Teligen data, 2008.

4.3.2 *Country features*

In all EU countries there is sustained substitution between fixed and mobile voice services, but there are also changes in the organization of the former. In the UK, there are more than 6 million Wholesale Line Rental (WLR) contracts, which have increased over time while the number of CPS (Carrier Pre-Selection) lines has decreased. There are more than 200 providers taking advantage of WLR new technologies. As mentioned above, the market share of British Telecom continues its long decline. At the end of 2008 it was 47.1% in terms of volume of calls, but still 54.7% in terms of retail call revenue (to this one should add the revenue from wholesale line rental to third parties). This is possibly the lowest market share for any incumbent in the EU, and the incumbent's termination charges on his network are kept low and closely scrutinized by the sector regulator, Ofcom, and the other competition authorities. The whole industry is in private hands.

The countries that have most liberalized their telephony services after the UK are Denmark and Sweden, perhaps among the most advanced telecom economies in the world. In contrast to the UK, in Sweden, market opening has not been accompanied by full divestiture of public ownership. Telia Sonera, one of the major global players, is the incumbent company, resulting from the merger of two government owned entities: Telia (Sweden) and Sonera (Finland). While listed in the stock exchange, the company, which is also very active in other Scandinavian countries and elsewhere, is still controlled by the Swedish and Finnish governments. Moreover, a number of municipal companies provide local telephony services and compete with private providers. In Sweden in around 2009, Telia had increased its market share in fixed telephony to around 58% in terms of number of subscribers, reversing the previous trend of decline. While the environment is competitive, entrants complain about the fact that the incumbent tends over time to increase access rates, while the incumbent promotes migration of voice from PSTN to VoIP, offering low prices to users. In fact, in both the UK and in Sweden, the sector regulators are dealing with a substantial flow of legal controversies and appeals, often related to aspects of the market position of the entrant.

France and Germany share the common feature that the incumbent is partly owned by the government. As for fixed telephony, in France the incumbent, as mentioned above, still enjoys a large market share, in spite of a very active competition by VoIP operators, who—by volume of traffic—control nearly 40% of the market. Competition is also increasing with one mobile operator recently entering the fixed market with a bundled service offer. In Germany, while there are more than 180 alternative operators that offer telephony service in competition (LLU or own infrastructure) with Deutsche Telekom, the incumbent still holds two thirds of the market in terms of number of subscribers. Most of the network competition is based on cable operators, while one third of alternative operators use LLU (and less than 9% shared access). Since 2009 the regulator has fully deregulated the fixed national call market and the wholesale market. As elsewhere there are controversies about aspects of the offers by the incumbent.

Finally, it is interesting to mention the case of Italy, where the fully privatized Telecom Italia still holds two thirds of the market in terms of revenues of fixed telephony, and 78% of users need its infrastructure to access the network. The incumbent has aggressively marketed its IP offers, bundled with other services, and this seems to make it difficult for entrants to offer better conditions on a national scale. The regulator, who has fully liberalized retail prices since 2009, is often involved in cases deferred to administrative tribunals. To sum-up the EU country histories, Figure 4.2 shows the trends of ECTR/OECD indicators for fixed telephony in selected countries: the UK, Sweden, Finland, France, Germany, and Italy.

See Section 4.11 for further reading on country patterns.

We turn our attention now to the price of the service, as our yardstick to measure the benefit of the reforms for the consumer.

4.4 Telephony prices

In Chapter 2 we offered a first overview of trends in fixed telephony prices for the EU-15 from Eurostat. In this section, we take advantage of more in depth data, collected by Teligen (2008) for the European Commission, validated by national regulators, and referring to different telecom services. In contrast to

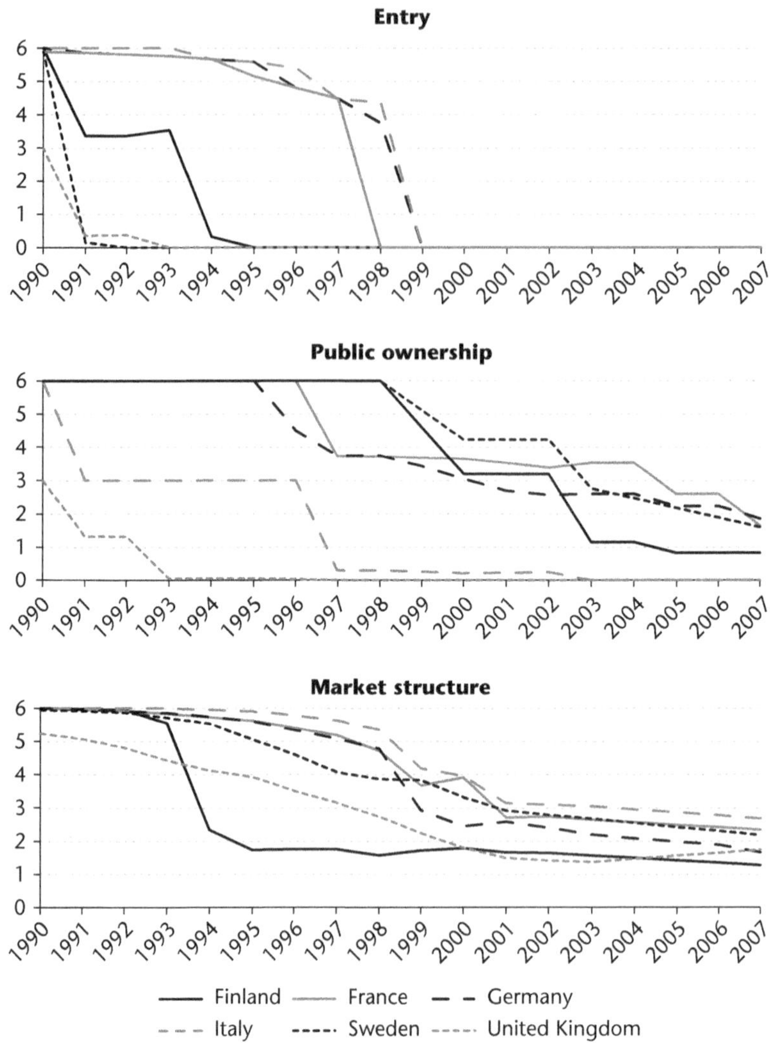

Figure 4.2. Telephony. Trends of ECTR reform indicators, 1990–2007, selected EU countries

Source: own elaboration of ECTR scores.

Eurostat, these data explicitly refer to information collected with the incumbent telecom company and its main competitor, at EU-27 level.[9]

Given the diversity of tariffs in telephony according to the type of user, the day of the week, the hour of the call during the day, the distance of the termination point, and the duration of the call, Teligen uses a composite

[9] In fact, it seems that Eurostat uses Teligen data as its main source of information.

basket based on weights given by the OECD in previous statistical studies (see Teligen 2008, pp. 11 and 20). Without entering into the details of this approach, it revolves around making certain assumptions about, for example, the distribution of business versus residential calls on a generic Wednesday at 11.00 or at other standard hours; or on Saturday or Sunday; according to distance, for example 3 km; in terms of duration, given the distance (for example a call on Sunday, at 15.00, lasting 7 minutes and to a correspondent distant 40 km). Clearly these assumptions should in principle vary across countries and periods of time. Moreover, standard tariffs, without considering discounts to certain categories of users, may alter the picture. Eventually, if the consumer picks up a provider who is neither the incumbent nor its main competitor, the data can again be misleading.

While in our empirical analysis in this chapter we are going to use Eurostat or ITU data (see Section 4.9), which offer a longer time span, it seems interesting to look at prices in the five countries we mentioned in the previous section taking advantage of this more detailed information. We also include in the discussion the EU-27 average that is influenced by the usually lower tariffs paid in the New Member States of the EU.

In order to see longer term trends, below we shall mainly discuss prices based on OECD baskets. We first show the cheapest package available to residential customers based on the 2006 OECD basket for the major EU countries mentioned in the previous section. The prices are distinguished between those for low, medium, and high usage customers (see Teligen, 2008 for the details). Figure 4.3 shows for each of these three types, the virtual price of the service for the appropriate basket, in terms of current Euros per month, VAT included, and combining the fixed and variable components of the bill.

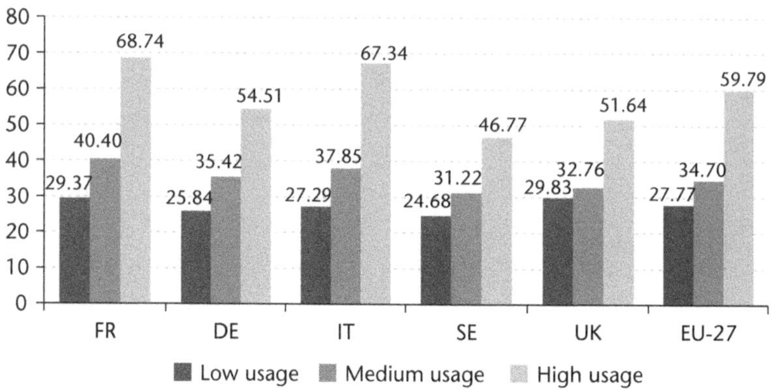

Figure 4.3. Results for low, medium, and high usage residential basket, selected EU countries, Euros per month, VAT included

Source: own elaboration of Teligen data, 2008.

Low use customers in 2008 paid 29.83 Euros per month for their basket of fixed telephony service in the UK, 29.37 in France, 27.29 in Italy, 25.84 in Germany, and 24.68 in Sweden. Turning to the medium user, the ranking of the countries in decreasing order of price is the following: France, Italy, Germany, the UK, and Sweden. The last two countries are also virtually the cheapest for high use residential consumers. The limitation of this approach is that it assumes the same basket across countries, which is obviously unrealistic. Moreover, consumption patterns, and hence the weights in the basket, are influenced by the prices of the components.

For this reason, it is perhaps better to look at specific prices (i.e. without attached weights). Table 4.1 presents three typical prices from the incumbents, taking 1998, a largely pre-reform year, as the starting point, and 2008 as the final year, again focusing on residential users only. The prices we consider here are for the following services: an international call (weighted by destination, time, etc); a 10 minute, 3 km call at weekdays at 11.00; a 10 minute call at same time, but at a 200 km distance.

The table shows the dramatic fall of prices in the relatively short time span of 11 years in all the five countries we consider, and for the EU average, on international calls and long distance calls, but not for local calls.

When focusing on these specific services, the UK (here represented by British Telecom standard offers) appears to be more expensive than the other countries (except Italy in one case). Sweden, with a liberalized market and a partly government-owned incumbent, offers by far the cheapest prices to its citizens. Germany and France, with a partly government-owned incumbent, and perhaps less market opening than the UK according to some indicators, offers prices that are lower than the fully privatized Telecom Italia.

Table 4.1. Incumbents' call price 1998–2008

Country	Incumbent	10 minute call* 3 km		10 minute call* 200 km		10 minute call to USA	
		1998	2008	1998	2008	1998	2008
Germany	Deutsche Telekom	0.437	0.400	2.933	0.510	4.316	0.290
France	France Telecom	0.423	0.353	1.747	0.775	3.437	2.320
Italy	Telecom Italia	0.236	0.222	2.164	1.148	4.988	2.116
Sweden	Telia	0.268	0.278	0.546	0.278	4.592	1.006
UK	BT	0.501	0.508	1.004	0.508	2.998	2.291
EU-27		0.3889	0.3782	1.8505	0.6300	4.432	1.878

* at weekdays 11:00.
Note: all charges are in current Euro including VAT; EU-27 simple average.
Source: Adapted from Teligen, 2008.

As here we are not controlling for possible confounding factors, we turn now to a more systematic analysis of the relationship between reforms and prices in fixed telephony.

4.5 Using the conceptual model

Modelling the behaviour of agents in regulated telecom markets is difficult, and is dealt with by an extensive theoretical literature (see e.g. Laffont and Tirole, 2000 and the Further Reading Section). Our objective in this section is more limited. We simply show that prices can change according to different parameters that interact with possible reforms, in a non-linear way. What follows is no more than a stylized story, based, with adaptations, on the illustrative model of the previous chapter. This conceptual frame will be implicit in the next two chapters as well.

In any country and time there is one homogeneous fixed telephony service and one price for it (instead of a bundle of services and a combination of time-based tariffs, subscription rates, connection charges, rental prices of equipments, etc.).[10] The regulatory regimes are stylized as follows. First, a vertically integrated public telephone operator (PTO), fully owned by a governmental body. Second, a fully privatized, but still vertically integrated monopoly, *without* any price regulation (except the constraint of no price discrimination across users). Third, the government appoints a regulator, who imposes a price-cap to the privatized incumbent. Fourth, the regulator orders mandatory unbundling as a preliminary step to liberalization. Fifth, a duopoly market, without price regulation, is established. Sixth, the regulator allows full market entry, granting as many licences as demanded. In this setting, we are interested to see how consumer prices change following a reform, after assuming that technology and demand is exogenous.

Let us first consider a telecom operator under a vertically integrated public monopoly, as was common in the EU before the mid-1980s (e.g. Swedish Telecom, British Telecom, France Telecom, Telecom Italia, etc.). The monopolist faces an exogenous linear demand curve for residential users, for a given per capita GDP: $p^{PM} = a-bq$.

The representative telephone user responds to changes of per capita GDP over time simply by a shift in their demand. There are no substitution effects (e.g. with mobile telephony). This is unrealistic, but can be accommodated in

[10] This assumption should be seen as just a heuristic device to present a simple theoretical model. More detailed models would try to disentangle conceptual models for different policies, including wholesale/access pricing and whether it tends to have an impact on the interconnection between different types of services.

the conceptual model by thinking of a changing b over time. The demand is sensitive, however, to a set of country features, including consumer preferences. The government-owned PTO's returns to scale are assumed to be constant in each time period and for each country. This assumption reflects a combination of zero marginal costs of the fixed telephony network (main lines, switches, and local loop), that is, the natural monopoly segment of the industry, and of positive constant average and marginal costs of operation and sales. In general, earlier empirical literature suggests that the marginal cost c of fixed telephony can be represented by a function of four variables (plus noise): population density, number of lines, technology change, and productivity per employee.

We assume that the public firm's costs c^{PM} are higher than in a first-best economy benchmark c^*, a value that can be inferred by engineering data, so that simply: $c^{PM} = c^*(1 + \alpha)$

For any given country and time, population density is exogenous. The other cost components, however, may be partly endogenous. The number of switched lines, the speed of technology adoption (including investment and human capital accumulation), and productivity of the workforce (typically the number of switched lines per employee) may be (to some extent) hindered by lack of incentives offered by the management under government ownership to minimize costs.

The α parameter captures—for each country and time—both inefficiencies, because of lack of incentives to minimize costs offered by the managers, and rent extraction by policy-makers. Any mark-up over costs, that generates dividends for the Treasury, is included, as rents to other stake-holders who actually control the firm as a coalition (perhaps trade unions, managers, or some suppliers), see Chapter 1.

After allowing for this social cost of public provision, the managers are instructed by their principal to maximize consumer surplus, under strict budgetary constraint.

Hence, the firm follows the traditional marginal cost rule for fixing price:

$$p^{PM} = c^{PM} = c^*(1 + \alpha).$$

A demand shift over time under this assumption has no impact on equilibrium price, as only the quantity of telephone service provided will respond (through a change from a_{ft} to $a_{f,t+1}$ for the same slope of the demand curve). Prices, however, will respond to changes of c^* (technology innovation, e.g. changes of switching technology, adoption of optical fibres, etc.) and possibly of α, (for example less harmful interference by government or trade unions, better management, release of investment constraints, decrease of the 'fiscal' mark-up). Over time the process results in lower unit costs. Telephone prices will tend to decrease, more or less slowly according to specific country and the firm's features. It is important, therefore, to disentangle price changes related to exogenous factors and to policy reforms.

In a different state of the world (time or country) the government privatizes the PTO, without, however, imposing a price-cap or rate-of return regulation. The only regulatory rule is that price discrimination is forbidden. In fact this never happened in the EU, it is just a way to represent the fact that the national regulator needs time to learn and to be effective, or truly independent, and this provides a benchmark case.

Since the privately owned incumbent has an incentive to minimize costs, the management is instructed by the new principal to wipe away any inefficiency and to maximize profits. As a result, the quantity and the price of service change respectively, as shown in the previous chapter. An increase in the demand for telephone services will raise prices and not just quantities, while a decrease of costs will decrease prices and increase quantities. We ignore changes in quality, as we understand prices as quality-adjusted, see Chapter 8 on this issue, however.

By comparison with the price of the government-owned PTO and of the unregulated private monopoly, the reform will decrease telephony tariffs whenever the inefficiency of the former exceeds the allocative inefficiency of the latter. The indifference value of the initial inefficiency parameter of the incumbent is simply: $\alpha^{Upr,PM} = (a - c^{*})/2c$.

This is also exactly the value that makes the consumer surplus the same in the two ownership regimes. In general, however, the indifference conditions between two regulatory regimes will be different for prices and for consumer surplus, as we discussed in the previous chapter.

In another country or time, the government appoints a telecom regulator with the power to impose a price-cap on the privatized incumbent's retail activities. The regulator perfectly knows the technology of the regulated firm, and picks up a price constraint sufficient to offer the private investor 'adequate' profitability. Suppose that for any time period there is a simple (implicit) mark-up β over the first-best (or minimum) long-run marginal cost that is incentive compatible. The new quantity and price equilibrium is now:

$$q^{Rpr} = [a - c^{*}(1 + \beta)]/2b$$
$$p^{Rpr} = c^{*}(1 + \beta)$$

The optimal production plan of the telephone operator is exactly the same as that under the unregulated regime. Moreover, price is not affected by the change of ownership plus regulation whenever: $\beta = \alpha$.

Again, the inefficiency parameter of the government-owned TPO sets a threshold for price indifference after the reform. Moreover, and differently from the previous case, demand or technology shocks will not change the price. The adjustment is entirely through quantity. The firm will earn more or less profit simply as a result of these adjustments given the mark-up (in fact, and in a *'RPI-x'* setting, β will be gradually adjusted by the regulator for nominal price changes and available information on productivity increase under the 'RPI-x' capping formula).

In a new step, the regulator imposes on the incumbent the ownership or legal separation of the telecom network, as is possible under the new EU telecom reform package. This new reform is best seen as preparation for granting access to entrants. Let us start with the simplest case, the establishment of a legal duopoly, as existed in the UK in the 1980s, with BT as the incumbent and Mercury as the sole entrant. At the same time (and differently from the British example at that time) suppose that the price-cap on end user tariffs is lifted, because the regulator hopes that duopolistic competition will contain rents.

The unbundling of the telephone network (an example is the above mentioned Openreach division of BT in recent years) has some administration and coordination costs. Moreover, both the incumbent and the entrant will face additional costs as compared with the vertically integrated monopoly arrangement, because of the increasing marginal costs of operations and sales, as mentioned above. Let us describe this simply by: $C^D = c^*(1 + \delta)$, where δ is the combined effect of the extra costs of unbundling and loss of economies of scale. We assume here that the extra cost is entirely loaded to the entrant, through connection tariffs.

We adopt here again the Chamberlin 'small group' competition model. There is no possibility of collusion (the regulator or an anti-trust body is in charge of preventing this) or of actual strategic behaviour. The entrant just assumes that a share of the telephony market is left unexploited by the incumbent. Thus the entrant fixes its optimal quantity and price supply of the service based on the 'free' segment of the demand curve. The entrant's price is clearly less than for the incumbent's segment, and because the telephone service is homogeneous, the incumbent must fix the price at the new lower level. Simple algebraic manipulation of the above conditions for the unregulated monopoly now gives the new price and quantity equilibria.

This is:

$$p^D = [3a + c^*(3 - \delta)]/4.$$

In fact, this regulatory mechanism can be seen as an alternative to price-cap regulation, because the incumbent is forced to adopt the equilibrium price in the entrant's market, and to provide access to the network. To simplify, we assume that the access charge is zero. Technological progress, investment, labour productivity, and scale factors are represented by decrease of c^*. Hence, if in a given country, the elasticity of telephone demand to GDP is

1, and, relative to time t, real GDP increases by g, and minimum costs decrease by h, the new price at time $t + 1$ will be:

$$P^D = [a(1+g) + c^*(1-h)(3-2\delta)]/4.$$

After having unbundled the network, the regulator is now ready to offer a telephony licence to any entrant. The total quantity of service supplied will now be:

$$Q^{ME} = q_1 + q_2 + \ldots q_j + \ldots + q_n.$$

As we showed in the previous chapter, the ε/s_j ratio can be interpreted as the elasticity of the demand curve faced by the single entrant. The smaller the market share is, the more elastic the demand. In this way, when $s_j = 1$ we are back to the 'unregulated' monopoly market structure, while as the market share tends to zero, the condition is that of pure competition: price equals marginal cost. The latter, however, are again not equal to minimum cost c^*, because in a network industry such as fixed telephony, after unbundling, each firm will face decreasing returns, not matched by the increasing return of the natural monopoly component. As in the previous duopoly case, there will be an inefficiency factor δ, that is now an increasing function of the number of entrants, so that the marginal cost is: $cn = c^* (1 + \delta_n)$.

Equivalence between prices under government-owned TPO, vertically integrated, monopoly, regulated privatized monopoly, and market opening is now given by the simple condition: $\alpha = \beta = \delta_n$. In other words, given the demand and supply shifters, the price will be the same if the inefficiency of the PTO under government ownership equals the regulatory permitted extra profit under a price-cap, or both equal the cost of unbundling/TPO. Basically: if privatization and liberalization have no influence on c^*, that is on technology through a dynamic effect, x, the comparative statics of regulatory reforms do not predict necessarily decreasing prices for the consumer.

To sum up: telephone prices, in this simple framework, can vary across countries and over time because of forces that push in different directions. The change of regulatory regime, from vertical integration, either under public or private ownership, to unbundling and limited market opening, will lower prices only for some combinations of the key α, β, δ parameters, but not for all of them. It is difficult to predict which of the above stylized arrangements will deliver lower telephone prices to the consumer. When the vertically integrated government-owned PTO was very inefficient, paradoxically even an unregulated private monopoly could lower price. Price-cap regulation can mimic the cost plus rent of the public firm or allow a mark-up of an oligopoly. Market entry is an alternative to price regulation or public monopoly, but it needs costly access or network unbundling in some form, and cost duplications of several activities with increasing returns to scale.

In other words, telephony prices can respond directly to changes in market structures and demand, to changes in costs indirectly induced by regulatory reforms, or to exogenous trends of some costs. The empirical analysis should disentangle, as far as possible, these components of telephone price variation over time and across countries.

4.6 Earlier literature

Despite the large body of empirical literature on telecommunication and its changing institutional setting, relatively little attention has been paid to price determinants through cross-country analysis.

Li and Xu (2004) use a large set of 177 countries over the period 1990–2001 to estimate the impact of privatization and competition in telecommunication with ITU and World Bank data. They use only simple dummy variables to capture shifts in regulation and find positive effects of privatization on output, productivity, and resource allocation, but their estimates also show that full privatization increases the prices of local calls. The authors propose two possible explanations: a combination of strong network externalities and non-competitive behaviour among existing operators, or an increase in service quality (not captured by data) that drives up the cost of phone calls. They do not find a significant effect of competition on prices or a modest one in containing the adverse effect of privatization (interaction term).

Cave and Vogelsang (2003) point out that access price can play an important role in promoting competition and influencing final prices, while Bauer (2003) finds no statistical evidence of price increase eventually determined by the substantial licence fee paid by mobile operators.

Grzybowski (2008) examines, like us below, the impact of regulation on telecommunication prices estimating a reduced form model for the European Union from 1998 to 2002. Regulation and cost factors (including technology) explain 94% of price variation and the former has a significant negative impact on incumbent retail prices. Prices are taken from an earlier study by Teligen for the European Commission, and cover a period of four years only. As measures for regulation Grzybowski uses several micro variables taken from the different EC Implementation Reports like carrier pre-selection, number portability and local loop unbundling, and a dummy for market liberalization.

In a previous paper Grzybowski (2005) studied the mobile market, estimating a reduced form model using panel data for the EU countries from 1998 to 2002 and showing the relationship between mobile and fixed telephony markets. Liberalization of fixed telephone has a negative impact on the prices for mobile services and a positive effect on the demand, while competition enhancing measures like the introduction of number portability have a negative impact on mobile call prices.

Nicoletti (2001) estimate a panel with data on 23 OECD countries in 1991–7 and show that prospective competition (proxied by the number of years remaining to liberalization) and effective competition (proxied by the market share of new entrants) both reduce the price of all the telecommunication services. No clear evidence could be found concerning the effect of the government ownership of the industry on performance and price.

Estache, Goicochea, and Trujillo (2006), in a paper focused on the link between corruption and utilities reform, analyse a large sample of countries and find that privatization of the incumbent telecom operator increases prices while establishing an independent regulatory authority decreases them.

Gasmi et al. (2009) study fixed charges for telephone and cellular prices in 29 developing and 23 developed countries looking for the effect of different institutional governance and they find that privatization increases prices in developing countries while it has no effect in developed economies. Viani (2007) studied 67 countries for 7 years and finds that monopoly of telephony services is not associated with lower prices, increases residential local prices, and does not help universal provision. He finds, however, that mandatory vertical network separation reduces international call volume and the number of fixed lines. Edwards and Waverman (2006) study interconnection charges in 15 European countries for a few years (1997–2005) and find that public ownership increases access prices while regulator's independence decreases them. We comment on the latter paper in greater detail in Chapter 10.

Other papers in earlier literature usually focus on specific countries, and are often quite different in their objectives relative to our research question: Do consumers pay less in the countries that have adopted the policy reform paradigm? We mention in Section 4.11 several country studies.

4.7 Data

To answer the question about the relationship between reforms and prices, we report extensively on the findings by Bacchiocchi et al. (2011).

The investigation concerns the EU-15 countries for the period 1997–2003. Almost all the information concerning the telecommunication sector comes from the ITU World Telecommunication Indicators (2006) dataset. The database contains time series data, mainly collected from an annual questionnaire sent out by the Telecommunication Development Bureau (BDT) of ITU, for the years 1960, 1965, 1970, and annually from 1975–2005 for around 100 sets of telecommunication/ICT statistics covering telephone network size and dimension, mobile services, quality of services, traffic, staff, tariffs, revenue, and investment.

The price indicators for fixed telephone calls, instead, come from Eurostat, and distinguish between 10 minute local, national, and international calls.[11] Monthly telephone connection charges are also considered.

Among the explanatory variables, in the regression models discussed in the following section are included the number of fixed telecom lines (*Lines*), the amount of investment in the telecommunication sector (*Investment*), the number of ISDN channels (*ISDN channels*) as a measure for the technological progress, the population density (*Density*), and the number of subscribers in the mobile telephone sector (*Mobile subscribers*).

4.8 Explaining telephone price dynamics

The empirical analysis is based on the specification and estimation of equations for local, international, and national prices, and connection charges. For each of the dependent variables, the regression models include, among the explanatory variables, aggregate or detailed measures of the level of privatization and market opening of the sector. Given the structure of the data, the empirical analysis uses panel data techniques, with country fixed effects accounting for all the variability not explicitly modelled by the regressors. The general specification is the following

$$p_{it} = \beta_0 + \beta_{0i} + \beta_1{}' x_{it} + \beta_2{}' r_{it} + \varepsilon_{it}.$$

Where β_0 and β_{0i} are the constant term and the country fixed effect respectively, β_1 is the vector of parameters associated to the regressors x_{it}, while β_2 is the vector of parameters associated to the regulatory variables r_{it}, that is alternatively the OECD/ECTR, aggregate telecommunications *Regulatory Index* (see Chapter 2, Figure 2.1 and above Figure 4.2) or the *Public ownership, Entry regulation*, and *Market structure* indicators. For each dependent variable alternative specifications are proposed, see Appendix 4.A.1.

Although we aim at estimating the conditional correlation of prices and a set of controls, we attempted to correct for possible correlation of some regressors with the error term using instrumental variable methods. The instruments, due to the restricted amount of information from institutional

[11] In a previous version of our research, we concentrated our analysis on the single indicator for the 3-minute fixed telephone local call (peak rate) collected by the ITU, mainly because of the long time series collected for the EU-15 countries. The results, however, probably were not significant for an overall quantification of the reform process, in that considering the local call price only could underestimate the impact of the liberalization policies.

databases, are found by taking lagged values of all the variables used in the regressions.[12] For details, see Bacchiocchi et al. (2011).

4.8.1 *International call prices*

The results of the estimation when the dependent variable is the price of international calls are reported in Table 4.2. Columns A and B use the aggregate indicator *Regulation Index* for the liberalization policy interventions. The two specifications differ according to the presence, in column B, of the number of subscribers in the mobile telephony sector, *mob subscribers*. The third and fourth specifications, reported in column C and D, use the same regressors as A and B, but consider the disaggregated reform indicators,

Table 4.2. Telephony, international call price (log), estimation results

Variables	International call price			
	A	B	C	D
Regulation index[1]	0.404***	0.040		
	(0.151)	(0.175)		
Entry regulation			0.010	−0.024
			(0.023)	(0.023)
Public ownership[1]			0.002	0.008
			(0.023)	(0.021)
Market structure[1]			0.865***	0.287
			(0.315)	(0.322)
Lines[1]	−2.727**	−1.537	−2.138*	−0.917
	(1.232)	(1.197)	(1.263)	(1.194)
Investments[1]	−1.014***	−0.733**	−0.639*	−0.460
	(0.326)	(0.314)	(0.326)	(0.301)
ISDN channels[1]	−0.131	0.033	−0.191**	0.039
	(0.083)	(0.087)	(0.079)	(0.090)
Mobile subscribers[1]		−0.471***		−0.534***
		(0.132)		(0.132)
Density[1] (log)	−0.686	−1.248	−4.997	−3.854
	(3.299)	(3.073)	(3.255)	(2.980)
Constant	−4.792	0.269	14.470	12.565
	(15.672)	(14.324)	(15.448)	(13.693)
Observations	92	92	93	93
Countries	15	15	15	15

Standard errors in parentheses. For the regulatory variables see Table 4.A.1.
*** $p < 0.01$, ** $p < 0.05$, * $p < 0.1$
[1] Log.
Source: own elaboration on ITU data.

[12] The Sargan J-test statistic is available for each regression in order to check for the goodness of such instruments. The null hypothesis of exogenous instruments can never be rejected by the data.

Table 4.3. Telephony, national call price (log), estimation results

Variables	National call price			
	A	B	C	D
Regulation index[1]	0.343*** (0.099)	0.174 (0.119)		
Entry regulation			0.020 (0.015)	0.007 (0.015)
Public ownership[1]			0.003 (0.015)	0.004 (0.014)
Market structure[1]			0.579*** (0.198)	0.370* (0.221)
Lines[1]	−1.511* (0.807)	−0.865 (0.817)	−1.621* (0.813)	−0.950 (0.844)
Investments[1]	−0.582*** (0.214)	−0.450** (0.214)	−0.551** (0.216)	−0.441** (0.218)
ISDN channels[1]	−0.154*** (0.054)	−0.081 (0.060)	−0.180*** (0.050)	−0.102* (0.061)
Mobile subscribers[1]		−0.216** (0.090)		−0.193** (0.094)
Density[1]	−3.240 (2.161)	−3.437 (2.099)	−3.860* (2.161)	−3.870* (2.117)
Constant	6.780 (10.257)	10.415 (9.775)	8.876 (10.249)	12.148 (9.799)
Observations	91	91	91	91
Countries	14	14	14	14

Standard errors in parentheses. For the regulatory variables see Table 4.A.1.
*** p<0.01,** p<0.05,* p<0.1.
[1] Log.
Source: own elaboration on ITU data.

Entry regulation, Public ownership, and *Market structure.* When the aggregate indicator of market structure is included, the results in column A show that it is significant and positive, indicating that an increase in the liberalization process induces a reduction in the prices of international calls (the original market share variable is coded as '100—share of entrants', hence the share of the incumbent is positively associated with price). The result, however, becomes less clear when the estimation is controlled for the diffusion of mobile telephony, measured as the number of mobile subscribers. In this second case, in fact, the coefficient continues to be positive, but becomes not significantly different from zero. A very similar pattern appears for the distinct aspects of liberalization.

In particular, in regression C, the only significant variable is the market structure, indicating that a more liberalized market reduces the levels of prices. This evidence, as before, disappears when introducing the number of mobile subscribers. The mobile sector thus clearly plays the role of 'competitor' and the rapid diffusion of the latter contributes to reducing the prices of fixed telephone international calls. The prices of international calls are also reduced with investments in the telecom sector, while other technological development seems to play a limited role.

4.8.2 *National call prices*

Table 4.3 reports the results for national calls prices. The results highlight a situation that is very similar to that observed for the international call prices. Among the regulatory variables, the only one that matters is *Market Structure* (the share of the incumbent), and even in this case, with an impact that decreases (but is still significant) when controlling for the diffusion of the mobile telephony market. The amount of investment in the telecom sector is extremely important in explaining the level of national calls prices, as well as the technological progress, proxied by the ratio: number ISDN channels/ number of lines.

4.8.3 *Local call prices*

The results for the dynamics of local call prices are shown in Table 4.4. As in the previous cases, column A and B consider the aggregate regulatory indicator, while in column C and D the detailed aspects of the regulator interventions are included.

The aggregate indicator is positive and significantly different from zero. This result does not substantially change when we control for the mobile market diffusion. When we include the disaggregated indicators, *Market structure* and *Public ownership* are significant and have the expected positive sign. Moreover, this result is robust to the two alternative specifications in C and D, that is, with and without taking into account the role of the mobile telephone sector. The policies of privatization and liberalization together contributed towards reducing the prices of local calls, but the role of ownership is a very minor one. The number of mobile subscribers, even in this case, plays a relevant role, though less important than for international and national calls prices.

4.8.4 *Connection charge*

The last model considers the dynamics of the connection charge, and the results are provided in Table 4.5. The aggregate index is significant only when we exclude the mobile diffusion indicator. The coefficient of the number of fixed telephone lines is always significant and positive, as well as the *ISDN channels* variable, probably an indication that connection charges are used to recoup fixed investment. A remarkable result instead concerns the role played by the *Entry regulation* variable, which enters negatively in the explanation of

Table 4.4. Telephony, local call price (log), estimation results

Variables	Local call price			
	A	B	C	D
Regulation index[1]	0.221***	0.209***		
	(0.047)	(0.059)		
Entry regulation			−0.009	−0.013*
			(0.007)	(0.007)
Public ownership[1]			0.019***	0.019***
			(0.007)	(0.007)
Market structure[1]			0.452***	0.361***
			(0.095)	(0.105)
Lines[1]	−0.601	−0.722*	−0.455	−0.420
	(0.386)	(0.405)	(0.380)	(0.388)
Investments[1]	−0.098	−0.087	0.028	0.055
	(0.102)	(0.106)	(0.098)	(0.098)
ISDN channels[1]	0.163***	0.167***	0.127***	0.158***
	(0.026)	(0.030)	(0.024)	(0.029)
Mobile subscribers[1]		−0.015		−0.078*
		(0.044)		(0.043)
Density[1]	−1.568	−1.639	−3.567***	−3.386***
	(1.034)	(1.039)	(0.980)	(0.969)
Constant	−1.086	−3.430	7.788*	4.919
	(4.911)	(4.844)	(4.649)	(4.451)
Observations	92	92	93	93
Countries	15	15	15	15

Standard errors in parentheses. For the regulatory variables see Table 4.A.1.
*** $p<0.01$, ** $p<0.05$, * $p<0.1$.
[1] Log.
Source: own elaboration on ITU data.

the connection charge. More precisely, moving towards less restrictive legal conditions for entry into the fixed telephone market leads to an increase in the connection charge. In other words, the gains obtained in the decreasing levels of prices for calls, are partially counteracted by the increasing of the connection charges.

Finally, we have performed some additional estimates, using ITU data (1990–2007) and using a dynamic panel model with generalized method of moments (GMM).[13] We find that telephony prices in all the models are positively and significantly dependent upon lagged prices, showing substantial inertia in the price dynamics. The coefficient of the main demand shifter, GDP per capita, is significant, and positive, as expected. The substitution effect of mobile telephony has the expected sign, but is usually not significant. The aggregate and disaggregated OECD reform indicators are never significant.

[13] Not reported here and available from the authors.

Table 4.5. Telephony, connection charge (log), estimation results

Variables	Connection charge			
	A	B	C	D
Regulation index[1]	0.211**	−0.057		
	(0.105)	(0.138)		
Entry regulation			−0.019	−0.058***
			(0.018)	(0.017)
Public ownership[1]			−0.001	−0.001
			(0.014)	(0.013)
Market structure[1]			0.278	−0.232
			(0.245)	(0.241)
Lines[1]	2.380**	3.174***	2.417*	2.491**
	(1.129)	(1.115)	(1.249)	(1.044)
Investments[1]	−0.289	−0.154	−0.222	−0.062
	(0.271)	(0.263)	(0.273)	(0.247)
ISDN channels[1]	−0.043***	−0.037***	−0.124***	−0.033***
	(0.009)	(0.009)	(0.027)	(0.009)
Mobile subscribers[1]		−0.215***		−0.333***
		(0.075)		(0.069)
Density[1]	−3.975	−3.519	−6.255**	−3.692
	(2.798)	(2.687)	(2.513)	(2.283)
Constant	20.150	19.990	30.531**	21.091*
	(13.319)	(12.734)	(11.924)	(10.716)
Observations	100	100	101	101
Countries	15	15	15	15

Standard errors in parentheses. For the regulatory variables see Table 4.A.1.
*** $p<0.01$, ** $p<0.05$, * $p<0.1$.
[1] Log.
Source: own elaboration on ITU data.

4.9 Mobile telephony prices

As already mentioned, mobile market penetration in most developed countries is now wider than for fixed telephony, and this is the case in some developing countries too. The oligopolistic nature of the market here is determined by the allocation of radio frequencies, which are again a natural monopoly element. According to Melody (2011, p. 117) all regulators failed to anticipate the growth of the market and this created a spectrum shortage: 'The liberalization policy has been managed oligopoly, not liberalized entry. The industry has been enormously profitable everywhere.' How far, however, has this arrangement been beneficial to consumers?

In this section we focus on the mobile telephony market and propose a descriptive and econometric analysis similar to the one performed for the fixed telephony prices reported in the previous sections. As for the fixed telephony market, attention is paid to the impact of policy reforms on the dynamics of prices and the other costs sustained by final users. A linear regression model for each of these price components is estimated as a function of the policy variables indicating the different steps of the reform process towards market liberalization.

These policy variables are summarized in the three ECTR/OECD indicators for mobile telephony: entry regulation, public ownership, and market structure. The dynamic paths of the market structure and public ownership indicators are reported in Figure 4.4 for selected countries: the UK, Sweden, Finland, France, Germany, and Italy. As can be seen from Figure 4.4, the legal conditions for the liberalization of the market were accomplished well before the other points of the entire reform package, at least for the main EU countries reported in the figure. Concerning public ownership, with the exception of Italy and the UK, for which the process of denationalization started at the beginning of the 1990s and was completed in a few years, for all the other countries privatization was introduced only in the second half of the 1990s and progressed relatively slowly, see Figure 4.4b. The liberalization of the market, instead, as shown in Figure 4.4c, can be described as a very slow process, with almost constant improvements year after year, reaching only half of its final aim, however, at the end of the observed sample in 2007.

The second point worth discussing concerns the fragmentation of contracts proposed to consumers by the different operators. This clearly poses serious problems in the collection of an indicator for the dynamics of price calls for the mobile telephony service. As is well known, mobile operators propose different contracts in terms of pre- and post-paid packages, and for each of these, different charging methods with initial and additional charging increments. To give an idea of the complexity of the problem, Table 4.6 shows the number of contracts for the two main operators in each of the EU-15 countries, together with the range of minimum charges for calls to PSTN and to on-net mobile. To complicate the picture, the operators have different methods and charges for each package, such as the fraction of time for which the costs are calculated (minutes or fractions of it, e.g. 30 seconds or 1 second).

As said before, this rather complicated aspect of mobile pricing translates into the difficulty in finding reliable time series data describing the dynamic paths of the price sustained by final users. In the following econometric analysis, we use the data on prices of 3-minute mobile calls (peak and off-peak), on the connection charges, and on monthly subscriptions, over the period 1990–2005, although with many missing values for almost all countries at the beginning of the sample.

The econometric analysis is similar to that performed for the fixed telephony and consists in estimating linear regression models for the price components as a function of the regulatory indicators and other possible explanatory

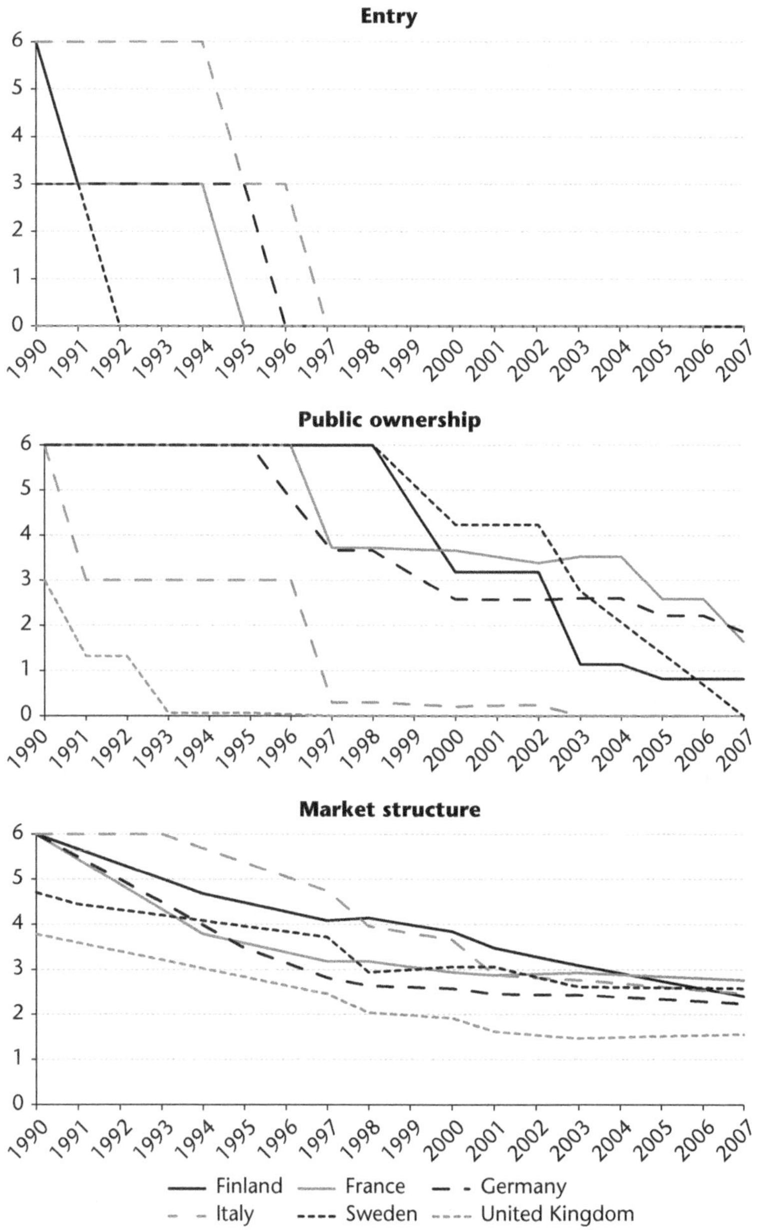

Figure 4.4. Mobile telephony. Trends of ECTR reform indicators, 1990–2007, selected EU countries

Source: own elaboration of ECTR scores.

Table 4.6. Mobile telephony, number and prices of different contracts for the two main operators in the EU-15 countries, around 2008. Current Euro

		Charge to PSTN		Charge to on-net mobile	
	No. contracts	min	max	min	max
Austria	23	0.04	0.40	0.03	0.20
Belgium	23	0.11	0.50	0.11	0.50
Denmark	28	0.04	0.07	0.04	0.10
Finland	29	0.08	0.08	0.08	0.08
France	58	–	–	–	–
Germany	32	0.09	0.45	0.05	0.45
Greece	49	0.10	0.18	0.03	0.18
Ireland	23	0.03	0.05	0.03	0.05
Italy	30	0.15	0.19	0.15	0.19
Luxembourg	8	0.09	0.12	0.07	0.1
Netherlands	43	0.07	0.45	0.15	0.45
Portugal	38	0.03	0.51	0.03	0.35
Spain	54	0.14	0.58	0.14	0.58
Sweden	25	0.06	0.10	0.06	0.10
United Kingdom	41	0.06	0.06	0.06	0.06

Source: own elaboration of Teligen data.

variables. The model considers a panel for the EU-15 countries over the period 1990–2005, which, however, greatly shrinks given the presence of missing values. Moreover, due to the scarce availability of time series for the mobile telephone sector, and the consequent reduced number of possible control variables, a dynamic specification of the panel data model is preferred, in order to capture part of the variability of the price variables through their dynamic effects. Furthermore, concerning the regulatory variables, the entry regulation indicator does not show sufficient variability in the period of investigation, and thus has not been considered in the econometric analysis.

The results of the estimation procedure, performed through the Arellano–Bond GMM technique, are reported in Table 4.7. Each model is specified by using the general *Regulation Index* for the Telecom sector first, and subsequently by using the specific indicators for the mobile sector, that is *Public Ownership* and *Market Structure*. Other than the lagged endogenous variable, an indicator of the demand for the service (number of mobile subscribers standardized with the population of the country) and an indicator of the technological developments (number of ISDN channels standardized with the population), have been included as control variables. The standardized number of mobile subscribers has been considered as endogenous, and thus instrumented. The first two columns report the estimated results for the peak price, while the following pair of columns refer to the off-peak price. Interestingly, the estimated coefficient for the *Regulation Index* is not significant in either case, while when concentrating on the specific aspects of the

Table 4.7. Mobile telephony price (log), estimation results

Variables	Mobile price peak[1]		Mobile price off-peak[1]		Connection charge[1]		Monthly subscription[1]	
	(1)	(2)	(3)	(4)	(5)	(6)	(7)	(8)
Regulation index[1]	−0.086		0.101		2.468***		1.647***	
	(0.134)		(0.108)		−0.625		−0.522	
Public ownership[1]		−0.233		−0.097		0.418		−0.983*
		(0.149)		(0.112)		−0.513		−0.525
Market structure[1]		0.688**		0.390		8.421***		6.839***
		(0.338)		(0.292)		−1.653		−1.353
Lagged mobile price peak[1]	0.610***	0.402***						
	(0.110)	(0.141)						
Lagged mobile price off-peak[1]			0.514***	0.449***				
			(0.098)	(0.101)				
Lagged connection charge[1]					0.799***	0.741***		
					−0.04	−0.044		
Lagged monthly subscription[1]							0.787***	0.699***
							−0.04	−0.042
Mobile subscriber/pop[1]	0.041	0.017	0.020	−0.017	0.138	0.255***	0.247**	0.461***
	(0.057)	(0.077)	(0.048)	(0.064)	−0.09	−0.096	−0.097	−0.105
ISDN channels/pop[1]	0.009	0.012	0.003	0.002	−0.033	0.023	−0.024	0.016
	(0.008)	(0.008)	(0.005)	(0.005)	−0.033	−0.036	−0.027	−0.03
Constant	−0.067*	−0.050	−0.039	−0.022	0.038	0.051	−0.065	−0.082
	(0.036)	(0.034)	(0.025)	(0.025)	−0.069	−0.067	−0.065	−0.063
Observations	70	56	107	92	274	262	271	259
Countries	11	10	14	14	15	15	15	15

Standard errors in parentheses. For the regulatory variables see Table 4.A.1.
*** $p<0.01$, ** $p<0.05$, * $p<0.1$.
[1] Log.
Source: own elaboration on ITU data.

liberalization process only the *Market Structure* variable in the peak price equation is significantly different from zero and with the expected sign. From columns (5) to (8) the estimated coefficients for the connection charge and monthly subscription are reported. As for the price variables, the *Regulation Index* is always significant whereas only the *Market Structure* variable is significant and with expected sign. The common and important result, therefore, is that the *Public Ownership* variable does not play any role in influencing the price components. Moreover, concerning the monthly subscription equation, it enters with a negative sign and with slightly significant magnitude. All the autoregressive components are significant, indicating a strong persistence in the price components, and among the control variables the dimensionality indicator has a positive and significant effect while the technological proxy is not significantly different from zero in all regressions.

As for the fixed telephony sector, the only positive contribution to the reduction of prices seems to be provided by the decrease of the market share of the incumbent, while the ownership of the incumbent is not significant.

4.10 Summing up and a glimpse into the future

After a brief discussion on technology features, regulation, and market structure and a simple conceptualization of price determinants, we cited previous empirical analyses about the statistical association between fixed telephony prices and reforms in the EU-15. Our empirical findings, either qualitatively, by inspecting TELIGEN data, or with different models using both Eurostat and ITU data, and different econometric specifications, suggest that ownership change, from public to private, plays no role at all or a very limited one in explaining fixed telephony prices, possibly with a minor decrease of prices for the variable part of the local calls tariff. The market share of entrants seems to play a more beneficial role in some models, as one would expect in a more competitive environment, but the result is not very strong, and substitution by mobiles seems much more important. The indicator for entry regulation has very limited explanatory power (or even has the 'wrong' sign for connection charges and local call prices in some models).

Overall, it seems that technology, substitution with mobile telephony, and other demand factors, combined with unidentified country features, have much more explanatory power than any of the regulatory variables. In other words, to put it bluntly, the evidence that the average consumer benefits from privatization of PTOs is negligible, and perhaps less than hoped for from liberalization. For mobile telephony we find again no role of the public or private ownership, while liberalization is clearly beneficial.

Given our illustrative conceptual model, these empirical findings are consistent with a modest wedge between cost inefficiencies of the government owned and the privately owned PTOs in the first place, or with persistent allocative inefficiency in liberalized markets, or with a combination of both features. Moreover, as in fact vertical disintegration has so far been limited in telephony (except in the local loop), it is not to unbundling costs that one has to look in the first place to understand what has happened. It seems likely that both modest (or negligible) inefficiencies, if any, of government-owned PTOs were balanced by modest gains from competition outcomes, after discounting for technology and demand changes, including substitution with mobile telephony. Thus, from the perspective of the consumer, the privatization of fixed telephony in the EU-15—in spite of dramatic changes in market structure—has been less beneficial than expected. There may have been winners and losers across types of consumers, but this is something that we shall discuss later, in Chapter 9.

Moreover, the reform architecture of fixed telephony through privatization and market opening in the 1990s has perhaps underestimated the new challenges facing the industry in a broader socio-economic perspective. Historically, for over 100 years, telephony has been a successful example of connecting citizens and their activities through a powerful networking technology. The copper-wire line covered nations' territories, allowed voice communication between cities and remote areas, and eventually linked continents. Only a few decades ago, postal services took days or weeks to transfer letters and documents. Now we are disappointed if our intercontinental telephone call takes more than one second to get connected. The impact on growth and welfare of allowing everybody instant exchange of information through voice cannot be underestimated.

Governments have played a major role in pushing copper wire everywhere. Most of the infrastructure investment was performed by public (or private) vertically integrated enterprises, under franchised monopoly. Policy-makers, after certain experiences with private oligopolies in the first half of the twentieth century, opted for this arrangement as they considered competition wasteful when the core of the telephony business was linked to building new infrastructure under returns to scale of a fixed investment and network externalities. The essential nature of the service has been so appreciated by governments and citizens that being connected to telephony has been regarded as a core indicator of social progress, for example by the World Bank and other international development agencies. In Germany, art. 87 of the Constitution (Grundgestz für die Bundesrepublik Deutschland) states that every citizen has a fundamental right to a basic level of telecommunication service. The EU directives on universal service obligation should be seen against this background. From this perspective, during the last century, telephony acquired the same policy status as the delivery of water, sewerage, public transport, and energy services.

When the pro-competitive reforms were implemented, in the 1980s and 1990s, voice over fixed telephony was already in the industry's past. A major technological change was the dramatically shifting frontier of what today we understand as electronic communication. From the perspective of the information society, voice transmission on fixed copper lines is a declining service, while the new challenge is how to fully exploit the huge potential of generic technologies allowing any kind of electronic communication to everybody under the Internet protocol and its evolution.

From this angle, the policy issue of the future is that of building the broadband network needed to accommodate the necessary bandwidth for a range of information services. The investment effort involved in this technological shift is very large, and can be compared with the early building of the telephony network: a nation-wide long-term project involving every kind of territorial and social circumstances and, on an EU scale, tens of millions of households, firms, and organizations. This issue has been widely acknowledged by European institutions and governments since the adoption of the Lisbon Strategy. It is also a topic high on the agenda of the US government. Under the Technology Opportunities Program (US Congress, 2009) federal funds are allocated to provide broadband access to consumers in unserved or underserved areas. Holznagel et al. (2010) offer a review of some ambitious national plans to support broadband penetration, including China and Australia. See OECD (2007) for an analysis of the potential economic impact of the new technologies.

In terms of penetration, data from the 'Progress Report on the single European Electronic Communications Market 2008' (EC, 2009), show that in the EU-27 broadband lines (of any type) covered on average 22% of the population (subscribed lines per 100 inhabitants). The variability is high, from around one third in Sweden, Denmark, and the Netherlands to around 10% in Romania and Bulgaria. By comparison the penetration ratio is around 27% in the USA and 25% in Japan.

In this context there are two parallel policy concerns: first, the speed of investment while the fixed telephony incumbents still try to defend the rents extracted from the fully depreciated copper networks that they own; second, the sustainability of investment in 'rural' areas.

The latter term includes wide territorial contexts outside the densely populated cities. While in Europe these areas have been successfully reached by fixed telephony, usually thanks to cross-subsidies in the tariff structure managed by government-owned providers, or direct subsidies to households, it is apparent that broadband will not be supplied to them without some form of public engagement. According to Holznagel et al. (2010) there are three reasons for the massive undersupply of broadband outside the urban context: weak legal obligations for the incumbents, technological barriers, and lack of profitability.

First, the Universal Service Directive (EC, 2002) defines the obligation for fixed telephony incumbents to provide a 56 kilobits/second (kbps). This obligation was written having in mind the standard PSTN connection, while the ITU defines broadband as anything faster than the ISDN connection, typically 1.5–2.0 megabits/second (mbps). In the USA, the FCC considers basic broadband as offering service in the range of 768 kbps–1.5 mbps. Germany in its national strategy has defined the objective of at least 1 mbps, and Finland has a plan to provide the same service level to 99% of its population by 2015. Switzerland, outside the EU, has been a first mover in explicitly including broadband in the universal service obligation. It is apparent that the EU Universal Service Directive cannot contribute to spreading the information society beyond the urban context. Moreover, technological advances in content that can be supplied and exchanged (downstream and upstream the consumer terminal equipment) suggest that greater capacity will be needed in future.

Thus, the second obstacle to broadband penetration in sparsely populated areas, which in some countries may comprise 10–30% of households, is the intrinsic limitation of the old twisted pair copper wire to carry enough information. This network was designed to carry voice analogue signal, under a channel-based technology, and must be transformed to do the new job of carrying any type of data. As briefly mentioned in Section 4.2, this adaptation can be implemented by DSL technology, which in its ADSL version can carry up to 16 mbps. The investment needed is an upgrading of some equipment at the users' home and at the distribution cabinet. Thus, the incumbent operators, who own the local loop, have promoted ADSL, as to them this option is much more profitable than creating a new network. However, adapting the existing network does not work in sparsely populated areas, as the speed of the signal drops exponentially with distance in the loop, because of the intrinsic characteristics of the transmission of electrons through the copper wire. Coaxial wire lines are also not good enough for generic broadband because, in spite of very large capacity, it is shared by different users, and is not economical in sparsely populated areas. Other problems are faced by wireless technologies. Wireless local area network (WLAN), using the Wi-Fi standard, typically implies a dense and close net of connections (around 100 metres from the household's equipment and the base stations) and needs many base stations to cover the territory. WiMAX technology uses different frequencies which are licensed. It is more powerful than Wi-Fi and would need fewer base stations to cover the territory. It seems more promising as an alternative to the wire line local loop than as a standalone technology for the core network.

The most important mobile technology, the universal mobile telecommunication service (UMTS) also offers fast access to the Internet, but it has coverage problems. In general, wireless technologies have the disadvantage of using a shared medium, problems of reach, quality of transmission, and some concerns about acceptance of exposure of the population to waves. In spite of earlier hopes, they will not supplant the fixed network. But what will?

In practice, optical fibre technology seems to be the most promising generic solution for the future of broadband. It can supply up to 10 gigabits per second, the cable is not expensive to produce and maintain, and it is largely neutral to different contents and supporting technologies.

However, laying down a FTTH (fibre to the home) network, including sparsely populated areas, would not often be provided by either incumbents or their competitors. Competition will not offer long-term investment in rolling out an entirely new network where revenues are less attractive than elsewhere. This is similar to the reason why in the last century competition did not work in providing nationwide fixed telephony coverage.

While the issue of extending to sparsely populated areas the benefits of the information society may be controversial, or perceived as not a priority by some, the other concern that we have mentioned above is the overall speed of broadband technology adoption through competitive mechanisms. A privatized incumbent in fixed telephony faces several challenges. First, from mobile service providers. Second, from other fixed telephony providers, who under the EU directives and national legislation have the right to access the local loop or other network facilities owned by the incumbent. Third, and most important for the future, VOIP services. Given that the incumbents, even when they are partly owned by governments, as France Telecom or Deutsche Telekom, have no further well defined public interest objectives, their investment strategies are constrained by profitability. As they are listed companies in the stock exchange, value for their shareholders is influenced by short-term news. Hence, it is unrealistic to expect that these firms will push for scrapping their wire line network not just in less populated areas,[14] but also in those urban contexts where they can adopt different solutions to protect their rents.

As we have seen, competition in fixed telephony has done little, from this perspective, to change prices and to wipe away the market dominance of the incumbents. But, and perhaps much more important, it is not clear to what extent competition has promoted or hindered investment in the network of the future by most players. Regulators face a strong trade-off. As the

[14] Rural poverty may be less important than urban poverty in some countries, hence it is difficult to make general statements on this topic. See references in the Further Reading Section on the subject.

Table 4.8. Digital agenda targets

Targets	2009	2011	Note
Fast broadband (>30 MBPS) coverage for all	0	50%	(1)
50% of household taking up broadband subscription >100 MBPS	0	2%	
100% increase in ICT R&D public spending	52%	56%	
Roaming at national prices	45%	50%	(2)
33% of SMEs selling online	33%	37%	
50% of population buying online	74%	86%	
60% of disadvantaged people using Internet regularly	70%	85%	
75% of population using Internet regularly	81%	90%	
15% of population having never used the Internet	82%	89%	
50% of population using e-government	76%	82%	
25% of population using e-government and returning forms	72%	84%	
Broadband coverage for all	94%	95%	(3)

Source: EC Digital Agenda 2010–20.
(1) Target to be achieved in 2020
(2) Target to be achieved in 2015
(3) Target to be achieved in 2013

incumbent gradually and selectively moves to FTTH (or to fibre-to-the-building or fibre-to-the-curb), the entrants are cut off from their previous access to the local loop, hence natural monopoly will be re-established. Alternatively, if the regulator forces the incumbent to allow entry to its fibre investments, these will be delayed. In terms of social costs and benefits, some years of delay in adoption of broadband may be a substantial welfare loss.

The fact that governments and the EU[15] are now launching publicly funded plans to close the digital divide across regions and social groups is perhaps a signal that some of the problems encountered by telephony in its early days are still with us.[16] See also Table 4.8 on the digital agenda for Europe targets and progress. The key issue of past times, which was only solved by nationalization and a franchised monopoly, was that of enforcing the political view that universal access to voice communication was a national priority. The issue of the future is how to offer to citizens universal access to electronic communication, in spite of high fixed investment costs in new networks,

[15] Legal obligations and pricing regulation alone seem unhelpful for solving the lock-in situation of competitors in providing new investment. The EC is reconsidering its own state-aid legislation in this perspective, see the 'Commission's Guidelines on the application of EU state aid rules for public funding of broadband networks' (EC, 2009). These guidelines could open the way to public funding of projects by municipalities or governments, including PPPs. For instance, building common ducts for electricity, gas, water, and fibre are exempted from state-aid prohibition.

[16] This does not necessarily imply that the digital divide would not have happened in the absence of reform. In some countries under public ownership in the middle of the twentieth century there were long queues for connection, even if this may have reflected a booming demand and technical and financial constraints on the supply side. In any case, it is apparent that market solutions have their own problems, and public finance is still needed for rolling out the fibre.

coupled with very low marginal cost of traffic supply, decreasing revenues, and a huge beneficial network externality for the society.

4.11 Further reading

Melody (2011) offers a concise introduction to the monopolistic position of AT&T in the USA, after the initial two decades of network competition (1895–1915), telecom reforms (1968–84), further liberalization (1984–98), and the current global trends (1998–2010).

On the trade-offs faced by regulators when they move from price regulation to full liberalization, see Sidak and Spulber (1998) on the early debate in the USA on telecommunications and other networks. There are several papers on the empirical analysis of regulatory agencies in telecommunications. Estache, Goicochea, and Manacorda (2006) use cross-country panel data 1990–2003 and find different effects of private capital and independent regulations on several variables. They also provide a brief discussion of some of the papers we cite in this chapter, and also of Gual and Trillas (2006), Wallstin (2003, 2004), Gutierrez (2003), Fink et al. (2003), and Ross (1999, 2003). Most of these papers do not report direct estimates on prices, but on other performance variables. On investment see Friederiszick et al. (2008). Country case histories in the EU often show great variability of pattern. For example in Finland until the 1990s local telephone traffic was managed by cooperatives, while the long-distance traffic was managed by the post office. Subsequent reforms led to the establishment in 1998 of Sonera, a listed company partly owned by the government, and one of the early leaders in technology change. Hatternman and Murphy (1998) describe the issue of local monopolies, while Willner (2003) offers a more general analysis of privatization in Finland. On Sweden see Lindskog (2004). In the Netherlands the Dutch PTT was transformed in the royal KPM, then listed on the Amsterdam and New York stock exchanges in 1994, with complete privatization in 2006. KPM is still the dominant player, with a large number of competitors, however (16 in 2010). See Van Damme (2004) for privatization in the Netherlands. Servaes (1995) offers an account of some early reform in the Netherlands telecommunications.

An outlier in EU telecommunication is the small and rich Luxembourg where the state-owned incumbent still has a near market monopoly, a situation challenged by the legal infringement procedures of the EC.

On Denmark see NTA (2007) and OECD (2000). Here full liberalization happened in 1999, in advance of the schedule dictated by the EC. There are now 65 operators in this small country, many of them owned by foreign shareholders. Prices have fallen significantly and are now lower than in most EU-15 countries. On privatization and liberalization of telephony in Ireland see Palcic and Reeves (2010) and ODTR (2001). Palcic and Reeves (2011) critically discuss the role of employee shareholding schemes (ESOPs) in privatization of Telecom Éireann.

An interesting paper providing a perspective on the welfare impact of reforms in Portugal is Doganoglu and Pereira (2006). In spite of liberalization the privatized incumbent group PT still has a market share of 78% of the residential fixed telephony market, but liberalization has benefited consumers of regional/national and international calls, though not for local call prices, which have considerably increased.

Turning to the major players, see ARCEP (2011) for a recent analysis of fixed telephony in France and several previous reports by this regulator (established in 1997). Currently the share of the entrants is greater than 40%, having substantially constrained the dominance of France Telecom, which is now owned by mostly private shareholders (around 75% of equity).

On Telecom Italia see Florio (2007). On Great Britain several interesting reports are available on the regulator's website (OFCOM: www.ofcom.org.uk). Attemborough (1992) discusses price rebalancing in the first years after privatization and their welfare impact.

On early reform steps in Germany see Federal Ministry of Economy and Technology (1999) and for updates several reports of the OECD Communication Outlooks. On the dominant position of Deutsche Telekom, see European Court of Justice (2010).

On Spain see Jordana and Sancho (2004), Tejeiro and Ruiz Cañete (1999) and Arocena (2003).

There are some interesting issues raised by the comparison of regulatory issues in fixed telephony and mobile telephony. One of these is how to account for costs in the regulation of termination rates. The EC has taken the view that some costs should be excluded (coverage cost, recovery of fixed costs) and this would lead to very low termination rates (EC, 2009). Frontier Economics (2008) suggests that pushing down those prices in the EU (close to the US low level) could decrease consumer welfare because of a 'waterbed effect', that is tariff rebalancing of call prices, connection charges, and rentals. This seems to be a good example of the paradoxes of piecemeal price regulation. Demand-side network effects can be encountered in many industries and telephony is an important example. Koski and Kretschemer (2004) define them as the general property that utility of a product increases with the number of users. They are sometimes called 'demand-side' economies of scale, suggesting that network effects are similar in their outcomes to economies of scale on the supply side. They observe that TPA of local loops by law as in the

EU affects the R&D strategy of the incumbent, as there is a spill-over to entrants of benefits of innovation. This leads to potential disincentive to invest, if access charges do not counterbalance this effect. They review some earlier findings on the topic of R&D, network effects, and pricing issues. They also review some literature on other related issues, such as entry deterrence and the appropriate public policy to deal with demand-side economies of scale, abuse of dominant position, securing efficient entry, promoting generation charges, and battles on telephony standards, see also Economides and White (1994).

On the important issue of infrastructures in telecommunications, including broadband, some reading may include Aron and Burnstein (2003), Bourreau and Pinar (2004 and 2005), Garcia-Murillo (2005), Hausman and Sidak (2005), Jorde and Sidak (2000), Kotakorpi (2006), and Röller and Waverman (2001).

Some of the issues raised by the debate on the network infrastructure are the following: a) definition of the network, as in fact in telecommunications there are different platforms, which in turn are composed of different hardware and software; b) interplay between access regulation and investment incentives, given some apparent trade-offs for the owner of the infrastructure; c) types of unbundling, which can go from the milder form of accounting separation to functioning/management separate business units, until a strong ownership barrier prevents service providers owning the network(s); d) regulatory trade-offs between promoting service versus platform competition; e) public policy about the digital divide and other issues of access. The costs and benefits of different combinations of legislation and regulation are very uncertain. OECD (2001, 2003) acknowledged that there is inconclusive evidence in favour of separation. Röller and Waverman (2001) find evidence of causality between economic growth and telecom infrastructure. Others tend to deny this (Riaz, 1997). On broadband access see OECD (2007, 2008) and Preston et al. (2007).

Another interesting and controversial topic is network neutrality and pricing mechanisms that can discriminate among types of utilities (Cave and Crocioni, 2007; Ofcom, 2006; Thorngren, 2006).

Biglaiser and Ma (1999) explain why competition can deter long-term investment. Bourreau and Pinar (2005) argue in favour of platform competition, but Hoffler (2005) finds that social welfare can be decreased because of wasteful duplication of investments. De Bijl (2005) and Woroch (2002) discuss the alternatives of platform and service competition, see also Hausman and Sidak (2005) on the possible beneficial effects of allowing increasing prices to incentivize network expansion.

On the 'essential facility' nature of telecom bottlenecks see Buehler (2005). On discrimination see Mandy (2001) and Reiffen and Ward (2002). Doh et al. (2004) discuss public and private ownership of the network. On the role of public effects in broadband development and the digital divide see Holznagel et al. (2010).

On the more general topic of performance of telecom providers, some helpful reading would include Bortolotti et al. (2002), De Bijl and Peitz (2005), Doyle (2000), Parker (1999), Sidak (2006), Wolak (1996), and Woroch (2002).

Appendix: 4.A.1 ECTR scores

Table 4.A.1 Telephony. Sectoral indicator of regulatory reform

	Weights by theme (b_j)	Question weights (c_k)[1]	Coding of data		
Entry regulation:	1/3		Free entry	Franchised to 2 or more firms	Franchised to 1 firm
What are the legal conditions of entry into the trunk telephony market?		$1/4{*}w^t{*}(1-w^m)$	0	3	6
What are the legal conditions of entry into the international market?		$1/4{*}(1-w^t)(1-w^m)$	0	3	6
What are the legal conditions of entry into the mobile market?		$1/2{*}w^m$	0	3	6
Public ownership:	1/3				
What percentage of shares in the PTO are owned by government?[2]		$1-w^m$	% government ownership/100 * 6		
What percentage of shares in the largest firm in the mobile telecommunications sector are owned by government?		w^m	% government ownership/100 * 6		
Market structure:[3]	1/3				
What is the market share of new entrants in the trunk telephony market?		$1/4{*}w^t{*}(1-w^m)$	6-normalized market share[2]		
What is the market share of new entrants in the international telephony market?		$1/4{*}(1-w^t)(1-w^m)$	6-normalized market share		
What is the market share of new entrants in the mobile market?		$1/2{*}w^m$	6-normalized market share		
Country scores (0–6)			$\Sigma_j b_j \ \Sigma_k c_k$ answer$_{jk}$		

[1] The weight w^m is the OECD-wide revenue share from mobile telephony in total revenue from trunk, international, and mobile. The weight w^t is the OECD-wide revenue share of trunk in total revenue from trunk and international telephony.
[2] 'PTO' stands for 'public telecommunications operator'.
[3] For the purposes of calculating the indicator the market share of new entrants has been normalized to be between 0 and 6 with 6 being the smallest market share over all countries and time and 0 being the largest.
Source: OECD.

Appendix 4.A.2: The 2009 EU reform package for telecoms

Twelve reforms are included in the new package agreed between the European Parliament, the Council of Telecoms Ministers and the Commission on 5 November 2009:

1. Portability and contracts: right of European consumers to change, in one working day, their fixed or mobile operator while keeping their old phone number. The maximum initial duration of a contract signed by a consumer with an operator will be no longer than 24 months. Operators must also offer consumers the possibility of agreeing to a contract with a maximum duration of 12 months.
2. Information: Consumer contracts must specify, among other things, information on the minimum service quality levels, as well as on compensation and refunds if these levels are not met, subscriber's options to be listed in telephone directories and clear information on the qualifying criteria for promotional offers.
3. Internet: Any measures taken by Member States regarding access to or use of services and applications through telecoms networks must respect the fundamental rights and freedoms of citizens, as they are guaranteed by the European Convention for the Protection of Human Rights and Fundamental Freedoms and in general principles of EU law.
4. New guarantees for an open and more 'neutral' net: National telecom authorities will have the power to set minimum quality levels for network transmission services so as to promote 'net neutrality' and 'net freedoms' for European citizens. Consumers must be informed about the nature of the service to which they are subscribing, including traffic management techniques and their impact on service quality, as well as any other limitations (such as bandwidth caps or available connection speed). The Commission will report regularly on the state of play in net neutrality to the European Parliament and the Council of Ministers.
5. Consumer protection against personal data breaches and spam: Names, email addresses, and bank account information of the customers of telecoms and Internet service providers, and especially the data about every phone call and Internet session, need to be kept safe from accidentally or deliberately ending up in the wrong hands. Operators must respond to the responsibility that comes with processing and storing this information. Other provisions are related to spamming.
6. Emergency services, 112: Extending the access requirements from traditional telephony to new technologies, strengthening operators' obligation to pass information about caller location to emergency authorities, and improving general awareness of the European emergency number '112'. Provision on the availability of terminal equipment offering the requisite services and functions for users with disabilities.
7. National telecoms regulators' independence: Eliminating political interference in their day-to-day duties and adding protection against arbitrary dismissal for the heads of national regulators.

8. A new European Telecoms Authority: The new European Telecoms Authority 'BEREC' (Body of European Regulators for Electronic Communications) will replace the loose cooperation in the 'European Regulators Group'. BEREC decisions will be taken, as a rule, by majority of the heads of the 27 national telecoms regulators: by a simple majority when BEREC gives opinions in the context of the Commission's analysis of remedies notified by national regulators, and by a two thirds majority in other cases. Such BEREC decisions will be prepared by an independent supranational office with expert staff. BEREC will also advise, support, and complement the independent work of national telecoms regulators, especially when it comes to regulatory decisions with a cross-border relevance.
9. EC and competition remedies for the telecoms markets: The new EU telecoms rules will give the European Commission the power to oversee regulatory remedies proposed by national regulators (e.g. on the conditions of access to the network of a dominant operator; or on fixed or mobile termination rates). When the Commission, in close cooperation with BEREC, considers that a draft remedy notified by a national regulator would create a barrier to the single market, the Commission may issue a recommendation that requires the national regulator to amend or withdraw its planned remedy.
10. Functional separation: National telecoms regulators to be able to oblige telecom operators to separate communication networks from their service branches, as a last-resort remedy.
11. Broadband access for all Europeans: The reform will help overcome this 'digital divide' by better managing the radio spectrum and by making it effectively available for wireless broadband services in regions where building a new fibre infrastructure is too costly; and by allowing Member States to expand universal service provisions beyond narrow-band Internet access.
12. Next generation access networks (NGA): On the basis of the new rules, the Commission plans to issue a recommendation for the regulation of access to NGA networks. The rules governing the sharing of network elements, such as ducts or in-building wiring, between operators are also updated by the reform.

Source: Adapted from European Commission, DG Information Society, available at http://ec.europa.eu/information_society/policy/ecomm/tomorrow/reform/index_en.htm.

5

Testing the paradigm: Electricity

5.1 Introduction[1]

Electricity is a crucial commodity in an economy as it serves both as an important input in industrial production processes and as an essential final service to households. Data for the European Union show that in 2010 the electric power industry accounted for approximately 5.6% of the total consumption of fixed capital and 2.8% of total value added.[2] Between 2000 and 2010, total installed capacity in the EU-27, expressed in megawatts (MW), increased by approximately 2.6%, and reached the figure of 870 GW,[3] compared to around 1120 GW in the USA in 2009.[4]

As of 2010 in the EU-27, household energy consumption accounted for around 25.6% while industry consumption, excluding transport, accounted for 24% of total consumption.[5] However, the increased level of energy efficiency in advanced economies is causing a general decrease in the consumption of energy, with an average annual rate, at the EU level between 1995 and 2009, of–1.3%,[6] while in the service sector this trend has been accompanied by an increase in the consumption of electricity per employee[7] mainly due to increased use of IT and other electrical equipment (such as the increased use of air conditioning in warmer months in Southern European countries, changing consumer attitudes, and the widespread use of computers in offices). Overall, however, these contrasting trends highlight the importance of the electricity industry and the potential repercussions of price variations on economic growth and consumer well-being.

[1] With Chiara Del Bo and Carlo Fiorio. This chapter draws also in some sections on Fiorio, Florio, and Doronzo (2008) and on Fiorio and Florio (2010).

[2] Eurostat data, authors' elaboration.

[3] Eurelectric, 2011.

[4] Edison Electric Institute data.

[5] Eurostat, authors' elaboration.

[6] Eurostat, authors' elaboration.

[7] European Environment Agency, 2011.

Traditionally, vertically integrated monopolies were a response to the special features of electricity supply, such as the need to carefully standardize some technical parameters (voltage, frequency, and others), non-storability in the longer term, the physical laws requiring instant equation of supply and demand, and the need for a complex and integrated network which must be controlled by a system operator. The diffusion of generation from renewable sources, such as wind and solar photovoltaic, which are typically intermittent, not precisely predictable, unevenly distributed in space, and non-storable, highlights the need to increase the feasibility of storing electricity efficiently for longer periods of time, or managing the system by a mix of technologies. There are today several options for grid energy storage, which are, however, characterized by varying levels of efficiency. Traditionally, the most widely used technique is pumped-storage hydroelectricity, where energy is stored in the form of water, taking advantage of the height differences between bodies of water, and is considered cost-effective and energy-efficient, with 70% to 85% of the electricity used to pump water that can be regained.[8] It is estimated that the EU had, in 2008, 40.3 GW net capacity in this form, corresponding to almost 10% of total net electricity capacity.[9] This solution is, however, not available for many countries, and in any case covers the possibility of storage to only a limited extent.

Despite these special systemic features, however, a wave of reforms promoting competition in this industry has been initiated in most markets, including the USA and Europe, accompanied by regulation. This chapter offers an overview of the technological and economic features of electricity markets and provides an empirical analysis of the effects of privatization and liberalization on the residential prices of electricity.

Section 5.2 introduces some technological features of electricity supply. In Section 5.3 a brief description of the industry and the rationales for reform is presented. Some features of the electricity industry and its reform in selected EU countries are then reviewed, thus adding additional details to the initial remarks in Chapter 2 (Section 5.4). In Section 5.5 earlier empirical literature is mentioned, while Sections 5.6 and 5.7 present an empirical analysis of country panel data on residential prices in the EU-15. A discussion of new electricity policy in the UK and its implications follows in Section 5.8 while Section 5.9 concludes.

[8] Electricity Storage Association's website, http://energystorage.org/tech/technologies_technologies_pumpedhydro.htm, retrieved online 17 February 2012.

[9] Authors' elaboration on International Energy Statistics data, available at http://www.eia.gov/cfapps/ipdbproject/iedindex3.cfm?tid=2&pid=82&aid=7&cid=CG1,&syid=2004&eyid=2008&unit=MK, retrieved online 17 February 2012.

5.2 Technological and regulatory features

5.2.1 *The core principles*

A distinctive feature of the electricity industry is the use of multiple inputs and technologies (coal, natural gas, oil, nuclear power, geothermal, hydropower, solar, wind, biomass) to produce a highly standardized,[10] mostly non-storable product, with a variable demand over the day and year. At every point in time, production must exactly balance demand, making electricity a somewhat special commodity. In fact, the lack of a careful balance would disrupt the service. The industry can be described as including four different activities: generation, transmission (the high voltage network), distribution (the middle[11] and low voltage network), and retail (supply to final consumers). These four activities or industry segments present different technological and economic characteristics. Generation is often considered potentially competitive, because economies of scale in most types of production process are said to be negligible. Transmission and distribution are natural monopolies, at the cross-border, national, and regional level, because of the high network fixed sunk costs. Retail supply is often seen as potentially competitive, because trading and marketing activities do not necessarily imply high fixed costs. This conventional view of the industry, however, runs the risk of ignoring the industry's systemic nature, which in turn makes it unique. To re-assess this issue, it is perhaps helpful to go back to some basic technological notions.

Since the middle of the nineteenth century, the core technology of the electricity supply industry has been based on the conversion of mechanical energy into power. This applies to almost all the generation process, with the—until now—marginal exception of some forms of solar and chemical conversion processes. The basic physics involved has been known since Faraday (1791–1867), and Maxwell (1831–79) and establishes a relationship between magnetism and electricity, such that a changing magnetic field induces an electric field (Faraday's Law). Typically, any major generator in the industry is based on a stationary conductor, with a magnet connected to a rotating shaft. The rotation induces a current in the stationary conducting ring wrapped by a continuous wire. Thus, mechanical energy is needed to rotate the shaft and create an alternate current, which is then transported, transformed, and distributed elsewhere.

In fact, the key technological issue is the energy efficiency of this process, as part of the thermal and mechanical energy is unavoidably lost in the process of conversion to electricity. Most of the electricity supplied in the world is still based on some type of turbine or other related equipment, where the

[10] See, however, the discussion on the quality of electricity supply in Chapter 8.
[11] The middle voltage network may sometimes locally belong to the transmission network.

mechanical energy is derived from the potential and kinetic energy of a fluid, such as water, wind, and, most often, steam. In the latter case, water must be heated, by burning fuels such as coal, gas, diesel, or biomasses, and then the steam is forced to propel the blades mounted on a shaft. This, in turn, moves the magnetic field at the desired rotation speed, and the current is generated. Nuclear power plants, generating heat from fission, also produce steam which drives a turbine connected to a generator.

The relationship between this generic electricity generation technology and the origins of the industrial revolution itself, based on James Watt's steam engine, from this perspective is still very close (appropriately the Watt is the power unit). The initial ingredients are mostly the same: fuel, a furnace to burn it, a boiler with water to create steam, a condenser, a shaft to transmit motion, and a cooling medium. Motion in the nineteenth-century steam engine went directly into wheels or tools, while in the contemporary electricity generation process it is applied to the rotator of the magnetic field.

While declining over time, coal is still the most important fuel for energy generation around the world, with another fossil source, natural gas, having taken the second position. For example, in the USA, around 2010, coal was the source of nearly half of the generated power, gas (and nuclear) making around one fifth each, and the remaining shared by hydropower, wind, oil, geothermal, biomass, garbage, and others. As mentioned, only photovoltaic energy (and other very minor sources) is based on completely different principles that do not involve a rotating machine. We shall elaborate later in more detail on the mix of primary energy sources in different EU countries.

Thus, in spite of the wide variety of currently available energy sources, it should be acknowledged that most of these sources do not differ in the role they play in the fundamental electricity generation principles. The available technologies, however, do differ greatly, as we shall discuss below, in terms of technical efficiency, capital and operative costs, and externalities. The tension between this variability of generation techniques and the constant need to balance the system poses planning trade-offs which are, in turn, related to long-term investment decisions.

After being produced, power flows immediately through cables to transformers, which upgrade the voltage of the AC current in such a way that high voltage transmission is possible. With this operation, further efficiency in the system is ensured, since losses in transport are minimized. Then substations are needed, which reverse the process and downgrade the voltage to more manageable and less risky values, until it reaches end users at, typically, 220 V in Europe and 120 V in the USA. From the medium tension (typically 15–20 KV) substations this power is further transported and then transferred to low tension and distributed to the users' premises, where it runs our lamps, computers, and washing machines, or is stored in the batteries of our mobile phones.

At any time, all switched on generators of any technology and vintage, transformers, high voltage lines, substations, and any switched on electrical apparatus of the end user, form a closed circuit where charges flow from negative to positive. From this perspective, the system is physically integrated at any time, whoever the owners of segments of it are and whatever their contractual arrangements, and must be balanced, or it will burn out somewhere because of overload of its components. This makes the electricity supply industry unique from a technological perspective, and perhaps the best example of a network service. The important point here is that different generation processes and primary sources in fact often play a different role in an integrated system, and 'competition' at the generation level can take different forms, but is crucially linked to the overall functioning of the system, from provisioning of primary sources to users' demand.

Given that only a minor share of the electricity flow can be efficiently and cost-effectively stored, to balance demand and supply in a technical sense (not necessarily in an economic sense) one transmission system operator must be in charge of dispatch.[12] This is the activity required to correctly switch the generation plants on or off, monitor the flows, the transmission grid, and other activities. Dispatch is managed in real time by control centres, where the system operator monitors tens or hundreds of control screens, which visualize the flows of hundreds of main transmission lines, including interconnections with foreign countries. For example, Terna, the government-owned company in charge of this activity for the Italian electricity system, manages the system in one place, monitoring the flows by means of approximately 100 control screens, which visualize the flows of 293 main transmission lines, including 9 interconnections with foreign countries, and 3 submarine cables. No other industry, including for example the management of airlines, has such a special pervasive central control mechanism. This fact has important consequences.

Under a vertically integrated electricity system, the control centre of the system operator, supported by dedicated software, balances supply and demand by switching on the generators according to a 'merit' order, which implies knowledge of the incremental cost, and other functional aspects, of supplying electricity at that point in time from any specific point. When the owner of the control centre and the grid is the same, as for most or all the generators and distributors, the merit order should imply cost minimization in the short run for the vertically managed firm or organization.

In principle, this process is not different under an unbundled electricity industry, but is more complex in terms of information flows, transaction costs, and incentive mechanisms for investment. Under the reform paradigm

[12] There are important differences in possible roles of the TSO, see Glachant and Perez (2011) for a comparison of the UK and the USA.

of the last two decades, however, it was maintained that such costs would have been less than the benefits of competition, particularly in terms of cost minimization in the longer term, because of shedding excess labour and faster adoption of innovative technologies.

In fact, competition is possible mostly in generation, and to a much more limited extent in retail sales. It is virtually impossible on the network itself, as duplicating transmission cables, transformers, substations, and other equipment would be costly and the balancing of the loads much more difficult. Competition downstream in turn is mainly related to differences across firms in commercial activities, advertising, billing and metering, and so on and it does not usually offer great margins to independent retailers if they actually need to compete. We shall not discuss these aspects here, because they do not substantially differ from other utilities (except for innovation in metering).[13]

5.2.2 *Costs of generation*

As far as generation is concerned, the crucial issue, as mentioned, is the variability of production costs across firms and across technologies in the long run. We focus here on the latter issue, because this is at the forefront of the current energy policy debate. The short-run pricing equilibria in the markets are only to a limited extent reflecting the marginal social cost of production. Looking at the costs from the limited perspective of the generator fails to capture social costs.

As we mentioned at the beginning of this section, the key efficiency issue in electricity supply based on fuels is not related to advances in the fundamental principles and engineering (with the exception of nuclear generation, where investment and innovation in security is highly technology-specific) of the generation itself, but in the efficiency of converting different fuels into mechanical energy, then into electrical energy. This is an important point, as—in a sense—nuclear, coal, gas, and oil, all do the same job of creating movement through steam[14] moving a turbine. Wind, hydropower, and in part CCT (Clean Coal Technology) gas use a fluid directly, but they perform the same conversion between different types of energy. The differences, however, in the capital and operative costs and features of the different sources are substantial, and to a certain extent are not well understood or known, particularly because of interrelations with other parts of the system. Several indirect and external costs should be added to these to gain a full picture.

[13] Quality of electricity supply is discussed in Chapter 8.

[14] In the combined cycle gas turbine (CCGT) the gas turbine is coupled with a steam turbine to generate additional electricity.

In the rest of this section we briefly discuss three interrelated issues: direct costs of electricity generation by technology, environmental externalities—that is additional social costs, and possible cost savings related to the demand side of the industry.

An indicator which is often used by practitioners to compare unit generation costs is LEC (Levelized Energy Cost), which is defined as the price needed to break even when electricity is generated by a median plant using a given source of energy.

A simple formula for the computation of LEC is:

$$LEC = \frac{\sum_{t=1}^{n} \frac{(I_t+OM_t+F_t)}{(1+r)^t}}{\sum_{t=1}^{n} \frac{E_t}{(1+r)^t}}$$

where t from 1 to n are years of operation, with n the year of decommissioning (usually up to 40 years), I is yearly investment, OM yearly operation and maintenance costs, F are fuel costs, and r is the discount rate. It is very difficult to compare these data across technologies because of different assumptions, for example about incentives, taxes, boundaries of the object of assessment (for example system costs related to the connection to the network), discount rates, time horizons, whether base or peak load is considered, or an average.

Figure 5.1 gives some data for the USA, based on EIA (2010) and other national sources and combines different recent surveys, originally expressed in different currencies for megawatthour, in different countries, with a simple indicator; with the cost of coal generation, which as mentioned is still often

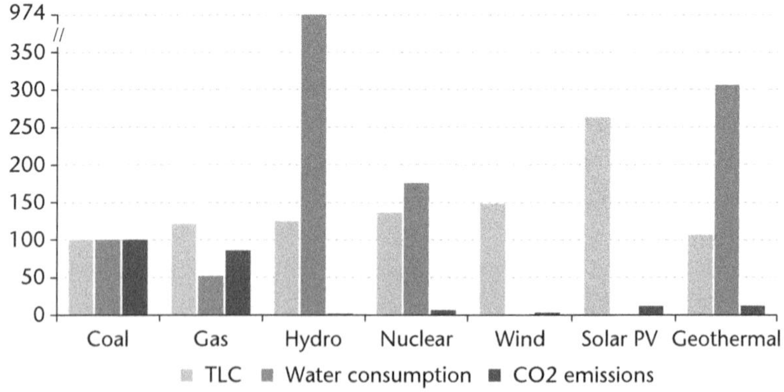

Figure 5.1. Total levelized costs (TLC), water consumption and CO2 emissions (base = 100 coal)

Sources: own elaboration based on Energy Information Administration (2010); Parsons Brinckerhoff (2010); Graham (2006); Nuclear Energy Institute factsheet (2007).

Table 5.1. Estimated levelized cost of new generation resources, 2016

Plant type	Capacity factor (%)	US average levelized costs (2009$/mWh) for plants entering service in 2016				
		Levelized capital cost	fixed O&M	Variable O&M (incl. fuel)	Transmission investment	Total system levelized cost
Conventional coal	85	65.3	3.9	24.3	1.2	94.8
Advanced coal	85	74.6	7.9	25.7	1.2	109.4
Advanced coal with CCS	85	92.7	9.2	33.1	1.2	136.2
Natural gas-fired						
Conventional combined cycle	87	17.5	19	45.6	1.2	66.1
Advanced combined cycle	87	17.9	1.9	42.1	1.2	63.1
Advanced CC with CCS	87	34.6	3.9	49.6	1.2	89.3
Conventional combustion turbine	30	45.8	3.7	71.5	3.5	124.5
Advanced combustion turbine	30	31.6	5.5	62.9	3.5	103.5
Advanced nuclear	90	90.1	11.1	11.7	1	113.9
Wind	34	83.9	9.6	0	3.5	97
Wind-offshore	34	209.3	28.1	0	5.9	243.2
Solar PV*	25	194.6	12.1	0	4	210.7
Solar thermal	18	259.4	46.6	0	5.8	311.8
Geothermal	92	79.3	11.9	9.5	1	101.7
Biomass	83	55.3	13.7	42.3	1.3	112.5
Hydro	52	74.5	3.8	6.3	1.9	86.4

* Costs are expressed in terms of net AC power available to the grid for the installed capacity.
Source: Energy Information Administration. Annual Energy Outlook 2011. December 2010, DOE/EIA-0383 (2010).

the most important fuel, taken as 100. Just to give an indicative money value to this normalized indicator, according to EIA (2010), the LEC in the USA, in 2009, for conventional coal was on average 94.8 USD per megawatthour, with a regional variation between 85.5–110.8 USD. The gas combined cycle levelized cost is one third lower than burning coal, while carbon capture and storage[15] is around one third more costly than the traditional, highly polluting process. Wind on-shore costs according to one British study (Parsons Brinckerhoff, 2010) are similar to new nuclear, while offshore wind generation and solar are more costly than nuclear.[16] Figure 5.1 also provides an indicative comparison of the amount of water used by each generating source and the average level of CO2 emissions, once again using coal as the base value (see below). As Table 5.1 shows, the proportion between capital and operating costs is also very different.[17]

As mentioned, the uncertainty surrounding these estimates is high as, for example, wind and sun conditions are site specific, but also because the precise design and vintage of conventional thermal plants have a large impact on costs. Moreover there may be costs which are not accounted for in company accounts, or not fully, and greatly vary across sources. These are, for example, the large differences in cost of disposal of waste, decommissioning costs, implicit costs of discontinuity of generation, and insurance obligations.

Important examples of system costs related to non-carbon technologies are those for wind and nuclear generation. In the case of wind, the load varies in an unpredictable way over time, even within an hour of operation. A grid that could accommodate this source of power should be able to respond very quickly to these changes, in order to avoid disequilibria in demand and supply. Thus wind will never be the elected choice for base load. On the opposite site of the spectrum, nuclear plants are designed to work under a narrow range of generation flexibility, and cannot be easily used to face peak load. In fact, a grid will need to accommodate different sources and generation technologies according to different needs, and this poses difficult issues to be solved if the right mix is not available. For example, if emergency load is provided by diesel generators (as in Italy during the exceptionally cold month of February 2012), the marginal cost of supply will be much higher than otherwise. Thus, which plants provide the reserve capacity needed to face peak load, or to exceptionally counteract a break in supply in some base load or medium load plants, is an important system cost factor. Another example of the hidden costs of a generation source, is provided by the fact

[15] The environmental impact of CCS is also highly uncertain.

[16] See also and Risto, T. and Aija, A. (2008).

[17] On this issue, see estimated costs of different generating technologies in Politecnico di Milano-Dipartimento di Energia (2010).

that some international conventions have exempted the nuclear operators from fully insuring themselves for third party liability, shifting the risk to taxpayers or third parties. The recent Fukushima accident suggests that the insurer of last resort is government, and that the cost per KWh of this shadow cost can be substantial.

Turning to the most important type of external costs, those related to environmental impacts, they can be classified according the type of natural resource involved.

a) As mentioned, most electricity generation plants are based on steam, including nuclear, coal, natural gas, geothermal, and solar thermal. Use of water is at a maximum with hydroelectric, much lower with gas, biomass, or coal, and negligible with wind. Nuclear power plants use water mainly to cool down the condenser, and this is why, for example in Japan, the plants (such as the one at Fukushima) are located near the coast. See again Figure 5.1.
b) The greatest concern with power technology is, however, related to emissions of different pollutants. Burning fossil fuels releases huge quantities of CO2 into the atmosphere, perhaps in the range of 10 billion tons per year for electricity generation (see Figure 5.1). The impact on climate change is a key aspect of the current international debate. The Stern Review (2006) was probably pivotal in precipitating the understanding of the large social costs involved in delaying the decarbonization of the economy, and the UK seems to have taken this concern seriously, with wide implications for the electricity industry (see again Figure 5.1 and below Section 5.8).[18] Costs of CO2 could be internalized if capture and storage techniques became more widely used.[19]
c) While all fossil fuels release CO2, other pollutants are found in different proportions according to the type of fuel. These include ozone, sulphur dioxide, nitrogen oxides, particulate, radioactive elements (which can be found in coal, for example), and other toxic metals, for example, mercury.

Only some of these costs are internalized because of tradable emission permits and other schemes, and the interaction between technology adoption or dismissal and environmental policy is still critical for the electricity supply

[18] See also World Nuclear Association (2011).

[19] A concrete example of Carbon Capture and Storage is Statoil's Sleipner Vest site in Norway, while Vattenfall in Germany operates the first generating facility with CO2 capture (De Leo et al., 2009), http://www.statoil.com/en/TechnologyInnovation/ProtectingTheEnvironment/CarboncaptureAndStorage/Pages/CarbonDioxideInjectionSleipnerVest.aspx [accessed 6 January 2013].

industry. This issue will be discussed in Section 5.8 below, when we report on the recent change in British energy policy.

Eventually, the demand side will be an additional factor of technological change, with potential implications for the industry pattern. For example, Chao (2011) reviews the impact of new standards in demand side management (DMS). These include changes in domestic electric appliances, ICT applied to optimize in real time demand and supply at the user level, and—perhaps most importantly—EU legislation and national directives on the energy performance of buildings. The new legislation sets out that any new building, or any building facing major renovation, should be certified to meet certain energy standards. Substitution of traditional lamps with LED or OLED technology, heat pumps, small solar sources, co-generation of heat and electricity, and so on, are only some examples of a more decentralized electricity system, which may lead to considerable savings, but also to obsolescence of part of the existing infrastructure.

Energy savings, sometimes labelled as Negawatts, that is negative power, represent perhaps the next technological revolution, and it is unlikely that this will see the traditional utilities, which after all sell power, as the core drivers. In fact, most of the change comes from either producers of appliances, or from government regulations. The possibility that the use of electric cars will spread in the future adds to the complexity of the upcoming electricity scenario, see Jamasb and Pollitt (2011) for an interdisciplinary set of contributions on this issue. In turn, local microgeneration systems may have a considerable impact on the design and function of the network.

5.3 Regulatory issues

As mentioned in Chapter 2, traditionally all or some of the activities in the electricity sector were vertically integrated in many European countries, with state or municipally owned enterprises playing an important role. The market was highly regulated, with very limited opportunities for users to switch to alternative suppliers. There was no third party access to the transmission grid. This integrated pattern was the deliberate result of policy reforms that consolidated the mostly private and fragmented European electricity industry in its earlier stages. The governments' view until the 1990s was that for economic, political, and social reasons the previous pattern, mainly based on regional private monopolies or collusive oligopoly, was either inefficient or undesirable (Millward, 2011). Despite some policy convergence in the interwar years, and further reforms in the 1950s, in the following half century there were persistent and significant variations in industry patterns across European countries, in terms of technologies, ownership, governance, per capita-consumption, spatial and vertical integration, market structure, and prices.

The reform wave that has characterized the electricity markets worldwide has been spurred by different policy goals, including especially pro-competitive arguments that link privatizations, liberalizations, and the creation of competitive markets as essential preconditions leading to higher efficiency and to an overall gain for final consumers. Other goals that are gaining prominence include environmental concerns related to pollution caused by electricity generation and the willingness to reduce exposure to price fluctuations and security of supply risks associated with imports of traditional fossil fuels. Later on in this chapter, we shall further discuss the new policy concerns surfacing in some European countries, most notably the UK.

5.3.1 *Liberalization*

Historically, the transition towards a more competitive environment in the electricity sector started with the creation of wholesale spot markets, which have been characterized by a certain degree of variability, across countries and times, in relation to the rules governing them (Sioshansi and Pfaffenberger, 2006). While the basis for price setting is related to marginal cost pricing, with special attention to the peak-load plant, markets can be designed as gross pools—where the price is set by the rules of supply and demand, and a net pool—where generators may enter into long term contracts with retailers.

In Latin America this process of reform, characterized by regulation and privatizations, started in Chile in the late 1970s, followed by Argentina. In Europe, the process started in the 1990s with the creation of the Pool (England and Wales) and the Nordic Pool. Similar processes were taking place in Australia and New Zealand and some US states, notably California. In the former, the reform process started in the mid-1980s with the adoption of incentive regulation, followed by full-fledged restructuring in the 1990s. The (in)famous Californian crisis spurred a vast academic and policy-oriented literature on the lessons learned from reform failure and possible re-entry of the public bodies into electricity generation. Joskow (2001, 2005) suggests that the main culprit is not deregulation and restructuring per se, but is related to errors (and bad luck) in implementation. Haney and Pollitt (2010) cite California, alongside South Africa, Turkey, and Ukraine, as examples of reforms which have stalled and have not delivered the promised benefits to the economy and society. Others, particularly Woo (2001), suggest that the liberalization design was flawed in the first place as it ultimately led to the emergence of a small number of generators which abused their market power. The key issue was thus poor market design in the liberalization package.

From an economic perspective, following privatization and liberalization in the UK in the 1990s and the EU directives over the last ten years (see Chapter 2), a paradigm has emerged, or 'a measure of consensus over some generic measures for achieving a well-functioning market-oriented industry' (Jamasb and Pollitt, 2005, p. 2). For the first time in the history of the electricity industry in Europe, a unique cross-country policy reform pattern has been advocated by international organizations, notably the European Commission, the OECD, and the World Bank for the transition economies.

To briefly sum up the main reform steps in Europe (and to complement the introductory material of Chapter 2), the three most relevant EU electricity directives are presented and progress towards an integrated single energy market is examined. The overall aim of this reform process has been to enhance competition in the electricity market, through the entry of new generating and retailing firms, and to extend the relevance of national markets through the creation of a European-wide market. The underlying social goal has been to benefit from the efficiency gains of competition in terms of lower collusion and abuse of the dominant position, which should lead to benefits to consumers and to society as a whole, including better international cooperation to increase security of supply (Pollitt, 2009).

The EU directive of 1996 (Directive 96/92/EC) set the stage for the creation of a European internal market for electricity, but was lacking concrete provisions in terms of market design. To this end, the 2003 directive (Directive 03/54/EC) provided specific indications on unbundling of transmission and distributions systems, established an independent national regulatory agency, and required free entry into the generation segment of the industry. Finally, the 2009 directive (Directive 09//EC) was prepared; taking stock of the progress needed to achieve the goal of the internal electricity market and including further requirements for unbundling of transmission assets and the creation of transmission system operators within integrated groups. The core ingredients of this latest directive are the following:

a) the unbundling of the grid is restated, but an alternative to ownership separation is introduced through the Independent System Operator model. This model allows the grid to be owned by the generator(s), but the management of the grid is divested to a third party. The only alternative is ownership separation;
b) the national regulators are given wider powers;
c) an Agency for the Cooperation of Energy Regulator (ACER) is established, with some powers related to trans-boundary electricity exchanges and grid investment, and the role of coordinating national regulators.

For additional details see Appendix 5.A.3. The overall European situation is one of increasing compliance with the directives and market opening is ongoing in most of the EU countries. However, assessment of the market impact of this reform process is not so clear cut (see below). For details see Appendix 9.A.1.

The electricity paradigm is usually simplified as the suggestion of three parallel reforms: privatization (sale of existing publicly owned electricity firms and licensing of private generators and suppliers), unbundling (associated with incentive regulation of the networks, third-party-access, establishing and independent regulator), and liberalization (i.e. allowing entry and competition in both generation and retail).

The World Bank (2007) suggests a more detailed list of reform ingredients:

a) De-monopolization and regulation: unbundling vertically integrated monopolies to foster competition in generation and supply; privatizing and shifting the role of the state from owner to regulator; promoting entry by foreign investors; establishing transparent energy markets; building regulatory capacity.
b) Prices and fiscal policy: promoting fully cost-reflective prices; eliminating production subsidies; taxation based on externality correction; enforcing metering and collection of bills; closing uneconomic plants.
c) Foreign trade: opening markets to imports; eliminating taxes on fuels and electricity; strengthening regional trading arrangements; expanding trans-boundary energy connections.
d) Investment policy: relying on energy companies to sustain investment, not on the public sector budget; supporting energy efficiency; increasing flows of foreign capital with appropriate measures.
e) Social protection: safety nets for the redundant staff; social service functions to be transferred to local governments, not to companies; supporting the poor through lifeline tariffs or means-tested subsidies (abolishing cross-subsidies).
f) Environmental protection: supporting environmental assessment; introducing emission norms; mainstreaming new environmentally friendly technologies.

Looking into the details of this comprehensive package of electricity reforms, it seems apparent that many items are only loosely interrelated, and can be implemented under a variety of industry organizations and policy patterns. Thus the degrees of freedom in the reform design are higher than is sometimes suggested.[20] For example, opening a market to imports may be compatible with different ownership regimes. In principle, there is nothing that prevents

[20] See Glachant and Perez (2011). They explain the origins of this large variability.

the import of electricity from abroad when the generation or the transmission network is government-owned. Vested interests against international trade can be strong under either public or private ownership. Unbundling can be legal, accounting, or ownership separation, with quite different implications. Promoting the entry of foreign investors is compatible with weak promotion of trans-boundary physical networks: in fact it would amount to offering rents, due to distorted domestic prices, to a coalition of investors larger than the national one. Vertical integration can be combined with liberalization under effective third party access. Liberalization of generation can be combined with constraints to retail competition. It is thus difficult to establish a compelling case for strongly linking together all the items of the standard reform 'package'. The case for implementing the comprehensive paradigm rests on a mixture of conjecture and evidence, with only limited cross-country empirical research based on standard econometric approaches.

Without empirical testing, however, some of the tenets of the reform paradigm are questionable, or at least depend upon a large number of preconditions. For example, while sometimes the reform paradigm has been justified by dramatic technological changes that were assumed to reshape the electricity industry, for example the new generation process using a combined cycle gas turbine and the decrease of capital costs and of economies of scale, this argument seems to be exaggerated. The mix of energy sources in Europe, as discussed in Section 5.2, is too diverse to provide the basis for a paradigm shift. Nothing of importance in the EU has changed in the last 20 years in most standard generation processes, from nuclear to hydropower, to justify such an overwhelming technological explanation for the paradigm shift.[21]

It seems more reasonable to look at the paradigm as a set of policy reforms based on increased confidence in market forces and private ownership, against the decline in confidence in planning and public ownership, for a number of reasons that we shall discuss in the concluding chapter. Moreover, the ambition by the EC to create an 'internal' market for electricity may contribute to the explanation of the policy reversal more than any compelling technological shift.

From this perspective, the driving force behind privatization of electricity companies is the idea that public ownership is intrinsically less efficient than private ownership, because of an incentive argument. In turn, the rationale for unbundling is to separate the potentially competitive stages from those with natural monopoly characteristics, which may require some public regulation. Eventually, liberalization will bring market forces in the industry, and competition will deliver production and allocative efficiency, hence lower prices, or

[21] Turbogas technology has been mentioned as one factor of change, but it has only been adopted in some countries.

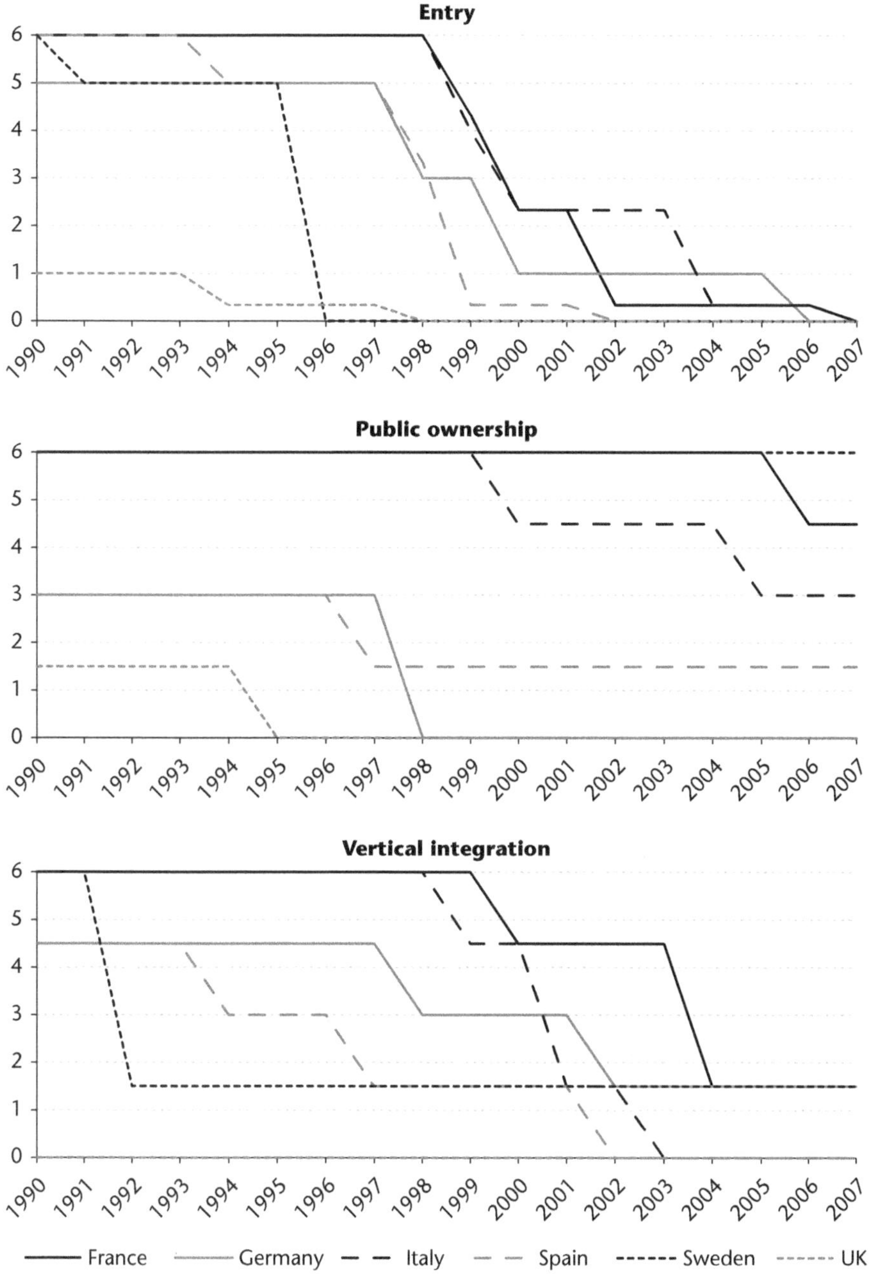

Figure 5.2. Electricity trends of ECTR reform indicators, 1990–2007, selected EU countries

Sources: own elaboration of ECTR scores.

lower mark-up over costs, for users (on this issue, see the previous discussion in Chapter 1).

Figure 5.2 offers a synopsis of reform trends in selected EU countries that will be discussed in the next section.

5.4 Country features

To show how variable the European electricity panorama is, in this section we present some general information on the EU-27 and focus in more detail on five major countries in the EU-15: UK, France, Germany, Italy and Spain, and three smaller Nordic countries: Sweden, Denmark, and Finland. In contrast to telephony, the patterns in technology adoption are important and we thus summarize them below, showing the countries' different energy production mixes, their trade patterns, and how this all translates into very different market structures.

5.4.1 *Sources of energy*

As mentioned in Section 5.2, an important aspect of electricity is that it may be produced with a variety of inputs (see Table 5.2) and the actual fuel mix used to produce a country's stock of electricity may have important effects on several economic variables such as prices to final users, firms' productivity, and emissions. This naturally leads to issues of environmental sustainability and is an important target of policy goals and actions. The following table provides the evolution of the fuel mix of electricity generation in EU-27 in the years 2008 to 2010 compared to the situation in the USA in 2010.

It can be noted that, on average, the share of renewables in the EU-27 is increasing, but the fuel mix is still unbalanced towards fossil fuels, both in Europe and in the USA (see Table 5.3). However, nuclear power and hydropower are more important in Europe than in the USA, especially in countries

Table 5.2. Fuel mix in the EU and USA, 2008–2010 (%)

	2008	2009	2010	%
Nuclear	27.7	27.8	26.5	19.6
Fossil fuel fired	55.2	52.5	52.1	68.7
Hydro	11.0	11.6	11.7	6.1
Other renewables	7.0	8.1	8.8	4.1
—of which wind	3.7	4.3	4.6	n.a.
Other	0.1	0.1	0.9	0.6

Source: Eurelectrics 2011 and Edison Electricity Institute. Own elaboration.

Table 5.3. Share of renewable energy (%)

	1990	1995	2000	2005	2006	2007	2008
EU-27	11.9	13.0	13.6	13.6	14.2	15.1	16.4
EU-15	12.9	13.8	14.6	14.5	15.3	16.5	17.7
NMS (EU-10)	4.1	5.4	5.4	6.4	6.2	6.5	7.2

Source: Eurostat. Own elaboration.

such as France or Italy, respectively. Table 5.3 summarizes the increase in the share of renewable sources between 1990 and 2008 in the EU-27 considering Old Member States (EU-15) and the ten New Member States. Renewable energy sources include wind power, solar power, hydroelectric power, tidal power, geothermal energy, solid biomass, and biogas.

The increasing attention paid to renewable resources as a replacement for fossil fuels is mainly related to environmental and security of supply considerations. Renewables may significantly help in reducing greenhouse emissions, a goal that is explicitly considered by the European Union (see for example Directive 2009/28/EC which has set an explicit quantitative goal of 20% of energy consumption to be produced from renewables by 2020) and may reduce the dependency on energy imports which in turn may reduce risks related to price volatility and producing country political risk, causing potential energy interruptions (Neuhoff, 2007). The main renewable source in the EU, as of 2009, is hydroelectric power, which accounts for 57.6% of the total of renewables, followed by wind (21.3%), biomass (17.7%), and solar (2.4%), with the highest growth rates associated solar and wind (Observ'ER, 2010).

The increase in the share of renewable energy to total electricity consumption is particularly relevant in the NMS, with an average annual growth rate of 4.2%, compared to the 2.1% of the older member countries, but the absolute figure is still significantly lower (7.2% compared to 17%). The highest share of renewables in 2008 is in Austria and Sweden, with the highest growth rates in Hungary and Denmark.

Looking in more detail at the situation in selected EU countries, it is clear that geography and national strategies have shaped a very uneven landscape in the EU. The productive mix in the UK[22] in 2010 comprises roughly 76% fossil fuels (gas 47% and coal 28%, very modest role of oil) with the remaining part being provided by renewables (7%) and nuclear power (16%).[23]

[22] Following the revised Electricity Market Directive (2003/54/EC), electricity suppliers must provide information on the fuel mix necessary to produce electricity sold to consumers. Implementation dates vary across countries, but most information can be accessed at http://www.reliable-disclosure.org/ [accessed 6 January 2013].

[23] Department of Energy and Climate Change (2011).

Table 5.4. Import dependency in EU-27 (Mtoe)

	1990	1995	2000	2005	2006	2007	2008	2009
Imports	1105.89	1095.54	1265.67	1455.54	1494.14	1464.61	1498.16	1407.18
Exports	350.96	377.60	440.57	472.09	485.07	478.16	483.88	463.58
Net imports	754.93	736.08	825.09	983.45	1009.07	986.46	1014.28	943.60

Source: Eurostat.

This pattern can be compared with the situation in France:[24] in 2008 approximately 76% of the electricity produced was generated by nuclear plants, only 10% from fossil fuels, 12% from hydroelectric, and 13% from all renewable sources. This situation leads possibly to much lower direct (short run) production costs in France than in the other EU-15 countries.

The Spanish productive mix is more balanced.[25] In 2008, there was a prevalence of fossil fuels (60%, most of it being gas) followed by solar (10%), hydroelectric (8%), with renewables in total accounting for 19%. The pattern in Italy is very different, where the current productive mix is the result of a political choice (following two referendums) to decommission nuclear power stations. In 2008, 80% of the electricity was derived from fossil fuels, and 15% was of hydroelectric origin, while only 3.4% was from other renewable sources. Germany has a productive mix more similar to Spain's, which includes 61% fossil fuels (mainly coal), 23% nuclear, and the remaining 11% produced by hydroelectric and other renewable sources.

Turning to the Nordic countries, differences in their energy sources for electricity are also impressive. In 2010 the energy in Sweden[26] was produced mainly by hydropower (45.7%) and nuclear (38.3%) with only 13.5% coming from fossil fuels. In Finland, the largest source of electricity was fossil fuels (48%) but a relevant part is also derived from nuclear (29%), with the remaining from hydroelectric and other renewable sources. As regards Denmark, the productive mix is dominated by fossil fuels (71%, mainly coal) with the remaining electricity entirely derived from renewable sources other than hydroelectric. In recent years, Denmark has adopted a bold public policy in favour of wind generation which has led to the result of 20% of national electricity production from this energy source (Haney and Pollitt, 2010).

[24] Data source: IEA Statistics and Balances.

[25] Data source: IEA Statistics and Balances.

[26] Data source: http://www.svenskenergi.se/upload/Om%20el/El%C3%A5ret/Filer/Electricity-Year-%202010.pdf, retrieved online January 20, 2012.

5.4.2 *Trade*

International trade in energy is very small in most EU countries, but not in all of them, and has been growing in recent years (see Table 5.4).[27] This increase is related to an overarching EU goal to create an internal electricity market, as exemplified by the various directives previously discussed. In fact, an implication of the adoption of Regulation (EC) 663/2009 has been the selection for EC funding of 43 additional projects for gas and electricity infrastructure, with a specific focus on electricity interconnectors.

By examining the experience of selected EU countries, the UK is marginally a net importer of electricity since 2004, after 30 years of being an exporter (Bolton, 2010), with net imports contributing 0.7% of electricity supply in 2010 (Department of Energy and Climate Change, 2011). The UK mainly imports from France and exports to Ireland.[28] In 2010 its imports satisfied around 2% of demand. Spain in 2010 was instead a marginal net electricity exporter.[29] France, in contrast, is the most important European net exporter of electricity, with 46 electrical interconnectors linking it to other countries in Europe. Roughly 12% of the electricity produced in the country is sold to neighbouring EU countries such as Belgium, Germany, Switzerland, Italy, Spain, and the UK.[30] Germany is also a net exporter of electricity with data from 2008 showing that the main countries to which it exports are the Netherlands, Austria, Luxembourg, and Switzerland, while a significant inflow comes from the Czech Republic, where one of the country's main producers, EON, has control of several production plants.[31] Italy is a substantial net importer of electricity: roughly 16% of demand in 2010 was satisfied by imports from abroad, mainly from Switzerland, France, Slovenia, Austria, and Greece.[32] Looking at the rather heterogeneous Nordic countries, Sweden imports only a small fraction of its internal demand for electricity (being a net importer in 2009, in contrast with 2008 where the balance was such that the country was a net exporter[33]), while Finland imported almost 15% of its demand in 2009,[34] and Denmark exports a substantial fraction of its production (almost 19% in 2010[35]). Other EU countries with a substantial

[27] With the exception of 2009, possibly due to the financial and economic crisis.

[28] European Energy Regulators (2011).

[29] European Energy Regulators (2011).

[30] RTE Electrical Energy Statistics for France (2010), authors' elaboration.

[31] http://rwecom.online-report.eu/factbook/en/marketdata/electricity/grid/germanyimportandexportofelectricity.html [accessed 6 January 2013].

[32] http://www.terna.it/LinkClick.aspx?fileticket=n%2b631NS8NM0%3d&tabid=418&mid=2501 [accessed 6 January 2013].

[33] http://www.energimyndigheten.se/en/About-us/Press-/Press-releases/Sweden-became-net-importer-of-electricity-in-2009/.

[34] http://www.stat.fi/til/ehkh/2009/04/ehkh_2009_04_2010-03-24_tie_001_en.html.

[35] http://energinet.dk/SiteCollectionDocuments/Engelske%20dokumenter/El/Market%20report%20-%20January%202011.pdf [accessed 6 January 2013].

international trade in electricity are Luxembourg, Netherlands, and Belgium, as imports account for more than 20% of their consumption. Austria and, outside the EU, Switzerland, have established themselves as active electricity cross-border traders.

5.4.3 *Ownership*

An interesting issue to be noted is the presence, among the core shareholders, of state or governmental bodies in four out of the seven major European electricity companies.[36]

By analysing the ETCR sub-indicator for public ownership in electricity (Table 5.5, rows), it appears that in 2007 many European countries were still characterized by a high degree of public ownership, despite the widespread

Table 5.5. Public ownership of largest company and wholesale market liberalization, selected OECD countries

	1980		2007	
	Non-liberalized	Liberalized	Non-liberalized	Liberalized
Private	Belgium		Belgium	Germany Spain United Kingdom
Mixed	Germany Spain			Finland Hungary Italy Luxembourg Norway Portugal
Public	Austria Czech Republic Denmark Finland France Greece Hungary Ireland Italy Luxembourg Netherlands Norway Poland Portugal Slovak Republic Sweden United Kingdom		Poland Slovak Republic	Austria Czech Republic Denmark France Greece Ireland Netherlands Sweden

Source: ETCR. Own elaboration.

[36] Namely: E.ON, GDF Suez, EDF, ENEL, RWE (Thomas, 2010).

process of reform that took place with the reception of the EU Electricity Directives. This fact should be stressed as it demonstrates that liberalized markets (verifiable through the indicator on the existence of a liberalized wholesale market, Table 5.5, columns) do not wipe away the possibility of the direct involvement of public bodies in electricity generation, transmission, distribution, and supply of electricity (Haney and Pollitt, 2010).

Considering the country specific situations, a certain degree of heterogeneity regarding the private or public nature of the largest firms in the national markets is present. Starting with the UK, the firm with the largest market share, NPower RWE, formerly Innogy, is a division of RWE AG, a privately owned listed German multi-utility company. The situation is radically different in France where EDF (Electricité de France) is 84% owned by the French government through the Caisse des dépots et consignations, while the German market is dominated by the privately owned RWE and E.ON. It should be noted, however, that electricity distribution in Germany is also largely provided by state-controlled municipal companies. Moving to the Mediterranean countries, the largest players in Spain, Endesa and Iberdrola, are both listed private companies while ENEL, the dominant player in Italy, is 34.1% government-owned. The situation in the Nordic countries is quite homogeneous, with Fortum in Finland, Vattefall in Sweden, and DONG Energy in Denmark all being state-controlled firms.

5.4.4 *Market structures*

Following the reform process aimed at the creation of an integrated and competitive European electricity market, the market structure in electricity generation has distinctly moved away from integrated local monopolies towards a more fragmented market structure. This trend can be seen by looking at the evolution over time of the market share of the largest generator in the EU countries, between 1990 and 2009 (Figure 5.3).

Looking at the overall variation between 1990 and 2009, Ireland has experienced a dramatic market opening, with the share of the largest generator dropping from 97% to 37%. In general—excluding Malta and Cyprus, two small islands—the average market share of the largest generating company in the EU is 52.4%, with an overall decreasing trend. While some countries are still characterized by a limited number of competitors, the general tendency is to expand the number of active firms, such as in the Netherlands, Italy, Austria, and Portugal.

In spite of market opening, however, the European electricity market is still deeply influenced by the presence of a small set of large companies, the 'Seven Brothers' (Thomas, 2003 and 2010) which, after a process of mergers and acquisitions are now made up of five truly global European companies and

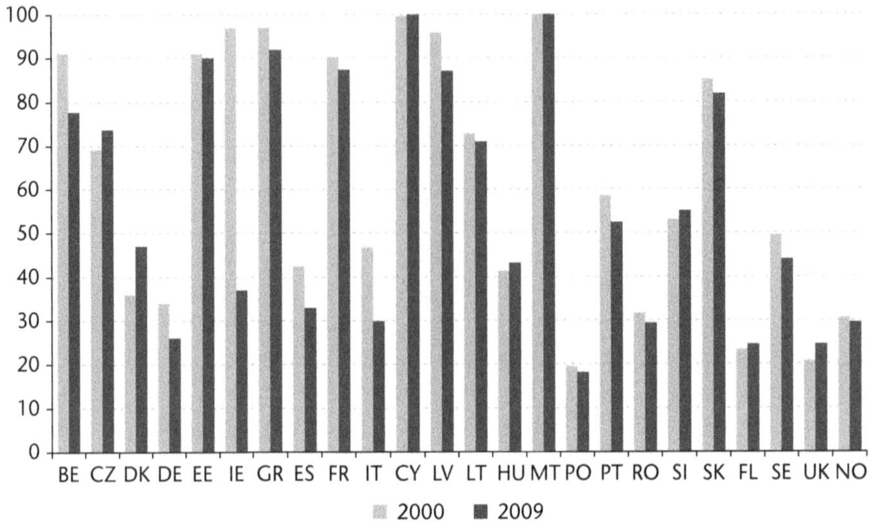

Figure 5.3. Market share of the largest generator in electricity market
Note: data for Romania and Slovenia in the first column refers to 2004.
Sources: Eurostat.

Table 5.6. The Seven Brothers: Major electricity companies in the EU, 2009

	Shareholding	Turnover	Net income	Worldwide workforce
		€bn	€m	
GDF Suez	35.6% state owned	79.9	4,477	242,714
EON	listed	81.8	8,645	88,227
EDF	84.7% state owned	66.3	4,088	164,250
ENEL	31.4% state owned	64.0	5,395	81,208
RWE	listed	47.7	3,571	70,726
Ibersdrola	listed	24.6	2,939	32,711
Vattenfall	100% state owned	19.8	1,300	40,026

Source: adapted from *Thomas (2010)*.

two relatively smaller firms with significant operations in more than one country. Table 5.6 provides financial statistics for 2009 for these companies.

There is thus no evidence that market structures are converging to a unique pattern in the EU, even if there are some common trends driven by the policy initiative of the European Commission which are related to the presence of high market shares and concentration. It should be noted, however, that the definition of market power in the electricity generation sector is quite different from the standard industrial organization framework. Given the merit order pricing structure, an operator is said to have market power when it owns a production facility, which is essential in the sense that its generating capacity is pivotal for meeting demand. If this is the case, this supplier may decide to withhold capacity in order to increase prices and thus exert market power.

Zooming in on the situation in the UK after the reform package, Sweeting (2007) finds that in the England and Wales Electricity Pool, an initial high degree of market concentration was in fact followed, after the restructuring and reform process, by the exercise of market power of the main firms in the market, even though indices of concentration were decreasing. The reading of this empirical result leads the author to conclude that the generators' behaviour was related to tacit collusion or price raising strategies. Newbery (2007) stresses the continued importance of reducing concentration as well, in order to ensure lower price-cost margins.

After the process of liberalization the number of companies producing electricity in the UK increased considerably, passing from 6 to 47 in two years, however there have been processes of re-integration, mergers, and acquisitions that have increased the concentration of the market. Generation comprises approximately 30 private entities (Department of Energy and Climate Change, 2011), many of which are joint ventures, with only 7 having a market share above 5%. In particular the market is dominated by three foreign-owned firms: Npower (RWE) (18.4%), E.ON (15.8%), and EDF Energy[37] (15.7%).[38] The retail market is characterized by the presence of six big suppliers[39] with an individual market share above 5% and which account for, on aggregate, 99% of supply (HM Treasury, 2010). The ownership of the transmission grid is divided between four different private companies. In the UK there are also many 'broker' type services which act to help customers choose the best contract structure offered by the main suppliers.

The main development in recent years is the greater degree of vertical reintegration between the main producers and supply companies,[40] and this fact poses difficult regulatory issues.

In France the structure of the industry is very different with respect to the UK. The French national government has, traditionally, a strong relationship with its energy industry. The primary electric utility *Electricité de France* (EDF), is a vertically integrated public company formed through a process of nationalization of the electric energy industry by the law of 1946. As regards unbundling, the government decided to introduce only a form of accounting separation between transmission and generation, essentially maintaining the vertical integration of the electric sector. Another characteristic of the process

[37] As of February 2009, British Energy is part of EDF.
[38] http://www.nationwideutilities.com/uk-energy-market-share.html [accessed 6 January 2013].
[39] E.ON, RWE Npower, SSE, EDF, Centrica, and Scottish Power (HM Treasury, 2010).
[40] Serrales (2006) and Commission of European Community (d).

was the reluctance to privatize EDF: the state in 2011 controlled 84% of EDF capital. Although new private generators have been given access to the market, EDF still owns and controls the majority of generation as well as the transmission and distribution networks. EDF was responsible for approximately 87% of the generation capacity in 2005, while the *Compagnie nationale du Rhone* (CNR), owned by Electrabel (50%) and by Caisse des dépôts et consignations (public) (33.2%), Elecrabel-Suez, and SNET (group E.ON) control almost entirely the remaining share. In 2001, CNR created *Energie du Rhone*, as a joint venture with Electrabel, the former Belgian national utility, to distribute and retail the electricity CNR produces in France. RTE, the subsidiary company of EDF created in 2005, is the unique administrator of the transmission system.[41]

In Spain, the most relevant outcome of electricity policy so far has been the emergence of a few big firms that have consolidated their generation assets. The market is overwhelmingly controlled by the three largest companies Endesa (24.3% of energy share), Iberdrola (19.6%), and GasNatural Fenosa (15%), which together account for 60% of the generation installed capacity (European Energy Regulators, 2011). The competition in the retail market is, in contrast, wide, with almost 200 suppliers, and only three had a market share above 5% in 2009.[42] In July 2006, the Iberian Electricity Market, MIBEL, started operating. It aims to create an integrated electricity wholesale market with Portugal, notably by creating a single market operator for the wholesale Iberian pool market; but, despite the start of MIBEL, it is too soon to consider that a unified Iberian market exists.[43] Recent developments show the entry of hedge funds, investment banks, and other European utilities into the market, but liquidity levels remain low, albeit increasing.

After a long history of private oligopoly, Italy changed to a vertically integrated public monopoly in 1963. ENEL, the listed public sector company, became a public limited company in 1992. After the privatization process, which started at the end of the 1990s, the government now controls 31.4% of the company. In generation, a decree allowed ENEL to retain a maximum market share of 50% after 2003. As a consequence, three of its generation companies were sold on the market: the sale of Elettrogen, Eurogen, and Interpower was completed by the end of 2003. In terms of amount of electricity generated, five operators had a market share over 5% in 2010: ENEL Group (27.8%), Edison group (10.7%), ENI Group (9.8%), E.ON (formerly Endesa) (5.5%), and Edipower (5.5%).[44] The distribution network is virtually

[41] Serrales (2006) and Commission of European Community (d).

[42] Eurostat data, http://epp.eurostat.ec.europa.eu/statistics_explained/index.php/Electricity_market_indicators, table 6.

[43] Serrales (2006) and Commission of European Community (d).

[44] Source: Autorità per l'energia e il gas. http://www.autorita.energia.it/it/dati/eem30.htm [accessed 6 January 2013].

controlled by ENEL, with the exception of few cities where the local municipalities own the distribution companies.[45]

In Germany, prior to the recent reform, the law of 1935 led to the existence of a de facto regime of private regional monopoly with nine vertically integrated supra-regional companies, which in 2000 still controlled 80% of production, 40% of distribution, and all transmission. Below this oligopolistic level that has been static for decades, there is a regional level, formed by about 80 firms, whose principal activity consists of acquiring energy from the large distributors and reselling it at a local level. About 800 entities operate at a local level, most of them publicly owned. With respect to generation, today approximately 84% of the total electricity in Germany is produced by just four large groups: ENBW, Vattenfall Europe, RWE, and E. ON. In 2008, looking at capacity distribution, RWE accounted for 33% of the market, E.ON for 23%, Vattenfall for 16%, and EnBW for 12% (Bundeskartellamt, 2011). It should be noted, however, that E.ON has recently given up capacity and shareholdings in municipal utilities to a considerable extent, although this has not changed the country's overall picture in a significant manner.

As for the separation between the transmission and the generation process, despite the two stages being officially divided since 2002, the European Commission argues that there is still insufficient unbundling, since the interest structures of the transmission system operator seems to be influenced by the supply interest of incumbent companies. The transmission assets are split between four companies while the number of distributors is particularly high. The ownership structure of the industry continues to be mixed, with the contemporaneous presence of public and private operators. To sum up, the German electricity market is still characterized by a high degree of vertical and horizontal integration dominated by a very few large companies and is characterized by an oligopolistic structure (Bundeskartellamt, 2011). This structure, in combination with congestion at interconnectors as well as some specific problems related to network access, is thought to prevent effective competition from developing and to increase barriers for new entrants and independent investments in new power plants.[46]

The three Nordic countries, despite being well integrated in one regional market, do not show structural similarities. A few companies dominate the production of electricity in Sweden. Two thirds of the production is provided by five companies: Vattenfall, E. ON, Fortum, Statkraft, and Dong.[47] In 2010, the former monopolist Vattenfall Fortum, still 100% government owned,

[45] Ferrari, Giulietti (2004) and Commission of the European Community (d).

[46] Reisch-Micklitz (2007) and Commission of the European Community (d).

[47] Vattenfall, company website http://www.vattenfall.com/en/core-market-sweden.htm [accessed 6 January 2013].

accounted for 50% of the Swedish electricity generation and had a share of the retail market of approximately 15–20%.[48] Supply companies and big end customers are able to buy electricity from the wholesale market. Customer switching is very common in the Swedish retail electricity market. In total, 54% of electricity customers have either renegotiated their contracts or switched supplier since the market reform in 1996, while, in 2011, 34% of consumers actively switched (The Swedish Energy Markets Inspectorate, 2011). Generally speaking competition for final customers is considered to work.[49]

In Finland the generation market is dominated by two big players, namely Fortum (majority state owned) and PVO. There are, moreover, several smaller producers on the market, including municipal companies, often publicly owned. The overall degree of competition in the country after the reform process is high: there are five companies with a market share above 5% in the generation stage and three firms that have a market share above 5% in the retail market.[50] Customer switching is reasonably widespread in Finland and has been allowed since 1998. The switching rate in 2010 was 7.6% (European Energy Regulators, 2011). Among small and medium sized industrial and business customers, it is estimated that 82% of the volume and 50% of electricity bought by customers was based on contracts that the customers had either renegotiated with their local supplier or made with a new supplier.[51]

After the reorganization of the energy sector, the newly restructured state-owned electricity and gas company DONG Energy now dominates the generation market in Denmark. This former gas company has acquired most of the electricity generation sector. In the reorganization, a few large power plants were sold to Vattenfall, the Swedish state-owned electricity company. The generation market thus remains oligopolistic with two main generating companies and numerous smaller generation companies with distributed generation assets. There is some competition from Nordic and German generators through the substantial interconnection capacity. In the retail segment of the industry, there is a great number of suppliers (55 in 2010) with five having a market share greater than 5%. Customer switching is relatively common in Denmark for industrial customers: more than 50% of customers are estimated to have either renegotiated their contracts or switched suppliers.[52]

[48] Vattenfall, company website http://www.vattenfall.com/en/core-market-sweden.htm [accessed 6 January 2013].

[49] Thomas (2006) and Commission of the European Community (d).

[50] Eurostat data, http://epp.eurostat.ec.europa.eu/statistics_explained/index.php/Electricity_market_indicators, tables 2 and 6 [accessed 6 January 2013].

[51] Thomas (2006) and Commission of the European Community (d).

[52] Thomas (2006) and Commission of the European Community (d).

In particular, approximately 17% of the large consumers and 6% of small consumers switched in 2009.[53]

All the Nordic Countries are part of a joint wholesale electricity market. The Nord Pool market was created in 1996 by Norway and Sweden and afterwards was completed by the entry of Finland (1998) and Denmark (1999). So the profile of the electricity industry of all these countries has to be considered jointly. If we look at the entire Nord Pool we can discern a competitive context: there are ten generating companies with a market share over 5%, while the aggregate share of the three largest generators is under 40%, and this seems to be compatible with mainly public ownership of generation and of the networks.

From the previous analysis, it appears that, in general in the EU, concentration in the industry remains high, with the largest three generation firms or the three largest retailers controlling more than 60% of the market in the large majority of countries, whatever the extent of privatization, with two polar exceptions: the UK and the Nordic countries, the former with no public ownership left, the latter with mostly public sector firms.

This brief overview of the electricity industry in some EU countries shows a striking variety of patterns. As mentioned, some common trends can be discerned, but it is not apparent that a unique European pattern is emerging. This, in a sense, is good news for empirical analysis, because it allows us to test whether structural diversity is associated with different performances. We try in the next sections to use available comparative data to test the reform paradigm in the EU-15. We consider this exploratory research, because we are fully aware of the difficulty of capturing the significant structural and legal diversity that we have mentioned in this section. There are, however, very few attempts in the literature to evaluate the electricity reforms across the EU, and it seems useful to contribute to the policy reform debate on solid empirical grounds.

5.5 Earlier empirical findings

As mentioned, while the consensus on the reform paradigm in the EU is rather high, the empirical evidence that supports it is more limited. After more than ten years of experience with its implementation, it seems appropriate to move from speculation on its merits to testing its impact on empirical grounds. Particularly, it would be interesting to check the differential impact of

[53] NordReg, 2010 https://www.nordicenergyregulators.org/upload/Reports/Nordic_Market_Report 2010.pdf.

privatization against other reforms, since, admittedly, their association is far from granted, see Jamasb and Pollitt (2005).

The overall effect of the reform package in the electricity industry on performance measures, prices, and general consumer welfare is difficult to ascertain, due to the variety of other factors that may influence the outcome. While some studies point in the direction of the negligible, albeit positive, effect of increased liberalization on industry performance (Erdogdu, 2011), other studies document the importance of the market power of electricity producers in increasing consumer prices, a result which could be wrongly attributed to the liberalization process per se (Percebois, 2008). In what follows we provide an overview of the earlier empirical findings on the effects of specific aspects of the reform package on price levels and investment and then focus on the general price effects of liberalizations in comparative cross-country and a few national studies.

One specific step of the reform package, the effects of ownership unbundling of transmission networks, has been extensively studied and some conclusions may thus be drawn with respect to it. The guiding principle behind this reform item is the acknowledgment of the nature of a monopoly of the high voltage transmission network, which should be separated from the potentially competitive activities, namely generation and retail supply. Failure to unbundle may lead the transmission network's monopolist to influence the mix of power stations in order to meet demand, thus providing a barrier to entry in the generation segment of the market. The monopolist may also influence the investment in network expansion and may make use of commercially sensitive information on final consumers to undercut entrants' offers (Frontier Economics, 2007). In order to reap the benefits of increased competition in the generation and supply sub-sectors, the EC has set the rules in the 2009 Electricity Directive (2009/72/EC) involving functional, legal, operation, and full ownership unbundling. Small firms are exempt from the full-fledged ownership unbundling and a set of unbundling requirements are set. The other key issue is the creation of Independent System Operators (ISO) and the separation of network ownership and its operation.

Some empirical evidence on the effects of ownership unbundling between the system operator of the transmission network and the rest of the electricity sector highlights the fact that ownership unbundling leads to higher investments in the electricity sector, along with third party access, while legal unbundling is not enough. On the other hand, unbundling appears to have no significant effect on consumer prices (Schmitt and Rammerstorfer, 2010). The effect on the quality of service may also be difficult to ascertain (Brunekreeft, 2008) and the overall welfare effects are still unclear (Nardi, 2010).

Empirical research on the impact of electricity reform at a national level is available in some EU countries, but a major problem in this area is often the lack of appropriate analysis of counterfactuals, that is comparing scenarios with and without reforms, or with a different mix of reforms (Newbery and Pollitt, 1997). The data needed to perform this analysis are often not readily available. The large variations in progress in the implementation of the reform paradigm across the EU countries, however, offer a proxy for such counterfactuals. By observing different industry characteristics in different countries we can try to infer the differential impact of reforms.

Looking at empirical research based on national analyses, the main result is in the direction of a positive effect of liberalization on firm performance and generation costs. When looking specifically at the price effect in the USA, Joskow (2006) reports a 5–10% decrease in retail prices for households and 5% for industrial consumers. For the UK, Waddams Price (2005) reports price reduction for households after liberalization. Looking at time series for the Nordic countries (Denmark, Sweden, Norway, and Finland), Johnsen and Olsen (2011) report that consumers pay the highest price in Sweden and the lowest in Finland and Norway. This result is linked to market structure and to the regulatory design of contracts, suggesting that liberalization in a market as complex as electricity is not sufficient to ensure price benefits for residential consumers. Abbott (2010) reports that in New Zealand, where the reform process started in the mid 1980s, after an initial decrease, prices for residential users have experienced an upward trend since the 1990s, with a contemporaneous decrease for industrial consumers.

Moving on to comparative studies, we briefly mention some earlier research based on cross-country analysis of the impact of regulatory reforms on prices, keeping in mind that the price effects may also have been heavily influenced by the countries' fuel mix, technological features, and overall market structure, all factors which should be accounted for in the empirical analyses. Hattori and Tsutsui (2004) look at 19 OECD countries (1987–99) and consider both industrial prices and the ratio between industrial/household prices. They find some support for privatization and liberalization determining price decreases. Steiner (2000), however, in a study of 19 OECD countries (1986–96), finds that privatization (and the time to it) increases prices, while unbundling and liberalization have the opposite effect. Zhang et al. (2002) studied electricity residential prices in 51 developing countries (1985–2000) and found no effects of the reforms. Martin and Vansteenkiste (2001) do not find an impact of liberalization on prices, while finding that public ownership increases prices in the EU-15 in the very short period they consider (1995–2000). More recently, Gassner et al. (2009) conducted a detailed empirical study for the World Bank on 1200 utilities in 71 developing and transition countries over ten years, including publicly owned and private sector participated ones, and, with a different regulatory index, use differences-in-differences econometric techniques. They find that privatization does not have an impact on prices and investment. Nagayama (2009) provides a panel data analysis between 1985 and 2003 linking prices and reforms in 78 developing countries in Latin America, Eastern Europe, Asia, and the former Soviet Union. The main result is a positive correlation between the liberalization measures and international prices, suggesting an increasing trend.

To sum up, out of the above mentioned cross-country empirical studies, several conclude that privatization increases consumer's prices, and a minority find that the prices decrease. The opposite is true for liberalization, while on the impact of unbundling the evidence is limited and mixed. This brief overview leads us to our empirical exercise, where the aim is to understand the effect of different elements of the reform package on consumer prices in the EU, while controlling for technological and country-specific features.

5.6 Data

The primary source for the average price variables is the International Energy Agency (IEA). The only alternative data source available for a EU-15-wide analysis would be the Eurostat, which, however, is available for a much shorter time series as it starts at best in 1991. The time series of IEA data starts in 1978, providing over 30 time series. The IEA net-of-tax electricity prices for households are expressed in €/unit and present a correlation with household net-of-tax electricity prices (yearly consumption of 3500 KWh of which at night 1300).[54]

As for the regulatory reform variables, including measures of entry regulation, public ownership, market structure, and vertical integration, at varying levels of detail, we used data provided by the ETCR dataset (see Chapter 2).

5.7 Results

Table 5.7 presents the results of the estimation of various models for electricity prices. The first column provides a test of the reform as a whole, estimating a model where the reform variable is the ETCR sector score ranging from 0 to 6. In the following columns, we use the set of dichotomous dummy variables as reform variables. All models are estimated using GMM.[55]

[54] IEA monetary variables, which are expressed in US$, were converted into Euros using the Eurostat Euro/US$ exchange rate.

[55] Due to the small dimension of the whole panel, and following Blundell and Bond (2000) who discuss the possibility that the error term might have a small degree of autocorrelation due to measurement errors, we estimate these models using as instruments the three most recent lags of the dependent variable starting from period *t-3*. All estimated models pass the key identification

Table 5.7. Net-of-tax electricity price (log), estimation results

	(1)	(2)	(3)	(4)	(5)
	GMM	GMM	GMM	GMM	GMM
Regulation index	0.003				
	(0.690)				
Public ownership[1]		−0.376***	−0.356***	−0.350***	−0.366***
		(0.000)	(0.000)	(0.000)	(0.000)
Entry regulation[1]		−0.106*			
		(0.072)			
Vertical integration[1]		0.099	0.079	0.037	0.066
		(0.233)	(0.393)	(0.598)	(0.462)
Entry regulation[1]: third party access			−0.043		
			(0.483)		
Entry regulation[1]: liberalized wholesale pool			−0.015		−0.024
			(0.741)		(0.596)
Entry regulation[1]: minimum threshold for consumers' choice			−0.098*	−0.100**	−0.092*
			(0.073)	(0.049)	(0.088)
Average price (period *t*−1)	0.854***	0.414***	0.407***	0.428***	0.414***
	(0.000)	(0.000)	(0.000)	(0.000)	(0.000)
Total combustible fuels	0.025**	0.016	0.021	0.023	0.021
	(0.011)	(0.517)	(0.408)	(0.354)	(0.409)
Electricity import[2]	0.000	−0.000	−0.000	−0.000	−0.000
	(0.720)	(0.913)	(0.900)	(0.936)	(0.917)
Electricity residential consumtion[2]	−0.022**	0.002	−0.008	−0.007	−0.005
	(0.012)	(0.948)	(0.805)	(0.820)	(0.870)
Per capita GDP[2]	0.016	0.231***	0.215***	0.225***	0.227***
	(0.274)	(0.000)	(0.000)	(0.000)	(0.000)
Year fixed-effects	yes	yes	yes	yes	yes
Constant	−0.292**	−0.676**	−0.683**	−0.612**	−0.647**
	(0.026)	(0.011)	(0.011)	(0.015)	(0.014)
Observations	402	402	402	402	402
Countries	15	15	15	15	15

Standard errors in parentheses. For the regulatory variables see Table 5.A.1. For details, see Fiorio and Florio (2010).
*** p<0.01,** p<0.05,* p<0.1.
[1] Dummy.
[2] Log.

The highly significant lagged dependent variable sheds doubt over similar panel data models in earlier literature without a dynamic specification, whose estimates are likely to be strongly affected by omitted variable bias.

It is interesting to note that the lagged dependent variable is highly significant also after the inclusion of year fixed-effects, regulatory variables, and other controls. If the aggregate index, ranging from 0 (full privatization, unbundling, and liberalization) to 6 (no privatization, vertical integration,

test allowing for zero second-order autocorrelation of residuals, as well as the Sargan test for over-identifying restrictions. All models are estimated using robust estimation as tests of homoscedasticity of residuals always reject the null.

and no free entry in the industry), is included in the regression, testing whether the reform package as a whole has any effect on average prices, one could conclude that no statistically significant effect is found. If, instead of the aggregate index, a dummy variable for each of the three dimensions of the electricity industry reform is included, it emerges that only the *Public Ownership* variable presents a consistently significant coefficient, reducing the average price by roughly 0.3 (log) points. Vertical integration, which is strongly correlated with public ownership, is never statistically significant. Entry regulation is significant at the 10% significance level, although it is mainly the freedom of choice of providers that drives this effect. Interestingly, the coefficient is negative, suggesting that allowing consumers to choose their provider is negatively correlated to the actual price paid. This could be interpreted as a consequence of more advertising expenditure by the producing firms that is reflected in average prices, or on the mistakes of the part of actual 'switches' as found in the UK (Wilson and Waddams Price, 2010), see Chapter 7.

Other control variables are not significantly correlated with average log prices, with the exception of per capita GDP, suggesting that electricity is a normal good.

5.8 A paradigm reversal? Electricity market reform in the UK

As the electricity reforms in the EU started in the UK, it is interesting to reconsider the situation and the outlook in this country.

The UK government recently announced, 20 years after privatization and liberalization of the electricity industry, a radical change of its agenda. The new policy was presented by the Department for Energy and Climate Change in 'Planning our Electric Future: a White Paper for Secure, Affordable and Low-carbon Electricity' (Department for Energy and Climate Change, July 2011). The previous consultation document, the White Paper, its accompanying studies, the parliamentary debates by the House of Commons, Energy and Climate Change Committee (House of Commons, 2011), the reports by the regulator and by other stake-holders, and the articles in the press on the announced reforms, have elicited a substantial amount of evidence and interpretation about the state of the electricity industry in the UK. The consultation document (December 2010) received 274 responses by companies and associations, most of them publicly available. The intensity of the debate, the range of opinions, and the widely different reform proposals are justified by a relatively high consensus about the difficult situation of the electricity industry in the UK, which is clearly stated in the White Paper. We briefly summarize below the UK government's diagnosis and the new policy proposals.

Despite the lip service paid to the previous arrangements ('Since privatisation in the 1980s, our competitive market and system of independent regulation has served us well, delivering reliable and affordable electricity' DECC (2011), p. 5), the Secretary of Energy identifies four unprecedented challenges:

a) In the next decade around 25% of generation capacity is expected to close and not be replaced in the UK. This is due to ageing of plants, low new investment, and to the fact that some polluting technologies are no longer in compliance with new environmental standards. According to some forecasts, the capacity margins will fall below 5% by around 2020, and for technical reasons such a narrow generation reserve implies that blackouts and voltage reductions may become highly likely. Moreover, some of the new technologies, such as wind, are intermittent. This fact has additional implications for security of supply, as the potential generation capacity will not be available in some weather conditions. Other non-carbon sources are inflexible, such as nuclear power, which cannot be started up or closed down at short notice to respond to peaks in demand.
b) The industry pattern and future investment trends, as forecasted, are not compatible with the UK government's objectives to achieve 15% renewable energy in 2020, and to cut by 80% carbon emissions by 2050. Without reforms, the industry is expected to pollute three times the government's target for decarbonization in 2030.
c) Electricity demand is expected to double in 2050 because of electrification of transport, heating, and other sectors, in spite of policies for household and industrial energy efficiency.
d) Prices to final users are expected to rise considerably, both because of increasing prices of energy inputs and of investment, and because of the impact of environmental policies. For households the electricity bill is expected to rise by around £200 per year (a 40% increase between 2010 and 2030). The impact is going to be felt particularly by the lowest income deciles (DECC (2011), p. 121), adding to the existing fuel poverty, but the rise could possibly be £160 with the reform (33% increase).

According to the White Paper there is 'broad consensus that current market arrangements will not deliver the scale of long-term investment needed, at the required pace, to meet these challenges. Nor they will give the consumers the best deal. This is partly because of the sheer scale of the investment required. Up to £110 million investment in electricity generation and transmission is likely to be required by 2020, more than double the current rate of investment' (DECC (2011), p. 6). While there are dissenting views on these forecasts, the investors in their response to the consultation seem in general to agree that, at the current prices and costs, they are unwilling to increase their investment in electricity generation, in transmission, and distribution (according to OFGEM, £35 billion will be needed for the latter sections of the industry).

Whatever the precise proportion of the two causes of the decay of capacity and network infrastructure, that is the lack of investment by the six major players and the un-profitability of the low carbon technologies, in both cases market prices have not provided the right incentives for long-term investment. Consequently, the new policy announced by the UK government will alter the pricing system:

> ...the social cost of carbon is not fully reflected in the market price as this does not take into account all the damage caused by climate change. The carbon price is also volatile and hard to predict,
>
> - making long term investment decisions more uncertain; and
> - the capacity and appetite of existing market participants to finance the unprecedented levels of investment needed is uncertain.
> - There are also likely to be insufficiently strong signals to invest in the level and type of capacity that we need in order to guarantee future security of supply. This is also due to the scale of investment and failures within the existing market. (DECC (2011), p. 7)

Hence, the UK government aims at radically changing the price signals by direct public intervention through the four instruments of the EMR. These are:

a) Long-term contracts between a 'central body' and low-carbon generators, a form of public procurement of electricity, based on Feed-in Tariffs with Contracts for Difference (FiT CfD). The central body will fix both a quantity of electricity (metered output) to be procured, a strike price (to be computed centrally) and a reference price. The latter could be different for different technologies, as it could be an average of price in the one-day ahead market (intermittent technologies), or in the one-year ahead forward market (baseload power plants). The strike price (indexed to inflation) will be calculated by the central organization in such a way that when the reference market price is below the strike price, the low carbon generators (including nuclear) will be paid the difference between the reference and the strike price. If the reference price exceeds the strike price, then the difference should be paid back by the generators. The cost of the subsidy or the claw back of the profit of low carbon will be passed to the consumer. Figure 5.4 shows the mechanism.

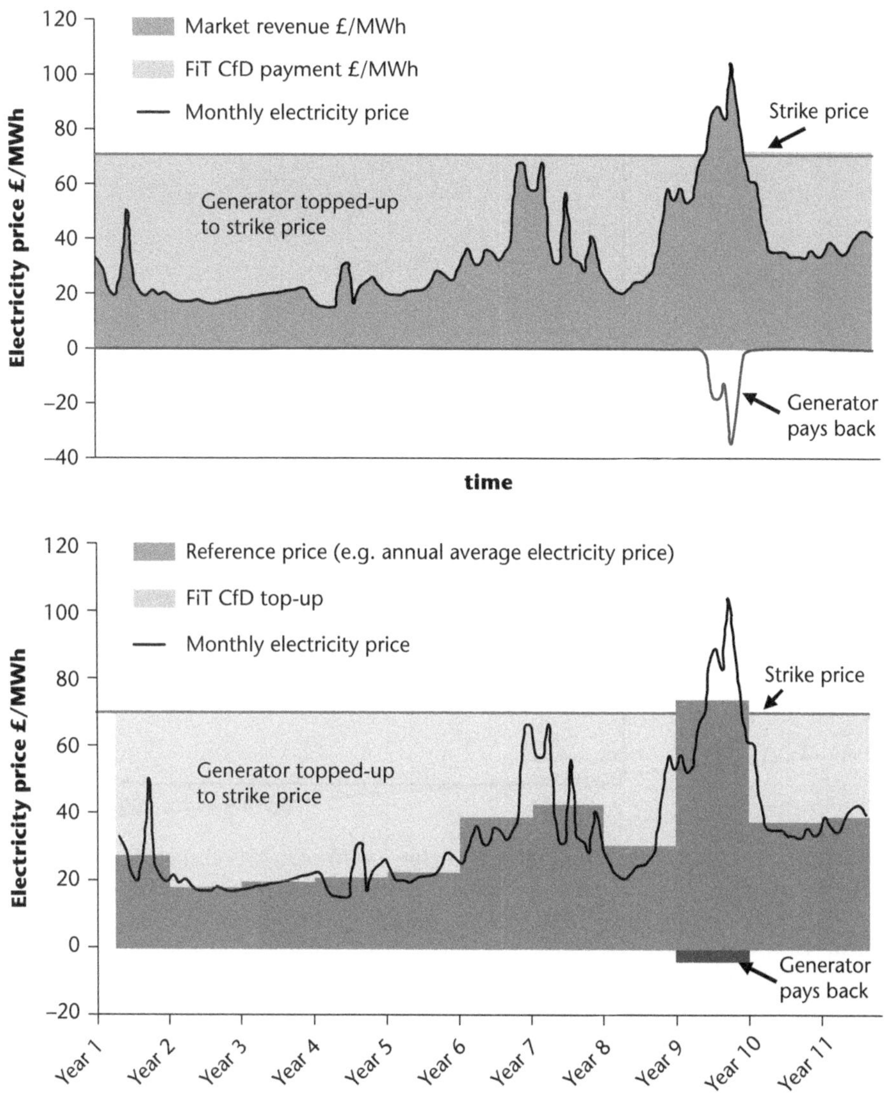

Figure 5.4. The proposed operation of a baseload Feed-in Tariff with Contract for Difference (top panel) and the operation of an intermittent Feed-in Tariff with Contract for Difference (bottom panel) in the UK
Source: DECC (2011).

b) A Capacity Mechanism, where again a 'central body' will offer contracts to electricity companies to increase their generation capacity. Two options are considered. The first is a targeted Strategic Reserve, where a centrally procured additional capacity is set aside and used only in certain extreme circumstances, that is when there is a risk of service

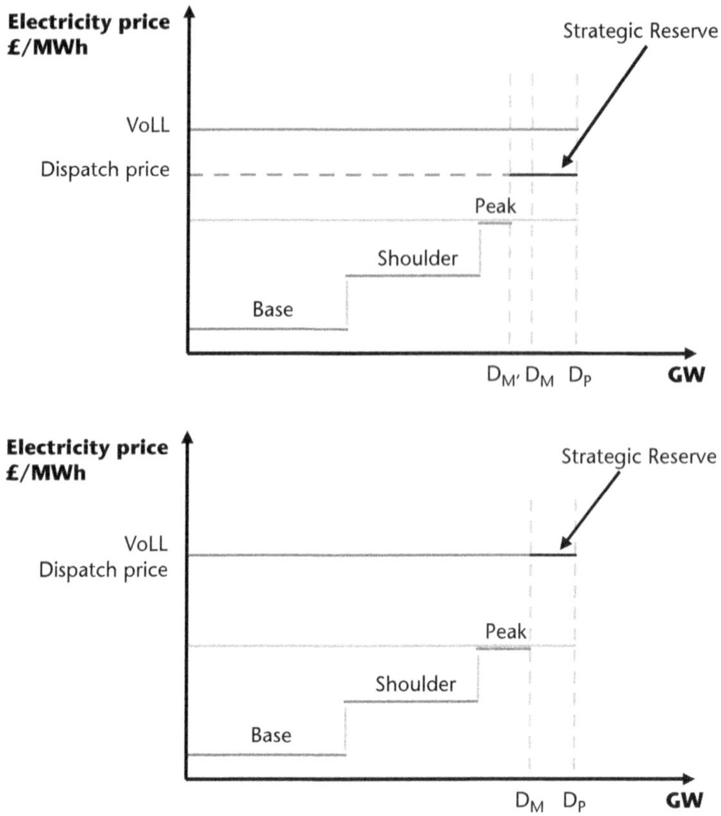

Figure 5.5. Proposed operation of Strategic Reserve with economic dispatch (top panel) and operation of a Strategic Reserve with last-resort dispatch (bottom panel) in the UK

Note: D_P is the maximum demand which can be served by a market which has a strategic reserve, and the capacity of the strategic reserve itself, combined. D_M is the maximum demand that could be served by a market with no capacity mechanism. Once a strategic reserve with a dispatch price lower than VoLL is introduced, it will replace some electricity market generation, since it effectively caps the revenues that can be earned from times of peak demand. $D_{M'}$ is the demand at which prices rice to the dispatch price. When $D_{M'}$ is reached the strategic reserve is dispatched.

Source: DECC (2011).

interruption. Alternatively, a market-wide mechanism based on a financial instrument (a 'call' option) would give an incentive to providers to invest in reliability of supply (or compensate users for service interruptions). Under both alternatives, demand side operators could also get the public incentive, for example by receiving an incentive to decrease their demand. Figure 5.5 shows the operation of the Strategic Reserve. System flexibility to accommodate the intermittent technologies, storage, and distributed energy (i.e. not using the main transmission lines) would also imply a 'fundamental transformation' in the functioning and network itself, and 'Without guidance from the Government, there is a risk that uncertainty over the rate of change could lead to insufficient or inappropriate investment, resulting in the network being unable to deal with future challenges' (DECC (2011), p. 103).Therefore the government will work with Ofgem to create guidelines for network companies (which are fully privately-owned in the UK). The government will also promote work on storage technologies, interconnections, and energy saving.

c) A Carbon Price Floor, starting in 2013, that is a stronger incentive for low-carbon technologies, additional to the EU transmissible certificates (Emissions Trading Systems), which price is considered too low and volatile to provide the right signal to investors. The level of support will vary over the future years, starting from around 35 t/CO2, based on a price floor targeting of £k30. The proposal is based on the Stern Review's (2006) position that it is socially desirable that polluters pay the full social cost of the externality associated with potential environmental damage, and that a carbon price would encourage investment in alternative technologies.

d) An Emissions Performance Standard, set at 450 g CO2/KWh per year, in addition to the requirement that any new coal-fired power plant must use Carbon Capture and Storage technologies. Grandfathering, which is an exemption, is allowed for existing plants.

These measures should complement other demand and supply side policies, including incentives for energy efficiency to households and firms, smart metering, a government-sponsored Green Investment Bank (www.bis.gov.uk/greeninvestmentbank), and others.

A crucial aspect of the EMR is that regulation by OFGEM will become in future only one element in a much more complex institutional framework:

> It is likely that an organisation at arm's length from the Government will administer the contracts. Other core functions to deliver the FiT and the Capacity Mechanism include: translating the policy objectives into technical requirements, delivering the contracts, data reconciliation, managing payments, and monitoring compliance and enforcement. The Government and the delivery organisation(s), working jointly, will periodically evaluate according to a planning cycle clearly laid out in advance, their future strategy in the light of possible change in costs, technological developments, and new challenges to the energy system. (DECC (2011), p. 10)

A five-year planning cycle is envisaged. The role of the regulator will be changed accordingly, as 'the current framework of broadly-scoped duties and weak guidance is very unlikely to be able to support a predictable regulatory environment that is coherent with the Government strategy...' (DECC (2011), p. 88).

Moreover, the UK government acknowledges that the electricity market is illiquid, and that Ofgem is trying (for the fourth time since the Pool) to achieve greater liquidity and remove barriers for independent generators. Market illiquidity is mainly due to the vertical re-integration of the market, with the Big Six who also own much of the distribution and supply, therefore selling to themselves in such a way that the unbundling of transmission and its regulation per se is not sufficient to establish a competitive market for generation.

Having briefly described some aspects of the current electricity policy debate in the UK, some comments from our perspective are in order.

The core idea behind privatization and liberalization in the UK and elsewhere in the 1990s was to achieve a competitive market mechanism which had to be able to balance demand and supply through efficient, cost-reflective prices, as discussed in Chapter 1. Regulation, initially in the form of a price-cap, then only in the form of supervision of access to the unbundled transmission network, was seen as set to gradually decrease its role. The regulator was designed as an umpire or a policeman on behalf of the different stakeholders, with the task of ensuring fairness and a level playing field in the market game.

The future landscape as set out by the White Paper and the EMR is radically different. As the White Paper's title says, it provides for a *planned* electricity market, where one or more government sponsored central bodies will take crucial long-term decisions.

By computing the target capacity margin relative to long-term capacity, the government, or an agent working for it, will actually make a decision on the total amount of investment needed. If the private sector does not deliver the investment, it will be the public sector that forces the desired level of investment simply by buying it on behalf of the consumers. At that stage, while this option is not mentioned in the White Paper, public procurement of generation capacity from the private investors may or may not be cost effective as compared to directly commissioning or acquiring some power plants, of the desired technology, by a Strategic Reserve agency owned or under the direct control of the Department of Energy.

In the public consultation some concern was expressed about the fact that a Capacity Mechanism can offer an incentive to some companies to set aside their capacity and be paid a long-term contract for this instead of actually generating electricity. In such a case, an even greater capacity would be needed, and it is not clear how great the intervention should be. In fact, there is a certain ambiguity in the White Paper as to the actual extent of the Capacity Mechanism. It seems that it is not intended simply to ensure the existence of an adequate margin (10% of capacity is mentioned) but in fact, combined with the FiT, to ensure that the total volume of generation and of investment is adequate. The margin, after all, given an increasing demand, depends crucially on how much the private sector will invest, and if it will not invest enough, either the reference market price will increase significantly or the government will need to intervene on a wider scale. Clearly a market approach would allow wholesale prices to increase until the profitability for investment is restored, but this in turn would not be compatible with an affordability objective, as declared by the government even in the title of the White Paper. In fact, the government wants the Strategic Reserve to be activated when market prices are close to the level that leaves users indifferent to paying or disconnecting themselves (Value of Lost Load). In other words, the policy is clearly intended to avoid disconnections and price spikes, hence it is equivalent to a price-cap.

A similar issue applies to the Contract for Difference approach. Here the government pays an incentive to top-up the revenues of low-carbon generators when the reference price is 'too low' as compared to a benchmark. Given that the low-carbon technologies are so diverse, spanning new nuclear plants to off-shore wind farms, the consultation has made clear that, in fact, different subsidy levels are needed for each type of technology. This obviously may create further effects and possibly distortions in technology adoption. In fact, by fixing a strike price, the central body in charge of administering the scheme will actually control the output volume by type of source. Thus there will be both an upper and a lower price that are centrally set, and quantity control of capacity. In turn, the Carbon Price Floor and the Electrical Power System (EPS) are the usual combination of environmental incentives and standards, and they will also alter market price equilibria in order to further force the government's environmental objectives.

These four instruments (contracts for difference, capacity mechanism, carbon price floor, and the EPS) may or may not work effectively, but their impact on the overall functioning of the electricity markets cannot be exaggerated. In fact, these markets are fragmented among a core segment under the control of a vertically re-integrated oligopoly, and an often illiquid set of other segments. Thus, if the UK government plans to cope with a considerable expected capacity gap in a decade, through central auctions for long term contracts targeted to capacity (set aside) and to low-carbon generation, it will become the most important player in the industry, with an effective control on output, investment, technological mix, and prices.

The existence of material market failures is fully acknowledged by the White Paper: 'These include: (a) reliability is a public good—consumers cannot buy reliability for themselves without providing it to everybody else ... (b) there are barriers to entry in the wholesale market ... and (c) prices in the electricity market may not send the correct signals to ensure optimal security of supply' (p. 66). The fact that under the proposed scheme the financing of new investments will not be provided by taxpayers (under public ownership), but by consumers through the cost of the subsidies within the bills they are going to pay (to private companies), is then more a distributive matter than a fundamental difference.

It seems that in any case the whole policy would need to be notified to the European Commission as it amounts actually to a form of state aid to private industry (including EDF, which is state owned abroad).

The British EMR goes well beyond the paradigm of the 1990s. This departure from the previous creed has two relatively independent drivers. One is the bold position of the UK government regarding decarbonization of the economy, which was also at the core of the Stern Review of Climate Change, by request of the New Labour government. However, the current Conservative–Liberal Coalition government, would have faced an investment gap in any case, simply because the privatized industry has eaten away part of the capital of the plants of the divested electricity industry, and does not have the 'appetite' for replacing such fixed capital. At the same time, the industry has eaten away much of the country's endowment of natural gas, which only 20 years ago seemed to justify the end of economies of scale in generation (we shall discuss natural gas in the next chapter).

The technological argument for privatization was in fact that the capital cost of a new gas plant was limited, and linking prices and variable cost through competition was possible and efficient with a large number of players. In the words of the White Paper:

> gas-fired power stations are a mature technology with low and predictable capital expenditure. They are quick to build and their fuel costs, which are a large proportion of operating costs, are naturally hedged because the price of electricity moves in line with the price of gas, since gas (or sometimes coal), is typically the price setting (or marginal) plant.... Each of the low-carbon technologies the Government is considering differs materially from this standard of investment choice...typically has high construction (capital costs) and low operating cost... It is therefore difficult to make an investment case for them in a market where wholesale electricity prices are predominantly set by the short-run marginal costs. (DECC (2011), p. 28)

In other words, the low carbon technologies, such as nuclear or in an extreme way wind, solar, and waves, are mostly based on fixed costs, with low or negligible marginal costs. When these features are combined with the empirical observation that major generation companies have preferred to vertically re-integrate with supply, not much is left of the original view that a competitive electricity market, with light independent regulation, would have solved the long term problems of secure supply, affordability, and environmental sustainability in the UK. If other European countries reconsider the British experiment and take a similar route in future, electricity will probably be the laboratory for a new paradigm shift (Rutledge and Wright, 2010).

5.9 Concluding remarks

After reviewing some technological and structural features of the industry, this chapter has offered a discussion of the impact of privatization and of liberalization of electricity supply on consumer prices in the EU. While the reform is often proposed as a policy package that includes privatization, vertical disintegration, and liberalization, we have disentangled the ownership effect after controlling for other reforms. We explicitly consider dynamics, use longer time series than in earlier literature, and data sources from international organizations, such as Eurostat, IEA, OECD, the World Bank, and Eurobarometer. These sources are widely available to researchers for further empirical analysis.

To conclude, we summarize our findings as follows. We find that the overall impact of the reform package on consumer prices, as summarized by the ETCR industry score, is never statistically significant for electricity prices. This is what one would expect if the adverse allocative effect of privatization is just counterbalanced by the beneficial effect of liberalization (see Vickers and Yarrow (1993) and our previous discussion in Chapter 3). An alternative interpretation is that the linear weights that the OECD uses for the indicator misrepresent the reforms. The results by Azmat et al. (2007), and by the earlier papers cited in Section 5.4, however, strongly suggest that the OECD scores work well for several other performance variables, including productivity and investment. Our results, compared with Azmat et al. (2007), who find declining labour shares, imply an increase of the profitability of reformed industries. This fact may well sustain investment, as in Alesina et al. (2005) but perhaps not in the long-run and nor for innovative environmentally-friendly technologies, as the current debate in the UK suggests. The combined evidence of lack of the overall reform impact on consumer prices, and of changes of the factor shares, implies a redistribution effect and a concern for affordability of tariffs that we shall discuss in Chapter 9.

Our bottom line is that while we have been able to identify a clear ownership effect, the overall evidence on the correlation of liberalization reforms with electricity prices is mixed. The variety of patterns in the EU, and the recent change of energy policy in the UK, suggest that the electricity reform paradigm of the 1990s, as adopted by EU regulation, needs a careful reconsideration.

5.10 Further reading

While this book deals with reforms in the EU countries, particularly in Western Europe, there is much to learn from the recent history of regulation of electricity in other regions. From a global perspective, the implementation of reforms in Asia is particularly important, given the booming demand for electricity in some key countries such as China and India. Particularly interesting and up-to-date are several chapters in the book edited by Jarvis et al. (2011). The case studies presented in the book's chapters, show to what extent the application of a 'one-size-fits-all' reform scheme, as often sponsored by the international financial institutions—particularly the World Bank and the Asian Development Bank—may create significant problems in practice. In the chapter on Thailand, Darryl Jarvis, after having discussed the fight among different stake-holders to control the regulatory process following the reforms, concludes that: 'Policy diffusion ... of regulatory models from vastly dissimilar environments is a problematic policy exercise. While international benchmarking and the tendency to emulate "best practice" standards is now part of the policy lexicon and raison d'ètre, the adoption of regulatory models and their transplantation into different institutional endowments can often produce unintended or even deleterious outcomes' (Jarvis et al. 2011, p. 147).[56] In comparing India with Brazil, Sunil Tankha shows that the heterodox Brazilian model of reform has been more successful. This model revolves around organizing 'competition for projects rather than competition in markets'. Here the government has created a state power planning agency to which power generators, public or private, submit offers for supply and 30-year contracts are adjudicated under a single buyer model. This stabilizes returns, pricing strategies, and avoids the need for costly financial hedging. In contrast, Junki Kim and Kyuhyuan Kim, in a simple but clear econometric analysis of the Korean electricity industry, fully privatized in 2001, show that profitability has increased, but that the restructuring had little impact on prices for the consumers. They conclude: 'The results ... show that electricity sector reform did not have a significant impact on the price of electricity. One possible interpretation is that the price regulation set by the government limited the impact of the electricity sector reform and the other is that the price could already be at an optimal level' (Jarvis et al., 2011, p. 349).

[56] This point is also well illustrated by the chapter on China by Kun-Chin Lin, Mika Purra, and Hui Lin, who document several decentralization attempts with the interesting outcome of a complex oligopolistic arrangement with state-owned enterprises, provincial governments, and different coordinating bodies at national level. This complexity emerges also from the chapter on Indonesia by Mika Purra, who points to the failure of the promise of neoliberal reforms. In India, Rajendra Kumar's chapter shows how three states (Andhra Preadesh, Gujarat, Tamil Nadu) despite a common framework for attracting independent power producers, have followed very different trajectories because of the variety of local circumstances.

Obviously, from a social welfare perspective, the increased profitability of firms may or may not be a good indicator of the advantage of privatization, but the lack of response of prices should be seen as a matter of concern, and is not far from our findings in the European context in spite of a longer history of reforms, and arguably a stronger institutional context (at least in some EU countries).

The interplay between technological evolution, electricity demand, and policy-driven changes is analysed, from different angles, in two recent books with a focus on the situation in the UK. The first, an edited volume by Tooraj Jamasb and Micheal Pollitt, *The Future of Electricity Demand*, offers an in-depth analysis of the challenges and implications for the industry of the changing nature of private domestic demand in the UK. The second book, edited by Ian Rutledge and Philip Wright, *UK Energy Policy and the end of Market Fundamentalism*, focuses instead on the social and economic consequences of the UK's energy policy of the past 30 years by highlighting its highly political and ideologically charged nature. Both volumes take into account the evolving nature of electricity demand, shaped by an increase in energy-intensive goods, a new array of energy sources, and growing concern over social (e.g. fuel poverty) and environmental (e.g. carbon emissions) concerns.

In the first book (Jamasb and Pollitt, 2011), the chapter by Platchkov and Pollitt (2011) provides an overview of the longer run trends of increasing global electricity demand and explains the potential impact in the UK of electrification of transport, water, and heating systems, now largely gas-supplied. Their main take is that demand will steadily increase and a possible coping mechanism will be to shift electricity demand through the use of demand-side management and smart grids and meters.[57] On the potential of demand-side participation, Torriti et al. (2011) provide evidence for possible price reductions but also warn against the constraints which may hamper the ability of consumers to manage electricity loads (thus limiting the potential price-reducing effects of demand-side participation). After presenting the technical aspects of demand side management and smart appliances for its integration, Jamasb and Pollitt (2011) focus on the social dimension of the evolution of domestic electricity demand. A major concern is related to fuel poverty, as discussed by Waddams Price (2011). The tension between policy targets such as carbon emission reduction and supply security, on the one hand, and affordability and fuel poverty on the other will exert opposite forces

[57] On the potential of demand-side participation, Torriti et al. (2011) provide evidence of possible price reductions but also warn against the constraints which may hamper the ability of consumers to manage electricity loads (thus limiting the potential price-reducing effects of demand-side participation). After presenting the technical aspects of demand side management and smart appliances for its integration, the book focuses on the social dimension of the evolution of domestic electricity demand.

on prices. If higher prices are important for security and environmental concerns, these will be detrimental in terms of affordability, especially for disadvantaged citizens.

The social implications are a key issue in the second book (Rutledge and Wright, 2010) with a focus on the impact that an extensively liberalized market, such as in the UK, may have on the adoption of renewable sources of energy (Gross and Heptonstall, 2010), the security and environmental protection (Keay, 2010) and fuel poverty (Boardman, 2010a,b). The book's overall aim is to analyse the unfolding of the UK's energy policy of the past 30 years and to provide a reflection on the lessons learned in the process. What emerges from the different contributions is that the effects of a highly ideologically-charged policy design has lead the UK to several problems in the energy sector, in particular with respect to security of supply and dependence on gas imports, a lack of liquidity and transparency especially in the wholesale market, issues related to the ageing of infrastructure and lack of incentives for future investment, and an overall '"portfolio power" over consumers rather than competition and effective choice' (Rutledge and Wright, 2010, p. 421).

The need for a 'new energy paradigm' is also advocated by Helm (2007), an edited book with several essays focusing on the long-term challenges and inadequacy of the short-term perspective of the policy and the regulatory arrangements that have emerged from some aspects of the reforms in the UK.

Appendix 5.A.1: ECTR scores

Table 5.A.1. Sectoral indicator of regulatory reform: Electricity

	Weights by theme (b_j)	Question weights (c_k)	Coding of data					
Entry regulation:	1/3							
How are the terms and conditions of third party access (TPA) to the electricity transmission grid determined?		1/3	Regulated TPA 0		Negotiated TPA 3		No TPA 6	
Is there a liberalized wholesale market for electricity (a wholesale pool)?		1/3	Yes 0		No 6			
What is the minimum consumption threshold that consumers must exceed in order to be able to choose their electricity supplier?		1/3	No threshold 0	<250 GW 1	Between 250 and 500 GW 2	Between 500 and 1000 GW 3	More than 1000 GW 4	No consumer choice 6
Public ownership:	1/3							
What is the ownership structure of the largest companies in the generation, transmission, distribution, and supply segments of the electricity industry?		1	Private 0	Mostly private 1.5	Mixed 3	Mostly public 4.5	Public 6	
Vertical integration:	1/3							
What is the degree of vertical separation between the transmission and generation segments of the electricity industry?		1/2	Separate companies 0	Accounting separation 3	Integrated 6			
What is the overall degree of vertical integration in the electricity industry?		1/2	Unbundled 0	Mixed 3	Integrated 6			
Country scores (0–6) (SEC)			$\sum_j b_j \sum_k c_k \text{ answer}_{jk}$					

Source: Conway and Nicoletti (2006).

Appendix 5.A.2: Estimating consumer prices using panel data models

Our approach for the empirical analysis of prices is based on the specification of a reduced form model of consumer prices using, as explanatory variables, aggregate or detailed measures of the level of privatization and liberalization of the sector, controlling for year fixed effects and other macroeconomic variables. The structural and evolutionary differences across the network industries considered justify why, while retaining a common empirical approach, our models have industry-specific ingredients in terms of explanatory variables and controls, in addition to a core set of indicators.

Following Blundell and Bond (1998, 2000), and Bond et al. (2003) we first study the autoregressive (AR) properties of the average consumer price for each industry. As we find evidence of prices following AR(1) processes, the econometric analysis of consumer prices is performed using dynamic panel data models, that is including among regressors the lagged dependent variable for explaining the strong persistence of prices measures.

For each network service sector, let P_{it} be a measure of current net-of-tax (log) prices for country i at time t, R_{it} the set of regulatory variables, which might include a score of the level of regulatory regime in each industry or measures of entry regulation, public ownership, market structure, vertical integration,[58] F and X_{it} a set of control variables, including demand variables such as an indicator of per capita income levels, sector specific variables such as type of energy source, input costs, and efficiency indicators, when available. The model for price levels is:

$$\begin{aligned} P_{it} &= \alpha P_{i,t-1} + R'_{it}\beta + X'_{it}\gamma + \delta_t + \varepsilon_{it} \\ \varepsilon_{it} &= (\eta_i + v_{it}) \text{ with } i = 1, \ldots, I;\; t = 1, \ldots, T. \end{aligned} \tag{5.A.1}$$

The year-specific intercept (δ_t) is included to account for common cyclical or trend components in prices, preventing a likely form of cross-country correlation. A key assumption of this kind of model is that of independence of the idiosyncratic disturbances (v_{it}) across countries. We treat the individual effects (η_i) as stochastic, which implies that they are correlated with the lagged dependent variable ($P_{i,t-1}$), unless the distribution of the η_i is degenerate. We also allow the control variables (X_{it}) and the regulatory variables (R_{it}) to be correlated with the individual effect η_i. Maintaining that the v_{it} component of the error term is serially uncorrelated, we then assume that X_{it} is strictly exogenous (i.e. uncorrelated with all past, present, and future realizations of v_{it}). As for the set of reform variables R_{it}, we assume that they might be correlated with the unobserved error term but that, due to the political and decisional process involved, they react with some lag to changes in v_{it}. In other words, we assume that R_{it} is predeterminate (i.e. R_{it} and v_{it} are uncorrelated, but R_{it} may be correlated with $v_{i,t-1}$ and earlier shocks).

[58] We also considered introducing as a regulatory variable the years spanned since the establishment of the sectoral independent authority (cf. Table 5.5) but in no regression was this variable statistically significant, hence it was omitted from estimations.

The assumption of stochastic individual effects implies that they are correlated by definition with P_{it}, and possibly also with X_{it} and R_{it}. If so, Ordinary Least Square (OLS) estimator of $\{\alpha, \beta, \gamma\}$ in the level equations of model (5.A.1) is inconsistent, since the RHS variables are positively correlated to the error term due to the presence of the individual effects, and this correlation remains even for $T, N \to \infty$. A Within estimator would eliminate the source of OLS inconsistency, η_i, by transforming the original observations as deviations from their own individual time mean, however it does not completely solve the problem. In fact, for a small T, the Within transformation induces a correlation between the transformed lagged dependent variable and the transformed error term, producing a biased estimator (Nickel, 1981). GMM, by using extra moment conditions, potentially produce consistent and efficient estimates, coping with endogeneity of the lagged dependent variable.

In fact, as no external instrument is usually available outside the immediate data set, an alternative to OLS and Within estimates is to either transform the data to eliminate the individual effects or find some instruments which are orthogonal with the error term but not with endogenous regressors. Arellano and Bond (1991) suggested[59] transforming the data in first-difference obtaining

$$\Delta P_{it} = \alpha \Delta P_{i,t-1} + \Delta R'_{it}\beta + \Delta X'_{it}\gamma + \Delta\delta_t + \Delta v_{it}. \qquad (5.A.2)$$

This transformation eliminates the fixed effect, although the lagged dependent variable remains potentially endogenous, as the $y_{i,t-1}$ term in $\Delta y_{i,t-1} = y_{i,t-1} - y_{i,t-2}$ is correlated with $v_{i,t-1}$ which is in $\Delta v_{it} = v_{it} - v_{i,t-1}$ and predetermined variables such as ΔR_{it} become potentially endogenous because they may also be related to $v_{i,t-1}$. However, longer lags of the endogenous regressors ($\Delta P_{i,t-s}, \Delta R_{i,t-r}$ with $s = 2, 3, \ldots, t+2; r = 1, 2, \ldots, t+2$) are orthogonal to the error and can provide additional moment conditions working in the GMM framework.[60] The Arellano and Bond estimator in this context is called the differenced-GMM method (henceforth, GMM-Dif).

Finally, we follow the practice of checking results for different numbers of instruments and use the Arellano and Bond (1991) autocorrelation test for testing whether autocorrelation in the idiosyncratic disturbance term v_{it} would render some lags invalid as instruments. In fact, a key identifying restriction in dynamic panel data models estimated using GMM is that there is no serial correlation in the v_{it} disturbances and this is tested using the first-differenced residuals. Since Δv_{it} is mathematically related to $\Delta v_{i,t-1}$ via the term $v_{i,t-1}$, negative first order serial correlation is expected in differences and evidence of it is uninformative. The test of r-order serial correlation is asymptotically normally distributed under the null of zero serial correlation.[61]

However, as our analysis is unavoidably constrained by the number of countries included in the EU-15, any estimation method in this context may lead to a biased estimate. As no analysis is feasible about which of them delivers the smallest bias, we

[59] See also Holtz-Eakin et al. (1988).

[60] Of course, also that the orthogonality of the exogenous variables with the transformed error term is exploited and corresponding moment conditions included in all GMM estimations.

[61] Arellano and Bond (1991) find that their test has greater power than the Sargan/Hansen test to detect lagged instruments being invalid due to autocorrelation.

interpret similar results using different estimation methods (OLS, Within estimation, and GMM) as evidence that true coefficients should not be significantly different from those estimated, see for details Fiorio and Florio (2010).

Appendix 5.A.3: The Third EU Package

The Third EU Package 2009 and Electricity market Directive 2009/72/EC repealing Directive 2003/54/EC provides for the following:

a) Rules for the organization of the sector. MS may impose on undertakings public service obligations related to security of supply, regularity and quality of service, price, environmental protection, and energy efficiency.

b) MS shall ensure that all customers have the right to choose their electricity supplier and to change supplier easily, within three weeks.

c) Electricity suppliers are obliged to inform final customers about:
   - the contribution of each energy source;
   - consumption data;
   - the environmental impact caused;
   - their rights in the event of a dispute.

d) MS shall put in place an independent mechanism (energy ombudsman or consumer body) to manage complaints or disputes efficiently.

e) MS must monitor the security of supply, define technical safety criteria to ensure the integration of their national markets at one or more regional levels. The NRA are to cooperate with the Agency for the Cooperation of Energy Regulators to guarantee the compatibility of regulatory frameworks between regions.

f) Generation: Member States shall define criteria for the construction of generating capacity in their territory taking account of aspects such as:
   - the security and safety of electricity networks;
   - the protection of health and public safety;
   - the contribution made towards the Commission's '20-20-20' objectives.

g) Transmission system operation: From 3 March 2012, MS must unbundle transmission systems and transmission system operators.

h) TSO are mainly responsible for:
   - ensuring the long-term ability of the system to meet demands for electricity;
   - ensuring adequate means to meet service obligations;
   - contributing to security of supply;
   - managing electricity flows on the system;
   - providing to the operator of any other system information related to the operation, development, and interoperability of the interconnected system;
   - ensuring non-discrimination between system users;

- providing system users with the information they need to access the system;
- collecting congestion rents and payments under the inter-transmission system operator compensation mechanism.

i) Distribution network operation: Member States shall designate DSO or require undertakings that those who own or are responsible for distribution systems do so. DSO are mainly responsible for:
   - ensuring long-term capacity of the system in terms of the distribution of electricity, operation, maintenance, development, and environmental protection;
   - ensuring transparency with respect to system users;
   - providing system users with information;
   - covering energy losses and maintaining reserve electricity capacity.

j) Unbundling and transparency of accounts: MS and the competent authorities have right of access to the accounts of electricity undertakings but shall preserve the confidentiality of certain information.

k) Electricity undertakings shall keep separate accounts for their transmission and distribution activities.

l) Organization of access to the system: MS shall organize a system of third party access to transmission and distribution systems. The tariffs based on that system shall be published.

m) Member States shall also lay down criteria for the granting of authorizations to construct direct lines in their territory, on an objective and non-discriminatory basis.

n) National regulatory authorities: MS shall designate a regulatory authority at national level. It shall be independent and exercise its powers impartially. It is mainly responsible for:
   - fixing transmission or distribution tariffs;
   - cooperating with regard to cross-border issues;
   - monitoring investment plans of the transmission system operators;
   - ensuring access to customer consumption data.

o) Retail markets: Member States shall ensure that contractual arrangements, commitment to customers, data exchange and settlement rules, data ownership, and metering responsibility are defined.

Non-household customers may contract simultaneously with several suppliers.

p) Derogatory measures: A MS may take the necessary safeguard measures in the event of a sudden crisis in the market or where the safety of persons is threatened. Derogations may also be obtained in the event of operating problems in isolated systems.

Source: adapted from http://europa.eu/legislation_summaries/energy/internal_energy_market/en0016_en.htm.

6

Testing the paradigm: Natural gas

6.1 Introduction[1]

The European Union relies heavily on natural gas as a primary energy source, both for final consumption by households and by service activities, and as an input in the production of electricity and other goods. Before the slump of 2008, the EU demand for natural gas was around 600 billion cubic metres, with a market value (at retail price, tax included) in the region of 180 billion Euros. In turn this value was worth nearly one third of the overall 620 billion Euros of energy expenditures by the EU. One half of natural gas is imported. Russia supplies one half of all imports, followed by Norway, and Algeria. Liquefied Natural Gas (LNG) is increasing its role and this will change the future scenarios of import in several countries. New non-conventional sources of gas, including shale deposits and coal-bed methane, could also have an impact on trade patterns. Domestic reserves in the EU, particularly in the UK and the Netherlands, are declining (Exxon Mobil, 2011), and some forecasts to 2030 place the import dependency of the EU at around 80% of demand (Bosmans, 2007). It seems apparent that this context of high dependency from imports makes the gas arena quite different from that of the electricity industry. The two industries are linked because of possible substitution of electricity and gas in some uses, and because of the demand for gas for electricity generation,[2] but their fundamental structural features are different.

Despite of their important technological and market differences, however, EU energy regulations have aimed to treat the electricity and natural gas sectors in a similar way. For instance, the recent (2009) Third Legislative Package on the EU Electricity and Gas markets states that both sectors require separation of production and supply from transmission networks. Ownership

[1] With Carlo Fiorio (Section 6.6).

[2] According to GME (2011, p. 5), out of 27.730 MWh, 14.568 MWh were produced by gas fired units.

unbundling is explicitly invoked by the Directive 2009/73/EC, even if there are softer alternatives in the form of legal/functional separation of the transmission system operator.[3]

One specific issue in the gas reforms has been the EC attempt to push the industry away from long-term contracts[4] in some market segments, particularly downstream, that is in distribution and supply to consumers. Related issues, which are specific to the gas industry, are that of security of supply and of trans-border interconnections.

The implementation of the paradigm poses difficult challenges to policy-makers and industrial players, given that it is well known that large and integrated firms in the gas industry often enjoy considerable economies of scope, scale, or coordination depending on the degree of conglomerate, horizontal, or vertical integration. This expected pursuit of efficiency may be at the expense of competition, in that large firms are likely to acquire a strong if not dominant market position, and productive and dynamic inefficiencies are likely to occur as a consequence. The gas sector displays considerable trade-offs of this type, especially since in most cases (but not everywhere) its current set-up has its origin in a long tradition of state monopolies, where vertical integration was the preferred arrangement.[5]

In order to assess how regulatory reforms have affected gas price dynamics, we combine again price and regulatory data sets as in the previous two chapters.[6]

The structure of the chapter is as follows: first, some technological features of the gas industry, and how they are likely to influence costs and prices are discussed (Section 6.2); then some country features, and aspects of the market structure, extending the discussion in Chapter 2 are reviewed (Section 6.3); earlier empirical findings on the effects of reforms on prices are briefly presented in Section 6.4; Section 6.5 presents the data, Section 6.6 the empirical analysis. Some concluding remarks and some open policy issues for the future are presented in Section 6.7.

[3] On this issue, see Jamasb and Pollitt (2005).

[4] For example, the EC (DG Competition) in December 2005 asked the German E.ON to release gas contracts as a condition for its acquisition of the Hungarian gas operator MOL.

[5] For a detailed history of nationalization and consolidation in Europe see Millward (2005).

[6] The strategy we adopt here is similar to the one used by Copenhagen Economics (2005) to estimate and forecast the impact of market opening policies on overall price and productivity changes and to use this forecast as input into their simulation model of the European economy. We use, however, a wider data set and a more robust econometric approach. We maintain that panel data techniques are able to get rid of of the national idiosyncratic aspects and common trends, so that the effects of changes in regulation regimes can be detected.

6.2 Technology, costs, and prices

6.2.1 *Technology and trade*

Natural gas is a composite commodity, mainly consisting of methane, but in the underground reservoirs it is found with other components: ethane, propane, butane, pentanes, carbon dioxide, hydrogen sulphide, and others. Natural gas can stand alone in the reservoirs in gaseous form or can be associated with oil, as a solution. While some components, particularly propane, increase the calorific energy of natural gas, others need to be separated out and purified. Thus, the first steps of the process, after exploration and discovery, are drilling the gas and processing it. Then, purified gas is pumped and transported through pipelines, or liquefied so it can be stored to allow flexibility on a daily or seasonal basis, or for strategic reasons.

The largest natural proven reserves in the world are in Russia, followed by Iran, Qatar, and other countries in North-Africa and the Middle East. According to Bosmans (2007), who cites BP statistics, the world proved reserves are around 180,000 billion cubic metres, which would be equivalent to 66 years of production before depletion (if no other reserves are discovered). Around 73 billion cubic metres of reserves are in the Middle East, 48 billion in Russia, 14 in Africa (mainly Algeria and Nigeria), and only 5 billion in Europe, and 7 billion in North America. These figures will change to some extent with new explorations and the development of non-conventional types of extraction technologies, particularly in the USA, such as shale gas, tight gas, and coal-bed methane.

The current geographical patterns of natural reserves, and of consumption, are such that a substantial share of production needs to be traded in distant developed or emerging economies, mostly through networks of high-pressure, transmission pipelines and pumping stations or through special ships for liquefied natural gas (LNG) and re-gasification plants. Japan, China, and India need to import substantial quantities of natural gas, and the exit of some nations (e.g. Germany) from nuclear power may also have an impact. From the main pipelines the gas is distributed through pipelines of smaller diameters until it reaches the final user. The typical investment pattern of the industry involves substantial lump-sum costs in exploration, development, and production, plus the cost of building and operating the networks.

From this point of view, a crucial difference as compared to the electricity industry is that economies of scale are related not only to the network component, but also to the whole process starting with exploration. The latter aspect also involves considerable risk-taking, and this justifies large enterprises as a way to pool exploration risks. The possibility that natural gas reservoirs are also combined with crude oil creates economies of scope. Development and production of gas includes drilling, field preparation, compression, and purification. All these activities are capital-intensive. From the processing plant the investment in the transmission pipeline is also a lump one, and it depends crucially on the distance between the production fields and the markets to be served. Eventually, even distribution involves economies of scale, related to the geographical pattern of final demand, which can be agglomerated or dispersed. Thus, the final cost of natural gas is determined by the size, quality, pressure of the fields, the rate of production for each point of take, their distance from demand, and the characteristics of the demand. Technology advances, which have contributed to decreasing development costs, include 3D and 4D seismic imaging of the site geology, changes in the pipes' material and design, measurement-while-drilling (MWD) during drilling, 'slimhole' drilling, offshore drilling (now possible at depths in excess of 3000 m), and hydraulic fracturing of gas-rich layers.

From the point of view of the investor, and of the social planner, there is in principle an additional cost to be considered. Natural gas is an exhaustible resource, and there is an option value in delaying its utilization. Moreover, the management of the network poses some dangerous issues. If the flow of gas is interrupted in the pipelines, and then is re-established, the mixture of air and gas can be explosive. Hence the system operator, the supplier, and the customer have to carefully manage interruptible contracts, or force-majeur stops of the gas flow. The swing between high and low level demand must be also managed carefully, and this contributes considerably to the cost of the service.

Storage, in contrast to electricity, is possible either in caverns or in depleted fields, and requires specific pumping stations. Security of supply in practice means that somebody has to a store sufficient quantity of gas to deal with an interruption of several days or weeks (some countries, e.g. Italy, consider that up to 100 days of storage is prudent). The repeated payment defaults of Ukraine to Russia in recent years, and interruption of service from Gazprom in retaliation—while only lasting a few days—has on different occasions provoked shortages in some EU countries that are connected to Russia through Ukraine (80% of Gazprom supply takes that route).[7] All these features, and the fact that in most cases different companies are involved in the investments and operations, make the determination of the cost of a cubic metre of gas a complex story, as we discuss below.

Turning to the demand side, natural gas is indeed very versatile. The value of benefits to the user greatly varies according the actual utilization, including substitution of a different fuel, typically coal or diesel. Clearly, net suppliers

[7] Professor Steve Thomas (in private correspondence) suggests that the fears of dependence on Russia have been excessive as, over 40 years, the record of reliability of Russia as a supplier of Western Europe is good, and the position of some importing countries would be to make pumps reversible and occasionally import from other EU countries.

and net importers have a different pattern of opportunity costs. Given our focus on household welfare, the typical demand for gas is for water and room heating and cooking. For each of these demands there are alternatives to gas. Customer related costs include connections, metering, billing and other sale expenditures, and safety controls.

In this context, the price paid by the final consumer is the result of a long chain of decisions taken perhaps many years ago in a very distant country, and of the complex interplay of current contracts, future markets, and regulatory mechanisms. In the words of Julius and Mashayekhi (1990, p. 116):

> Gas pricing can be a complex and controversial subject. For the most part, its complexities stem from the array of different considerations that come into play at the various stages where the gas changes ownership... Once gas has been discovered, potential buyers and potential sellers see the prospect of capturing substantial resource rents, but only at the expense of each other.

While this was written 20 years ago, it should still be remembered when discussing the European policy experiment with the gas industry.

6.3 Demand, supply, and country features

The evolution and maturity of the gas market in the EU differ considerably across countries. In some of them natural gas is steadily used as a primary source of energy, while in others, due to the limited availability of internal resources and/or to the scarcity of interconnections, its use is restrained. According to Honorè and Stern (2007), who cite IEA data and consider a broader set of 35 European countries (the 27 EU members, plus Turkey, Switzerland, and some small south-east countries), just three countries cover around 50% of the demand: the UK, Germany, and Italy. Other significant gas-fuelled economies in the EU are France, the Netherlands, and Spain. The highest level of consumption is recorded in the UK where natural gas has replaced oil as the main primary fuel, and the rate of diffusion among the household consumers is also large (35% of total consumption). The second market in the EU-15 is that of Germany where total consumption is slightly below the UK, but the share of household gas consumption is roughly the same as in the UK. Italy is the third largest natural gas consumer in Europe with a demand that has been steadily increasing in recent years. France, the Netherlands, and Spain represent intermediate market in terms of consumption. The French gas market is mature in age, but the share of natural gas in primary energy demand is small if compared to other mature markets and growth is not expected due to the dominance of nuclear power. In smaller European countries, the picture varies considerably depending on the geographic position and on the availability of internal resources. While Austria, Denmark, Belgium, Ireland, and Luxembourg show a non-negligible use of gas both in production stages and for household consumption, in Greece, Portugal, Finland, and Sweden the diffusion of this source of energy is small because of the limited interconnections and in some case because of specific policy choices.[8]

Obviously the use of natural gas and the dimension of national markets are also driven by the endowments of each country. The largest gas-fields in the EU are located in the UK and the Netherlands. Germany, Italy, and Denmark also have some gas-fields but these are more limited. Most EU countries do not have any gas-fields within their borders. A consequence of this heterogeneity of endowment is a situation of strong import dependency for most countries. Among the EU-15 countries we consider, only Denmark and the Netherlands are net gas exporters.

This general dependency from imports raises the question of the security of gas supply, which was evaluated in a European directive of 2004[9] and was recently reconsidered by a new directive (994/2010). As mentioned, presently the largest share of gas imports into the EU borders come from three countries: Russia (151 bcm per year), Norway (78 bcm), and Algeria (62 bcm), see Figure 6.1. Through the Nordstream pipeline (1224 km through the Baltic Sea) Russia will be able to supply an additional 55 bcm/year (around 2012), directly to Germany.[10] This new infrastructure will avoid gas travelling through pipelines in Ukraine, which, as mentioned, has proved to be a route at risk of service disruption because of the difficult economic and political relations of Russia with this former Soviet state. The Southstream project (ENI-Gazprom) will instead bring 63 bcm Russian gas to Italy through the Black Sea, while Nabucco—a 3300 km project endorsed by the EU[11]—will try to open the way to (mainly non-Russian) 31 bcm gas from the Caspian regions, Central Asia, and the Middle East, to Austria, across the Turkish–Georgian and Iran borders, and Bulgaria, Romania, and Hungary. Other priority corridors for gas have been identified by the EC.[12]

[8] Greece, Finland, and Portugal derogate from the provisions of the second European gas directive by virtue of their status as emerging or isolated markets. In Sweden nuclear power accounts for half of electricity supply. After the 1980 referendum to phase out all nuclear power plants by 2010, the government has given priority to renewable sources to fill the gap left by nuclear power and this choice has hampered the further development of the gas market.

[9] European Directive 2004/67/EC concerning measures to safeguard security of natural gas supply.

[10] The partners in the project are Gazprom, BASF, E.ON, Ruhrgas, Gasunie, and GDF-Suez.

[11] The project partners include Botas (Turkey), Bulgargaz (Bulgaria), Transgaz (Romania), MOL (Hungary), OMV (Austria), RWE (Germany), with financial support from the EU and the European Investment Bank also.

[12] Another submarine pipeline is under construction: the Galsi project between Algeria and Sardinia (http://www.galsi.it) with a capacity of around 8 bcm/year.

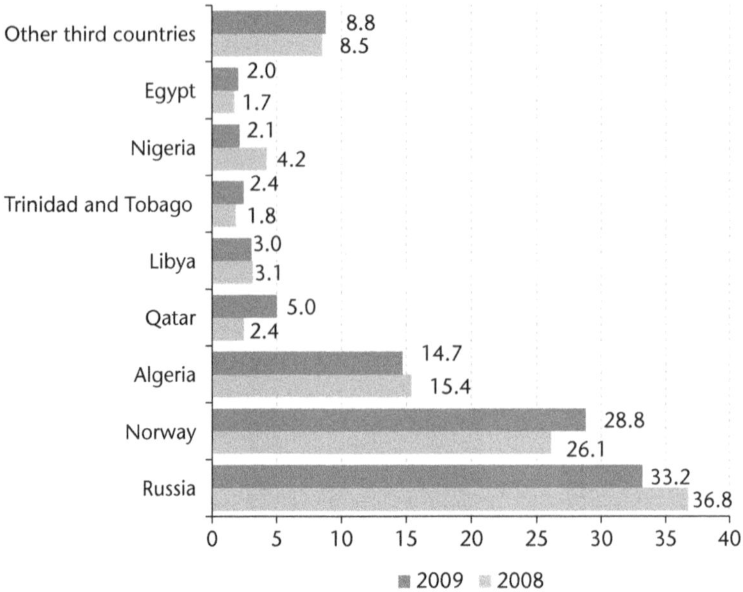

Figure 6.1. EU-27 imports of natural gas: Percentage of extra-EU imports by country of origin

Source: Eurostat 2010.

The capital investment involved in these infrastructure projects is huge. It has been estimated at 15 billion Euros for Southstream and 7.9 for Nabucco. Completion time is in general 10 to 15 years. There is still some doubt about the possibility of finding the finance and the natural gas flows necessary to sustain all the current investment plans, in spite of EC ambitions in this area.[13]

As a diversification strategy, some European countries have undertaken the construction of liquefied natural gas facilities (LNG). Spain is the European country with the highest number of LNG terminals in Europe: 37% of total gas imports are transported by ship from Nigeria and Middle East countries. The same approach is followed by France and Portugal where respectively 36% and 20% of total imports come from LNG terminals. The UK now has three LNG terminals and a significant share of its demand will be imported through those plants. Italy has also recently expanded its LNG operations, with imports from Qatar (Adriatic LNG). Another LNG grand new project for Europe involves a major liquefaction plant in Turkey, a new fleet of gas carriers, and several re-gasification facilities in the EU Mediterranean countries.

[13] See Communication 'Energy infrastructure projects for 2020 and beyond – a blueprint for an integrated energy network' (COM 2010/677).

To illustrate the diversity of the industry across countries, the rest of this section briefly sums up the main features in selected Member States. Some of the reforms implemented in these countries were presented in Chapter 2. The countries analysed are again the UK, the Netherlands, France, Germany, Italy, and Spain.

Despite the European Commission advocating a single cross-country policy reform pattern, there is still a large variability in the national natural gas market structure among the EU-15 countries. Tables 6.2 and 6.3 show the number of firms operating in the different segments of the gas industry and the type of unbundling for the TSOs. As we shall see, while natural gas is a relatively homogeneous commodity (after processing), in each of these countries the industry has pursued different structures and development paths.[14]

6.3.1 *Country features*

The starting point of the reform paradigm at the end of the last century was again the UK, see Newbery (2000), Thomas (2004), Florio (2004), and Parker (2009), which broadly paralleled the liberalization experiment introduced in the USA in the mid 1980s.

The privatization of British Gas,[15] however, was implemented initially (Gas Act, 1986) by government divestiture as a vertically integrated company, including significant control of domestic gas sources and a monopoly of sales to customers up to 23,000 tons for 25 years. This structure prevented new entries downstream for years, while there was free entry upstream. In 1997 British Gas PLC was divided into two companies: BG and Centrica, the former focusing on exploitation, production, and transport, the latter on supply. The network was spun off in 2001, then merged with the National Grid Company. The government required BG to progressively reduce its share in the industrial market, where it was replaced largely by oil companies, which diversified into the gas business. In the retail UK gas market there was instead a gradual process of merging with the electricity supply: all significant suppliers offer gas and electricity as a dual package. Centrica, still holds about 60% of the residential gas market, with the rest of the market going to the five major

[14] The EU pushes for harmonization of the national rules for gas specifications, in cooperation with the European Association for the Streamlining of Energy Exchange (EASEE-Gas), Gas Infrastructure Europe (GIE), and other stakeholders.

[15] The gas industry was nationalized by the Gas Act, 1948. The British Gas Corporation was created by the Gas Act, 1972. Natural gas in the North Sea was discovered in 1965. See Thomas (2003) for a concise and interesting history (1965–86) of the performance of British Gas before its privatization. British Gas acted as a monopsony upstream, but gas extraction was licensed to many private operators, typically under a combination of initial fees, a yearly rent, and production royalties to be paid to the British government by concessionaires. Thousands of 'blocks'(i.e. areas) were allocated over the years.

electricity companies (RWE/NPower, EON/Powergen, EDF Energy, Scottish and Southern, and Iberdrola/Scottish Power) and to smaller players.[16] It is not clear how far this development of horizontal integration was forecast by the policy-makers when they decided in favour of vertical disintegration of the industry. Moreover, the reform was implemented at a time when the UK could rely on an oversupply of its own gas reserves. Britain is, however, now facing a transition: from previous self-sufficiency in natural gas production (reservoirs in the North Sea were discovered in 1965 and are forecast to be near depletion after 2020) it is now a net importer due to declining internal production, see Table 6.1.To overcome this condition, the country created a trading connection with continental Europe via Belgium in 1998 and links to Norway and the Netherlands were reinforced,[17] including the creation of new LNG terminals in order to diversify the supply sources (Kent, Millford Heaven, Isle of Grain).

Given the recent history of gas availability, and the intention of the reformers to encourage competition, short-term contracting was the preferred option, in a manner that was quite different from the functioning of commercial arrangements in most of continental Europe. While European gas prices are usually linked to oil prices through specific formulae included in long-term contracts, in the UK the National Balancing Point acts as a virtual hub, based on short-term contracts.[18] Figure 6.2 shows nominal gas prices for residential users in the UK before and after privatization of British Gas. Thomas (2004) has suggested that the gas price stability in nominal terms, and its fall in real terms in the decade after privatization, should be attributed mainly to price regulation, while competition did little to contain gas rents. According to the regulator, prices in the UK are expected to increase in the future because of import dependency.

By way of comparison, the other gas-rich EU country, the Netherlands, has developed a different industry organization. In the Netherlands, the gas sector, while widely liberalized, is still largely controlled by the incumbent, which started out in 1963 as Gasunie, a joint-venture between the Dutch state (50% of equity), Royal Dutch Shell, and Esso, with the objective of developing Groningen and other fields discovered in 1959. The original legal arrangements were more complex.

[16] British Gas is now an exploitation and production company and has lost its dominant position.

[17] Since 1978 the Vesterled pipeline connects Norway to the St Fergus Gas Plant in Scotland. Since 1998 the Interconnector links Zeebrugge (Belgium) to Bacton Gas (UK). The latter has been connected to the Netherlands since 2006 (BBL Company). Another link with Norway is the Langeled pipeline, opened in 2003.

[18] The relationship between short-term and long-term prices is however a complex one, as the UK is no more insulated from continental oil-linked EU prices (see the Further Reading Section). The correlation of gas and oil prices in the UK is in fact still high.

Table 6.1. UK import and export of natural gas in thousand GWh, 2000–2009

	2000	2001	2002	2003	2004	2005	2006	2007	2008	2009
imports from:										
Belgium[1]	2.96	4.02	6.65	4.39	25.59	24.11	30.51	6.47	12.17	7.95
The Netherlands[2]	–	–	–	–	–	–	9.14	76.60	90.56	69.53
Norway[3]	11.28	12.73	37.89	71.75	95.36	127.90	157.04	225.76	283.72	260.44
Liquefied natural gas[4]	–	–	0.00	–	–	5.45	37.58	14.90	8.91	110.58
TOTAL IMPORTS	14.23	16.75	44.53	76.14	120.95	157.46	234.25	323.74	395.37	448.49
exports to:										
Belgium[1]	94.57	80.85	91.19	122.65	60.06	36.64	60.20	51.39	45.95	62.08
The Netherlands[5]	7.72	5.64	4.84	3.42	2.89	4.26	3.37	6.36	10.39	13.08
Norway[3]	–	–	–	–	–	–	–	0.15	0.39	0.28
Ireland	32.25	38.13	38.74	40.81	39.08	39.41	47.25	50.97	54.25	54.36
TOTAL EXPORTS	134.55	124.62	134.77	166.88	102.03	80.31	110.81	108.87	110.99	129.81
NET IMPORTS	–120.31	–107.87	–90.24	–90.74	18.92	77.15	123.44	214.87	284.38	318.68

[1] Bacton–Zeebrugge Interconnector.
[2] Bacton–Balgzand (BBL) pipeline.
[3] Langeled and Vesterled pipelines. Tampen Link (from Statfjord to FLAGS).
[4] From different deposits to Isle of Grain and Gasport Teesside.
[5] Dutch gas pipeline.
Source: Digest of United Kingdom statistics (DUKES), 2000–9.

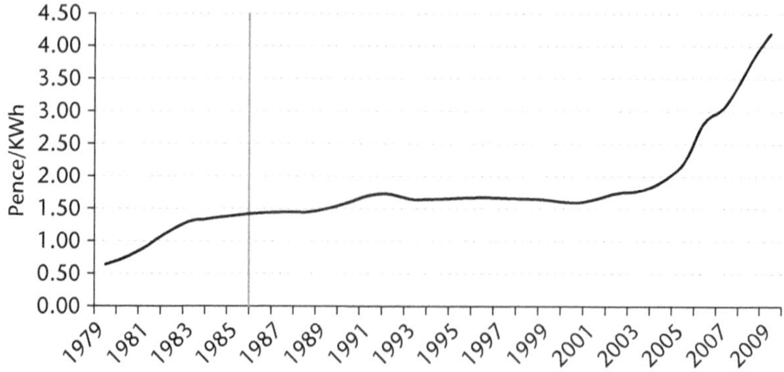

Figure 6.2. UK price of gas for households 1979–2009 (current pence/KWh)
Source: DUKES and IEA.

Table 6.2. The EU-15 gas sector: Production, imports, and retailing (around 2005)

Country	Production and imports		Retailing	
	Number of entities bringing gas into the country (production or imports)	Of which dealing with at least 5% of gas (imported and produced)	Total number of suppliers	Of which supplying at least 5% of the total
Austria	4	4	27	5
Belgium	4	2	32	2
Denmark	1	1	7	5
Finland	1	1	30	1
France	13	1	34	2
Germany	27	5	700	1
Greece	1	1	15	1
Ireland	7	5	2	2
Italy	26	3	389	5
Luxembourg	2	1	6	4
The Netherlands	n.a.	n.a.	25	4
Portugal	1	1	10	4
Spain	14	4	41	4
Sweden	1	1	7	5
UK	24	6	15	7

Source: Goerten and Clement (2006).

Exxon and Shell, through NAM (Nederlandse Maatschappij Aardolie) were awarded the extraction concession, and supplied the gas to NV Nederlandse Gasunie (25% Shell, 25% Exxon, 10% the Netherlands State, and 40% EBN—Energie Beheer Nederland, a government-owned entity). The company fully owns the 12,000 km long transportation network in the country. In 2005, Gasunie spun off GasTerra[19] that retained the original public–private equity

[19] Thus, Gasterra buys its gas from NAM and resells it to the distributors downstream.

Table 6.3. The EU-15 gas sector: Types of unbundling for the Transmission System Operator

Country	Type of unbundling	Country	Type of unbundling
Austria	Legal	Italy	Legal
Belgium	Legal	Luxembourg	Not implemented
Denmark	Ownership	The Netherlands	Ownership
France	Legal	Spain	Legal
Germany	Partly legal	Sweden	Ownership
UK	Ownership	Ireland	Not implemented

Note: Finland, Portugal, and Greece derogate from the provisions of the second European gas directive by virtue of their status as emerging or isolated markets.
Source: Commission of the European Community (b), 2006.

structure, and is in charge of gas trading. The Dutch state in the same year acquired the equity shares of its private partners in gas transportation activities, which are still under the name of Gasunie, now one hundred per cent state owned. This entity controls the transmission pipelines, storage facilities, plus a substantial network in Germany (2000 km) and an important connection pipeline with the UK. Gasunie is also a shareholder of APX-ENDEX, an energy exchange serving the Netherlands, Belgium, and the UK, and has an equity position in Nordstream and in other projects, including a LNG major project. A separate company (Gas Transport Services) acts as the national transmission system operator, following the Gas Act (2004) which introduced TPA and full market opening for final consumers.[20] Retail and distribution is carried out primarily by municipally owned companies which supply electricity as well. Thus, the current industry organization is mainly based on a concession to private producers,[21] on a re-nationalized infrastructure, legal but only partial ownership unbundling of the network and the TSO, and mostly municipal horizontally integrated local supply.

Comparing the UK and the Netherlands, the only two EU countries with significant gas reserves, it is apparent that—in spite of the common European legal framework—they have adopted very different gas policies. For a comparison of the government role in the two countries, see for example Mulder and Zwart (2006). See also Correljè (2008).

Turning to gas-deprived countries, geography and political circumstances have determined different paths here too.

The gas industry structure in France parallels that in the electricity sector with one large company, Gaz de France-Suez (GDF) created in 1946 by the French government, still dominating the market. It was fully public owned until July 2005[22] when 20% of the shares were sold by initial public offer. The

[20] In 2007 switching was around 6.8%.

[21] In addition to NAM other concessions have been granted to ELF, BP/Amoco, Wintershall, and VEBA

[22] Following law 83/2004, initially providing for a maximum of 30% of private shareholding.

Table 6.4. GDF-Suez's shareholder structure (2010)

Shareholders	% of share
Institutional investors	40%
Individual and other shareholders	11%
Sofina	1%
CNP Assurances	1%
Treasury stock	1%
CDC	2%
Employee shareholders	3%
Groupe Bruxelles Lambert	5%
French government	36%

Source: GDF-Suez.

merger between GDF and the private Company Suez[23] left around 36% of the shares in government hands, but there is little doubt that GDF is still under the effective control of the French government, see Table 6.4. There are two TSOs in the country: Gaz de France Reseau Transport (GDF-RT) and the privately owned Total Infrastructures Gaz de France (TIGF). Both operators have been unbundled in legal terms. In 2008 a separate company GrDF (Gas réseau distribution France) was established (by separating from the combined electricity and gas company EDF-GDF). The main DSO is also owned by GDF-Suez.

The main providers of natural gas are Norway (27% of total imports), Russia (21%), the Netherlands (20%), and Algeria (12%), with 95% of the long-term contracts controlled by GDF-Suez and Total. The share of alternative suppliers (such as Tegaz and local distributors) in imports is on the rise, while there are 12 foreign companies operating in the wholesale market. At the distribution level, there are 22 DSOs, with GDF again dominating this part of the market (96%). There are also 11 suppliers currently active in the retail market, which are independent of network companies.

A non-negligible share of French total import (about 20%) enters via LNG terminals. In the retail segment there is a limited competition: Gaz de France has not separated its retail activities and dominates the market for industrial and household consumers, while new foreign companies (particularly ENI with Altergaz, British Petroleum, and E.ON) have entered the market since 2005.

A more decentralized, but not necessarily more liberalized, industry structure can be observed in Germany. The structure of the German gas market is characterized by a multi-tier organization, with a small group of big players at the import and transmission level, another 24 regional companies at transmission level, and approximately 700 companies operating at the local distribution level. Ruhrgas, founded in 1926, served the Ruhr region with gas from coke, then in the 1960s shifted to gas supply. With more than 50% of the

[23] Following Law 1537/2006.

available gas, imported from Russia, Norway, and the Netherlands, it dominated the market. After some controversies involving different institutional actors, Rhurgas was taken over by E.ON in 2003, one of the two largest electricity companies in Europe.[24] The main competitors of E.on are Wingas, a company jointly owned by BASF and Gazprom, RWE the other dominant electricity company (through Thyssengas), VNG (Verbundnetz GAS), and BEB, see Table 6.5.

Table 6.6 shows the ownership pattern of the major pipelines.

Italy offers a very different example of a country that has adopted a partial liberalization strategy while keeping government control on the core player. The main participants in the Italian natural gas market are ENI, ENEL, Edison, Plurigas, Hera, E.ON, and Gaz de France-Suez (all listed companies except Plurigas). The biggest firm is the former integrated monopolist ENI, still partly owned by the government. In the wholesale market the share of the first three players is around 31% (Politecnico di Milano—Dipartimento di Energia for AEEG, 2010), ENI controls 84% of domestic production and 39% of imports

Table 6.5. Main gas import and transmission companies in Germany (2003)

Name	Ultimate parent	Gas supplied (bcm)	Transmission (high-pressure km)	Storage (bcm)
Ruhtgas AG	E.ON	50.6	10,750	<5
BEB Erdgas und Erdöl	Exxon Mobil (50%); Shell (50%)	16.4	3,439	>2.7
Verbundnetz Gas AG (VNG)	EWE (47.90 5); BASF (through Wintershall Erdgas Beteiligungs GmbH 15.79%); Gazprom (through ZGG-Zarubezhgas-Erdgashangdel-Gesellschaft mbH 5.26%); Gaz de France (through EEG-Erdgas transport GmbH 5.26%); Twelve municipalities in East Germany (25.79%)	15.8	7,300	>2
Wingas GmbH	BASF (65%); Gazprom (35%)	11.8	1,836	4.2
Thyssengas GmbH	RWE AG (100%)	6.7	2,500	>0.3
EWE	Administrative districts and towns in the Weser–Ems region, administrative districts and towns in the Weser–Elbe region	4	3,870	>1.1

Source: OECD 2004.

[24] The condition imposed by the German authority for allowing the take over included the sale of its stakes in a number of different gas companies and it was also required to auction a significant proportion of its gas import contracts to reduce its dominance in the wholesale market.

Table 6.6. Ownership pattern of the major pipelines in Germany (2003)

Pipeline	Commissioned	Capacity	Ownership
Megal (Czech Republic–France)	1980	22 bcm	Ruhrgas 50% GDF 43% OMV 5% Stichting Megal 2%
Tenp (the Netherlands–Switzerland)	1974	14.4 bcm	Ruhrgas 51% Snam International 49%
Netra (Etzel–Bernau)	1994	19.8 bcm	Ruhrgas 41.7% BEB 29.6% Statoil 21.5% Norsk Hydro 7.2%
Midal (Emden–Ludwingshafen)	1993	13 bcm	Wingas 100%
Stegal (Saxony–Thuringia)	1992	10 bcm	Wingas 100%
Wedal (Belgium–Midal)	1998	10 bcm	Wingas 100%
Jagal I/II (Malnow–Stegal)	1996/1999	24 bcm	Wingas 100%

Source: BMWi 2004.

through five main infrastructures under its direct or indirect control.[25] An ENI subsidiary, SNAM Rete Gas (52% owned by ENI, but to be divested in future), owns and operates 31.680 km of the total 33.768 km of the domestic regional and national natural gas transportation system (the remaining 2.088 km is managed by 9 different companies among which the Edison group). There are about 430 local distributors in the country, both private, public–private, or owned by municipal entities, while in the retail market at the end of 2006, 380 companies owned a gas licence. Most of them represent unbundled sales divisions of formerly integrated distribution companies. However the retail market is still strongly dominated by three largest groups: ENI with a market share of 40.3% of sales, ENEL (15.8%), the electricity incumbent, de facto controlled by the government, and Edison (7.9%). Moreover, ENI—to comply with the legislation—sells some of its Russian gas to other importers through transactions abroad, an issue that complicates the trade structure and possibly creates double marginalization.

Finally, a different case history again is offered by Spain, which is not interconnected with the main networks of other EU countries. In Spain the gas industry before liberalization was dominated by one integrated private company, Gas Natural. In 2002 the regulator forced the company to spin off 65% of the shares of Enagas, the private firm that controls Spain's natural gas transport system. The country imports all its gas mainly via pipelines from Norway and Algeria, but with a significant proportion coming via LNG terminals from countries such as Qatar, Oman, and Nigeria. Spain has developed

[25] TAG pipeline (mainly Russian gas), TENP pipeline (mainly Norwegian gas), Panigaglia LNG Terminal (mainly Nigerian and Algerian gas), TTPC pipeline (Algerian gas), Green Stream pipeline (Libyan gas).

a wholesale gas market in recent years, as the government introduced a gas release programme which operated from 2001 to 2004, and this resulted in six new entrants acquiring gas from the largest company, Gas Natural, which was forced to sell 25% of its contracted gas to new entrants to promote competition. After this process, the market share of the incumbent reduced from 100% in 2000 to 48% in 2005. Despite the fact that gas retail market has been fully open since January 2003 the overall setting is far from being considered competitive. The switching rate since market opening is only about 5% and Gas Natural still strongly dominates the market for households with a share of over 70%.

Table 6.7 provides some summary figures for the six major countries we have mentioned above and for the EU-27, while Figure 6.3 shows the change over time of the regulatory pattern.

This brief account of the structure of the natural gas industry in some Western European countries suggests the peculiar nature of the service, with a very small number of exporting countries, high dependence on imports, an intrinsically oligopolistic structure, and various combinations of horizontally and vertically integrated firms.

While the natural gas industry is a latecomer in the reform process, as compared with electricity or telecommunications, in fact there is has been considerable variability across countries for many years. As mentioned in the above discussion, for example, the extent of public ownership of the players is non-existent in the UK, limited in Spain and Germany, and substantial in the Netherlands, Italy, and France. Also the degree of vertical integration and of market opening is not homogeneous. Whatever the legal framework, and the formal need for the Member States to comply with the EU gas legislation, one can observe whether households pay different bills under different ownership and structural patterns of the industry. To this question we turn below.

Table 6.7. Natural gas balance sheet, 2009, in million tonnes of oil equivalent (GCV*)

	EU-27	DE	ES	FR	IT	NL	UK
a. Primary production	169.9	12.2	0.0	0.8	7.4	63.3	59.0
b. Total imports	372.4	84.8	35.3	45.2	63.0	19.5	38.6
c. Resources[1]	542.3	97.0	35.3	46	70.4	82.8	97.6
d. Stock change[2]	−5.5	−2.0	0.4	−1.8	0.8	0.0	−0.4
e. Total exports	76.9	10.1	1.0	1.8	0.1	44.2	11.9
f. Gross inland consumption[3]	459.9	84.9	34.7	42.4	71.1	38.6	85.3
g. Energy dependency[4]	64.3	88.0	98.9	102.3	88.5	−64.1	31.3

* GCV = Gross calorific value.
[1] $a + b$.
[2] + Decrease in stocks; increase in stocks.
[3] $a + b + d - e$.
[4] $(b - e) / f * 100$.
Source: Eurostat 2010.
Note: EU-27 figures comprise estimated 2009 data for Portugal.

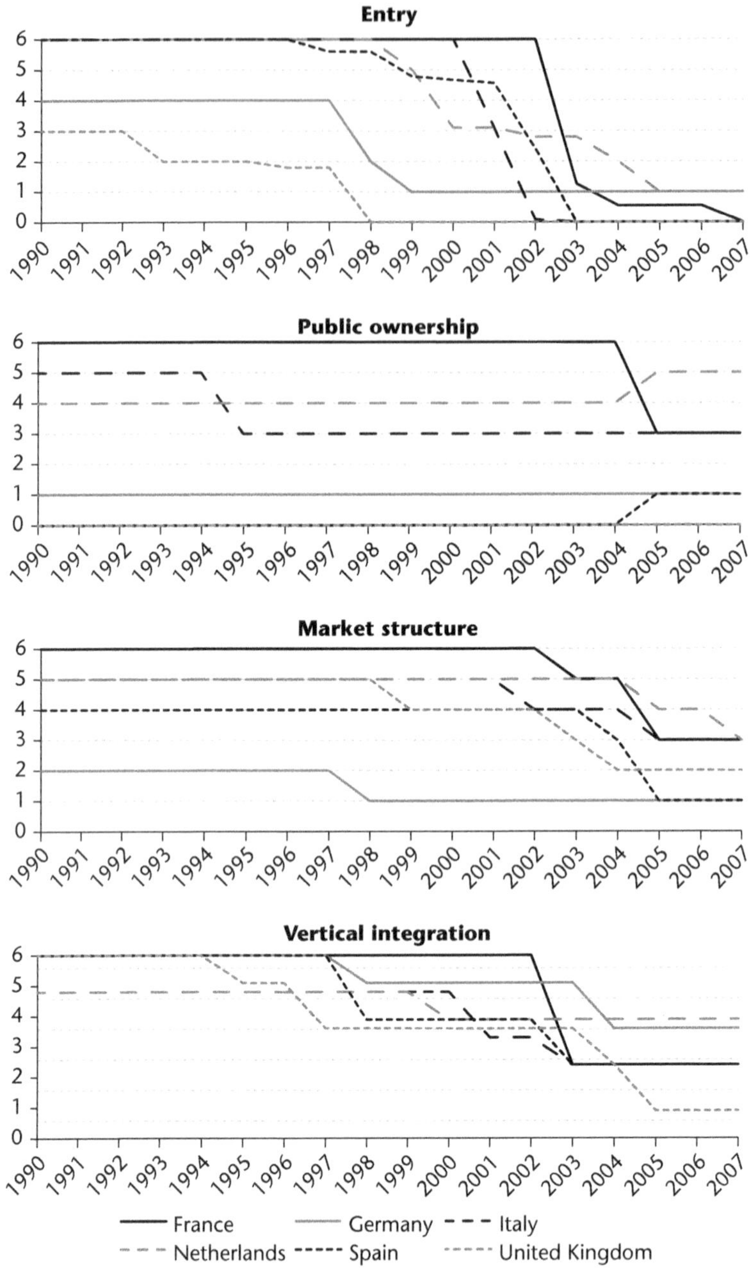

Figure 6.3. Trends of ECTR reform indicators, 1990–2007, selected EU countries

Source: own elaboration of ECTR scores.

6.4 Some earlier findings

A comparative analysis of the effects of a common policy reform on the European residential natural gas sector is a difficult task. As recently documented, for example by Asche, Nilsen, and Tveteras (2008), apart from the common feature of very small short-run demand elasticities to own price and a dramatic consumption increase at an aggregate level, large differences between European countries are easily detected in terms of shares on total energy consumption and natural gas grid coverage. Moreover, since in Europe long-term supply contracts from Russia, Algeria, or Norway usually employ price formulae based on the (time-lagged) prices of energy alternatives, such as crude oil, fuel oil, coal, and so on, the supply of oil and natural gas are strongly linked, and changes in crude oil prices have historically had a prominent role in shaping natural gas prices. Brown and Yücel (2008) confirm, for the USA, that movements in crude oil prices still explain gas prices, in spite of short-term markets.

Earlier empirical literature, looking at macroeconomic outcomes of reforms, provides some support to a negative view of vertically integrated monopolies.

Indeed, 'the devil could be in the details', and under some circumstances market liberalization could not bring significant positive welfare effects, for example see van Witteloostuijna, Brakmanb, and van Marrewijk, 2007, for an analysis of the Dutch natural gas industry.

Some concerns about the real effectiveness of the current patterns of regulatory reforms have been voiced by the Eurogas association: '[most firms] think that options based on legal and functional unbundling should be further considered, combined with an enhanced effective regulation and compliance system...The unbundling proposals contained in the EU Third Package do not address the issue of the required level of future investment' (Bosmans, 2007, p. 13). On this issue more articulated views such as that by the World Bank (2007) have appeared which, in comparison with the EU paradigm, leave more degrees of freedom to national policy-makers.

The lack of recognition of the very nature of the natural gas industry by the EU reform model has been pointed out by Jamasb et al. (2008). Natural gas transmission networks, unlike electricity and transport sectors, are often oligopolies rather than natural monopolies. The presence of companies with relevant market power could yield double marginalization processes in the case of unbundling of vertical stages (see also Davies and Waddams, 2007, and van Witteloostuijna et al., 2007). According to these views, reducing the market power of the incumbent at the distribution stage would be harmful for consumer welfare due to the loss of bargaining power vis-à-vis the upstream stage.

The literature is not univocal on this point. As remarked by Zwart (2009), in a simulation model (NATGAS) of the European gas market, in the presence of resource constraints the relationship between competition and elasticity of supply to own price is limited.[26] Honorè and Stern (2007) point to several external constraints for the future of European gas industry, while simulations carried out with the GASMOD (Holz, von Hirschhalsen, and Kemfert, 2008) and GASTALE (Boots et al., 2004) models lead to more optimistic conclusions, indicating that a liberalized downstream market (in accordance with the EU policy) would determine lower equilibrium prices even in the presence of an oligopolistic upstream market. In this view, downstream market liberalization could reduce the double marginalization effects, provided that a competitive market develops at this stage. Otherwise, consumers would benefit more from a vertically integrated industry. We shall further discuss these issues in the concluding section.

For a comparison with the impact of deregulation in the US gas sector, see Arano and Blair (2008) who find for the industrial user a benefit of deregulation policies in terms of prices and welfare. Growitsch et al. (2009) provide an empirical analysis of the separation of gas transport in 20 OECD countries 1989–2006, with panel data. They do not find any significant impact of unbundling on gas prices, while there is a very modest negative effect of liberalization. They use, as we do below, the OECD regulatory indicators. For a similar analysis, but with more optimistic results, see Copenhagen Economics (2005).

6.5 Data

As final gas prices often include substantial indirect taxation, we focus here on tax-exclusive prices. An EU-15 cross-country analysis of net-of-tax natural gas household prices can be performed using either the data provided by the Eurostat or those provided by the IEA. Brau et al. (2010) use both datasets, but as the Eurostat time series starts in 1991, they restrict their analysis to the period 1991–2007 for both datasets. The main differences in the two data sources reflect different aggregations of primary data at the national level, different average household definitions, and the exchange rate used.[27] Here, considering that the correlation of price variables coming from these two sources is high (about 0.75), we only use IEA data, allowing data to go back to 1978, the starting year of the IEA gas price series. Net-of-tax gas prices charged to domestic consumers are expressed in Euros per Gigajoule and represent our key price variable in this chapter. Figure 6.4 presents the price

[26] For a recent theoretical analysis of the contrasting economic effects in the case of vertical oligopolistic industries, see for example Reisinger and Schnitzer (2008).

[27] The price data provided by the IEA refer to the price paid by a non-better specified average household in US dollars, those provided by Eurostat refer to the price paid by households with yearly consumption equal to 83.70 GJ and is recorded in Euros. The IEA net-of-tax price data have been converted into Euros using the Eurostat exchange rate.

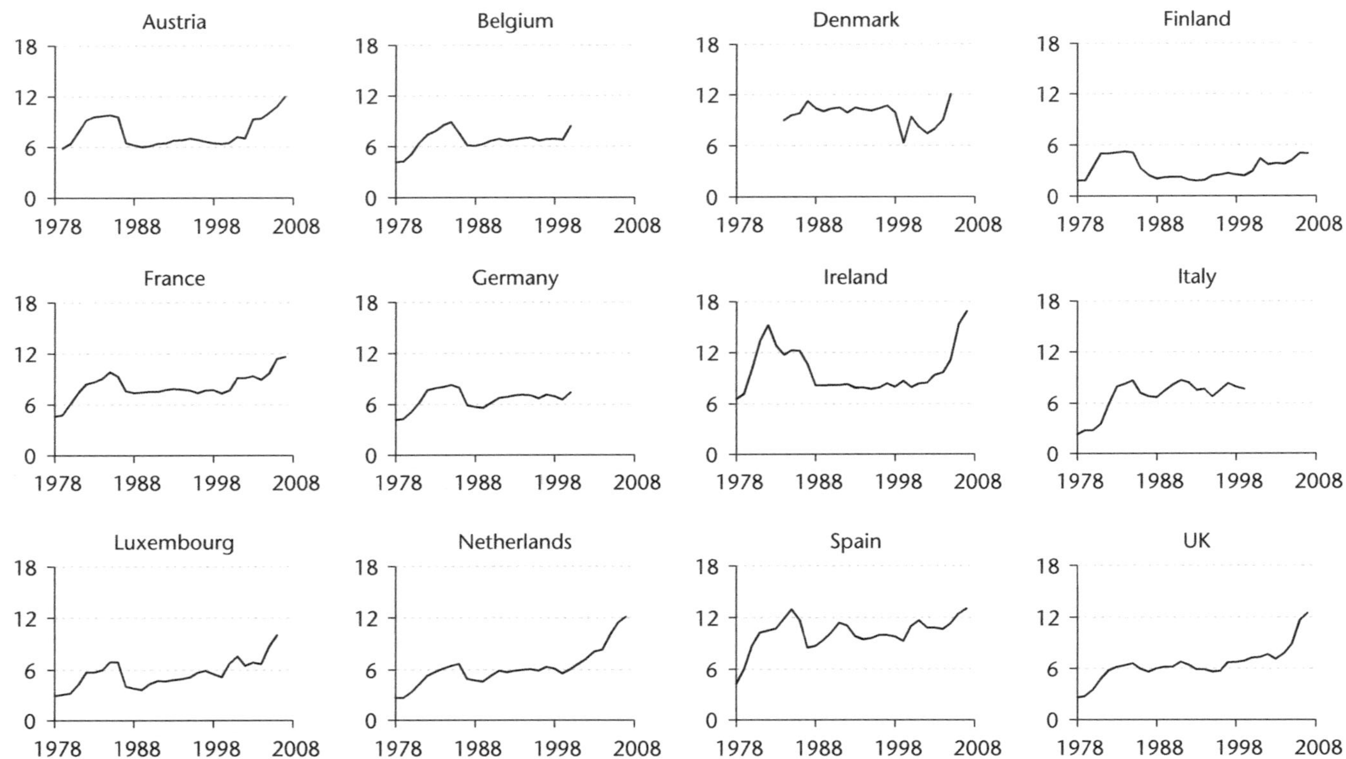

Figure 6.4. Evolution of net-of-tax average household gas prices in EU-15, price is in Euro/Gj

Source: own elaboration on IEA data.

dynamics in the main European national markets,[28] which is increasing, especially in the most recent years. It also shows that despite some differences in levels, the trend of domestic consumer price is roughly increasing across the EU-15 but with a lot of variation. During the early 1980s consumer prices increased in many countries. Some countries have also experienced a strong increasing dynamic in more recent years, for example see Ireland, the Netherlands, and the UK, in others the increase was weaker. However, these simple descriptive figures do not allow us to infer that specific combinations of industry reforms are systematically delivering lower or higher prices than the EU average.

In order to focus particularly on the countries we have briefly discussed in the previous section, Table 6.8 shows that in the UK, the country where reform started, in spite of the availability of natural gas domestic reserves, a fully privatized industry, a liquid spot market, and liberalized retail supply, the net-of-tax price of natural gas for the household[29] in 2007 is higher than the price in France, 12.50 in the UK and 11.63 Euro/GJ respectively. In the same year a household had to pay 11.95 Euros in Austria,12–15 in the Netherlands, and 11.79 in Italy. Only in Germany and Spain are prices higher than in the UK. As mentioned, in the 1980s, before the reform, gas prices for households in the UK were much lower than those in any EU country; then the relative advantage for the consumer gradually waned and it is now non-existent or reversed. According to the EC Market Observatory for Energy (2009), more recently (Eurostat) prices for households in the UK are still slightly lower than the average of the EU-27, but France still offers a slightly lower price to the average household, in spite of limited privatization, unbundling, and liberalization.

Interestingly, in the presumably more competitive market for industrial users of gas, according to the same source, the dispersion across countries and across bands of volume of consumption is less than for the household sector; and there is more price convergence. The authors of EC (2010)

Table 6.8. Natural gas prices for households in the EU, selected years and countries (Euro/GJ)

year	DE	IT	ES	FR	NL	UK	AT
1978	4.16	2.26	4.21	4.61	2.61	2.57	n.a.
1990	6.18	8.16	10.25	7.51	5.26	6.17	6.13
1999	6.52	7.66	9.28	7.30	5.52	6.85	6.32
2007	13.97[a]	11.79[a]	13.04	11.63	12.15	12.50	11.95

Source: IEA.
[a] Eurostat data.

[28] Notwithstanding the long time series provided by IEA data, this dataset does not consider household price for Sweden, a marginal consumer, Section 6.3.

[29] Household of group D3, with a yearly consumption of 83.70 GJ.

Observatory are uncertain about the interpretation: either convergence is a result of competition for the lowest price or: 'An alternative interpretation is that the majority of industrial users are paying according to an oil indexed formula. The use of a similar pricing mechanism produces a harmonization effect across the Member States' (EC, 2010, p. 45).

6.6 Empirical analysis

In order to further explore the relationship between reforms and prices, and to account for country-specific features, we estimated panel data models regressing the log of net-of-tax natural gas prices for domestic users against the ETCR reform indicators and tested the hypothesis that reforms had decreased prices, after controlling for country-fixed effects and crude oil prices and the 'Brent' series.

This estimation strategy is similar to the one used by Alesina et al. (2005) and Copenhagen Economics (2005) who estimate the impact of market opening policies respectively on investment in non-manufactory industries in OECD countries, and on electricity and gas prices for industrial users, and is very close to Brau et al. (2010), who also included additional controls for other macroeconomic variables.

Following the Blundell and Bond (2000) strategy, we estimated some simple autoregressive (AR) models of order 1 and 2 of log prices using OLS and fixed-effects. We include year time dummies to account for common trends. The (log of) natural gas price is highly correlated with its lagged value, calling for the estimation of panel models including the lagged dependent variable (dynamic panels) instead of static ones. Denoting with p_{it} the log yearly household natural gas prices for country i at time t, with R_{it} the vector of regulatory variables for country i at time t, with Z_{it} the Brent oil price, and with β a vector of time dummies, we specify the simple linear model:

$$p_{it} = \rho p_{i,t-1} + R_{it}'\gamma + Z_{it}'\delta + \alpha_i + \beta_t + \varepsilon_{it}, \qquad (6.1)$$

where $\alpha, \beta, \gamma, \delta, \rho$ are parameter vectors to be estimated and ε_{it} is an *iid* (over i and t) stochastic term.

In this model, time dummies account for common shocks on consumer prices and oil prices. Following Arellano and Bond (1991) they are considered strictly exogenous. In the case of feedstock prices, it is likely that true state dependence (e.g. habits and adjustment costs) is relatively small, but the dynamic specification is able to capture the effect of country-specific unobservable factors such as access to different pipelines with different import prices and the presence of *take-or-pay* contracts. Estimation was done by 5A2.

As information on natural gas prices is missing for some years and countries we have an unbalanced sample over 29 years.

Estimation results show that the lagged dependent variable is highly significant also after inclusion of year fixed-effects, regulatory variables, and other controls. Crude oil price—as expected—is positively correlated with natural gas prices. Given this pricing structure for wholesale gas, and the huge swings in oil prices in the period 1991–2003, it is hardly surprising that oil is an important determinant of domestic gas prices in Europe. After controlling for country-fixed effects, there is some evidence that the aggregate regulatory index (the ECTR score) is associated with lower prices, suggesting that a 1 point increase in the index is associated with 9.4% lower prices (Table 6.9, column 1). To disentangle which component of the reform set was driving the

Table 6.9. GMM estimation of dynamic panels for gas prices (log)

	(1)	(2)	(3)	(4)	(5)
	GMM	GMM	GMM	GMM	GMM
Regulation index	−0.094**				
	(0.027)				
Public ownership[1]		−0.240***	−0.233**	−0.227**	−0.245***
		(0.007)	(0.015)	(0.017)	(0.008)
Vertical integration[1]		−0.025	−0.016	−0.013	−0.030
		(0.691)	(0.811)	(0.840)	(0.614)
Entry regulation[1]		−0.028			
		(0.726)			
Entry regulation[1]: No third party access			0.047		
			(0.509)		
Entry regulation[1] dummy: No retail market open to consumers			−0.073	−0.043	
			(0.402)	(0.564)	
Restrictions by laws/regul. on no. of competitors No consumer choice of supplier			−0.019	−0.005	−0.020
			(0.900)	(0.974)	(0.890)
Average price (period $t-1$)	0.712***	0.806***	0.795***	0.797***	0.811***
	(0.000)	(0.000)	(0.000)	(0.000)	(0.000)
Brent oil price	0.025	0.076**	0.082**	0.071*	0.072*
	(0.591)	(0.028)	(0.041)	(0.053)	(0.052)
Year fixed-effects	yes	yes	yes	yes	yes
Constant	0.843**	0.220	0.231	0.268	0.229
	(0.048)	(0.240)	(0.417)	(0.340)	(0.405)
Observations	295	295	295	295	295
Countries	11	11	11	11	11

Source: own calculations using IEA, WDI source data.
For the regulatory variables see Tables 6.A.1.
Standard errors in parentheses. *** p<0.01, ** p<0.05, * p<0.1.
[1] dummy.

results, we used some dummies which identify the level of public ownership, of vertical integration, and entry regulation finding that public ownership has the stronger and the only significant effect, suggesting that public ownership is strongly associated with lower prices (Table 6.9, column 2). Hence, we tried to decompose further the entry regulation dummy to assess whether third party access, openness of the retail market to consumers, and restrictions on the number of competitors had any effect on average (log) prices but we found no significant coefficients, while the public ownership variable remained statistically different from zero.

Our results confirm those of Brau et al. (2010), who used a shorter time series, two different data sources, and tested a wide range of decompositions of the ECTR index for the gas industry.

6.7 Summing up and open energy policy issues

After a discussion of some features of the natural gas industry, this chapter has explored the impact of gas policy reforms on household prices in the core countries of the European Union over the 15 years to 2007. After controlling for macroeconomic factors, Brent oil price changes, own-price inertia, and country and year-specific effects, our findings suggest that the impact of the reforms on the household prices of the EU countries we have considered is—until now—at best negligible. Privatization per se does not lead to lower prices for consumers, and in fact there is some evidence pointing to public ownership of the utilities as a mechanism that caps prices. This effect is robust across different sources of price data. There is some limited evidence that softening entry legislation is beneficial to the consumer, as expected, but actual market shares of the entrants have no lowering effect or even tend to increase price in some models.

Our interpretation of these findings is simple. The threat of entry following legislation changes perhaps motivates the incumbent to some price restraint, but there is no actual further effect from competition. The market is in fact still intrinsically oligopolistic and mostly based on the long-term relationship between oil and gas prices. Privatization, unbundling, and formal legislation allowing entry do little to alter this feature. The OECD vertical separation indicators are never significant in the empirical estimates. The prices consumers pay for natural gas in Western Europe are still determined by what they have being paying in the past (own price inertia), with the well known linkage to crude oil price, plus country specific factors.

These findings suggest that until recent years the regulatory reform of the gas industry in the EU has had no impact on consumers' welfare, with privatization possibly a small adverse effect. High entry costs may explain why new players often have no incentives or no room for offering lower prices. Perhaps a more optimistic view could be gained by looking at prices for the business sector, but price convergence across countries with very different industrial organizations does not suggest such an outcome.

In order to further discuss our findings, it seems interesting to go back to Newbery (2000), who devoted a full chapter of his book to deregulating and restructuring the gas industry. He acknowledged that, differently from electricity, production costs of gas are opaque and ill defined, as they should include a resource rent that in turn depends upon the origin and destination of the commodity. The large scale of the investment, inflexibility, irreversibility, and the depletion premium have justified long-term contracts[30] as a way to share the risks among the different agents. The risk-sharing mechanism involves both—upstream—the owners of the reserves and those of the pipelines, and—downstream—the latter and the distributors and final suppliers. Governments tend to appropriate part of the rents, either directly as owners of the firms, or indirectly by taxation.

Thus, Newbery (2000, p. 345) suggested that:

> the benefits from liberalizing gas have very much to do with disturbing this equilibrium in which supply is restricted to raise prices above their competitive equilibrium level. If gas producers could be sure that any additional gas could be delivered and sold in the most valuable market, and if consumers could choose among different sources of supply, with confidence that enough gas was available at a reasonable price, then gas spot markets could develop and provide the assurance that previously could only be guaranteed by long term contracts and vertical integration.

According to Newbery, governments conspired with incumbents to share rents, and they cannot be trusted to provide efficient regulation. Hence, the final objective of gas reforms should be to have a competitive market, with a large number of competing suppliers, each of them enabled to access the network infrastructures. From this perspective, the EU energy policy, and the convergence between electricity and gas regulation, had to be appreciated, but it was deemed to be insufficient, and the experiences in the USA and the UK were discussed at length in order to draw lessons for Europe. Thus, Newbery concluded that while continental companies had resisted liberalization and unbundling, liberalizing access to the networks should have advantages in terms of less complex contractual negotiations, increase of supply, and more cost-reflective prices.

[30] For example in Italy around 60% of contracts have a more than 20 year duration. Long-term contracts and financial guarantees by governments are needed to secure huge infrastructure investment by all the players.

The argument of balancing risks of substantial investment through long-term contracts upstream and franchised monopolies downstream was then considered as slowly waning, and the conclusion was that:

> The decisive regulatory innovation is the creation of spot and futures markets for gas and gas storage... Provided that there are a sufficient number of competing producers (which can be assisted by interconnecting and extending the gas grid), competition can replace the system that had led to high prices and considerable inefficiency. Once that stage is reached, the remaining regulatory issues seem minor, and the sustainability of the new equilibrium looks as relatively unproblematic. But just because liberalization may be both irreversible and obviously superior for consumers, it is likely to be most fiercely resisted by incumbent gas companies. (Newbery, 2000, p. 383)

With the benefit of hindsight, one may observe that, indeed, it is easier to organize a liberalized market if the country has a substantial domestic production, but more difficult if you have to import most or all of your gas. Incumbents resisted the change (see the above citation by Bosmans); it is true that several EU governments were reluctant to disturb the existing equilibria; and it is also true that the European Commission, while clearly supportive of the British reform architecture, had to make compromises (particularly on unbundling). Then, the gas rent has not been curbed, and it is still in variable proportions shared by producers, owners of networks, traders and shippers, final suppliers, and by governments, (either through taxation or dividends from equity stakes). Therefore the consumers, and particularly households, still pay a relatively high bill for natural gas, as showed by the high profitability of gas.

The problem, however, with the slow progress of the proposed reforms in the gas industry is, perhaps, more fundamental than resistance and rent-seeking. In fact, the spot markets, as implemented in the UK and in the USA, have not yet been shown, beyond any reasonable doubt, to be able to solve the risk-sharing conundrum between gas producers and owners of the pipelines. After all, to break the resistance of the gas rentiers, the coalition of liberal policy-makers, industrial customers, and the consumer lobbies should be firmly convinced that they get a 'fair deal' in exchange for their support of the reforms. Clear advantages in terms of prices of one system against the others would be a crucial argument, with sustainability of investment being another one.

As Dieter Helm (2007) has noticed, the energy paradigm at the beginning of this century is again shifting dramatically. In his view, the market oriented reforms in the 1980 and 1990s, particularly in the UK, were supported by the abundant availability of gas, low international prices, and high demand because of the adoption of combined cycle gas technology by the electricity industry. According to Helm (2007, p. 10): 'One policy paradigm in energy has been provided by the set of ideas surrounding privatization, liberalization, and competition developed in the 1980s . . . It worked well in the 1980s and 1990s, but around the turn of the century a set of events began to take place in the energy markets that put the conventional wisdom under considerable stress.' Helm mentions also (in a footnote) that the change was not without problems on the side of distributive effects, an issue we discuss in Chapter 9. In his view, three types of events changed the priorities. First, the huge surge in oil prices around 1999; second, a number of spectacular power cuts in the USA, the UK, Italy, and elsewhere; third, the Iraq War, the Gazprom interruptions of supply to Ukraine, de facto to Europe, and the political interventions by Russian President Putin on Yukos and other major oil firms. These sorts of facts shifted the attention of European governments to security of supply and related long term-investment needs as the policy priorities.

These concerns for security of supply in a broad sense are now combined with the mounting feeling that energy policies need to also address the challenge posed by climate change. Natural gas has a clear advantage on the latter issue as compared with oil and coal, but not as compared with nuclear, renewable fuels, and energy saving. Gas has, instead, a clear disadvantage on the security of the supply side. It seems unlikely that new exploration will allow for a relevant number of producers. The supply side of the markets will remain oligopolistic. Perhaps LNG from Qatar, Trinidad and Tobago, or elsewhere, will be able to supplant some existing or proposed high-pressure pipelines with Russia. LNG implies, in any event, substantial fixed investment, which often takes more than ten years to be operative. In North America new sources of domestic supply will also change the scenario. Geography and technology, however, still seem to support the view of those who insist that securing long-term contracts upstream is needed, that increasing network trans-boundary interconnection is required as a priority, and that spot markets do not deliver the incentives to major risky investment projects. Banks need some form of guarantee to finance long-term projects, and currently only long-term contracts can play this role. Thus, declining domestic reserves of gas in the EU, the political risk of some non-EU suppliers (except Norway), the huge amount of finance involved in crucial pipeline networks or LNG ventures, strategic considerations, and the swings of price in spot oil markets in the recent two decades, all suggest that the natural gas industry of the future will retain some of its special features.

An increasing share of gas-related contracts will be negotiated in spot and future markets, but it is hard to see how far these markets can substitute for the long-term contracts so typical of the industry (except in the UK and USA). These contracts are not without their own problems. They are usually indexed to some basket of fuel prices, plus corrections for exchange rates and inflation, with averaging mechanisms over time, and have different types of take-or-pay clauses and different prices according to quantities and other circumstances.

Some contracts have destination conditions (not accepted by the EU), and other special foreclosure mechanisms. One may think that a core skill in the industry is in fact the ability to understand and negotiate these contracts. Getting rid of them, when the spot markets offer lower prices as during the world recession of 2008 and subsequent years, now seems attractive to some players. But, when prices in the spot markets are high or volatile, customers tend to regret not having enough gas at the somewhat smoothed prices offered by the bilateral contracts.

The opinions of experts and insiders are divided on the future of contractual relations in the gas industry, and it seems over-simplistic to say that the spot market design of the USA and the UK is the lesson that everybody needs to learn in continental Europe.

An example, among the many, of a swinging pendulum, is the fact that recently big US energy companies, such as Chesapeake Energy or Devon Energy, were pushing for long-term gas contracts (5–10 years or more) with the utilities. The US energy companies had for decades preferred to rely on short-term contracts in order to profit from soaring prices. However, the discovery of huge new gas fields in Texas, Louisiana, and Pennsylvania, plus the recession, halved gas prices in 2009, and this fact has changed the mind of the supply side of the industry. Customers were reported to be hesitant to move in that direction, which is new for the USA but common elsewhere (and in the USA itself for coal contracts). Core players are reconsidering their strategy. For example, it was reported that: 'Exxon Mobile Corp. agreed to buy U.S. natural gas producer XTO Energy Inc. in a USD 31 billion stock deal that was widely seen as an endorsement of the new gas fields' long-term viability. Exxon has experience with long-term sales contracts through its liquefied-natural-gas operations overseas, and many experts think the company will pursue them in the US.'[31]

On the other side of the long-term/short-term contract dilemma, there is some recent evidence that Gazprom, after the disruptions caused by the Ukraine defaults on payment, and the low price of oil during the beginning of recession in late 2008, declining demand of gas in Europe, and falling gas prices because of new proved reserves elsewhere, may change its approach to selling gas. If there is a partial decoupling of oil and gas prices, and Russia and Algeria feel the competition of LNG, Gazprom and Sonatrach may be compelled to change their contract strategy, and offer some gas on short-term markets, see Eprimc (2009). Thus, paradoxically, one possible scenario is with Exxon going towards more long-term contracts in the USA, and Gazprom

[31] See the *Wall Street Journal*, 30 December, 2009.

towards more short-term contracts in the EU. These are, however, just conjectures.

Zafirova (2010), a legal advisor with an insider view of the activities of several major players, suggests that the relationship between inflexible short-term prices and prices deriving from flexible long-term contracts are interrelated in various ways, and that a hybrid oil-based pricing and demand-supply pricing will tend to prevail in the future.

According to a very detailed study (Melling, 2010) Europe is currently a battleground between two contracting ideologies, with some chances for spot markets to win a larger share of the industry in the future, following the USA–UK models. While the study tends to argue in favour of the efficiency of spot markets, it also acknowledges that the balance in Europe will shift in that direction only if and when there is a reversal of the current situation of oversupply that makes spot prices attractive, with gas sold at a discount with oil. The reversal, however, could happen under declining domestic production in the EU, and expansion of demand from China and India, and this would determine high and volatile spot prices. The conclusion is that: 'Future rounds of the gas industry power struggle may play out increasingly between the EU and the producers' respective sovereign government. The prospect of a revolution in gas contracting practices has effectively been deferred by recent contractual changes. Whether or not the current dynamic precipitates a revolution or the more widely preferred managed evolution in contracting and trading of gas in Europe remains an open question . . . ' (Melling, 2010, p. 74).

If long-term contracts and oligopolistic arrangements are here to stay in Europe, that is the majority of trading over the next 20 years will be based on some evolution of the current mechanisms (while possibly allowing more room for spot markets), then much of the EU regulatory emphasis on mandatory ownership unbundling (with derogations) and on other liberalization reforms seems to be overshooting its target. A competitive component can be a welcome complement to a mainly oligopolistic-oligopsonistic industry, with a small number of dominant players on both sides of the market, but it seems realistic to say that there still will be, for decades ahead, the need to tax and redistribute gas rents in Europe (and elsewhere), and limited scope for squeezing them through mainly short-term competitive markets. Moreover, if security of supply, energy savings, and climate change are the core EU energy policy concerns of the future, it is difficult to see how European governments and institutions can actually stay hands off about industry, and treat them as just one mostly self-regulating market.

Any high-level policy decisions, for example about re-launching or abandoning nuclear generation, or supporting energy-saving in buildings, or supporting some transmission corridors instead of others, will probably have much more impact on the determination of the final price of gas to the burner-tip than forcing the industry to move to spot markets by EU legislation. Recent debates in the UK, following the Stern Review (2007), show that much more than competition is needed to address long-term energy challenges; see the previous chapter.

Moreover, in this complex 'upstream' context, the ability of households to search for alternative gas suppliers and to switch regularly to capture the best prices cannot do miracles in terms of promoting 'downstream' competition. The need to protect consumers of gas from un-competitive prices should possibly remain a permanent policy concern of governments and regulators for the future.

6.8 Further reading

Arentsen (2011) offers a global review of gas trends that enlarges to other regions the perspective of this chapter. There is a highly technical literature on gas prices, and particularly on formula indexing in long-term contracts, and on spot prices. We just cite three specimens of this literature. Holden et al. (2011), in a paper in the *Journal of Derivatives*, discuss the flexibility built into the long term gas contracts in Europe and probabilistic approaches to their evaluation. In fact, when the long term contact is properly analysed, it is far from being a simple fixed-price plus some index-linked correction mechanism, but is more similar to a structure of options, that poses complex risk valuation issues. To understand what the intrinsic welfare inefficiency of these contracts is one needs a benchmark: but where does one look for this? Simpson (2009), drawing from observation of Asia import prices, explores the possibility of building a benchmark global gas spot price index. He observes, however, that even in the USA, which liberalized gas some years before the UK, still around 80% of gas price changes cannot be explained by the fundamentals of gas demand and supply (Simpson, 2011). The paper offers a review of previous empirical research on the determinants of gas daily price in different regions, including at the National Balancing Point in the UK, and Henry Hub in the USA. The author opts for the latter benchmark, as HH is related to the currently most liquid market (worth 660 bcm in 2007, compared to 100 in the UK). Simpson finds that any gas spot pricing model should possibly include OPEC oil price (that Granger causes gas price, with some lag), coal, and he adds a global stock index as a proxy for demand and supply drivers (the stock index and coal price Granger causes oil price). He also admits that the oil market is, however, inefficient because of the OPEC cartel. Some readers, moreover, may also be puzzled about the 'efficiency' and information value of the global stock index after the global crisis of recent years.

Hence, one may conclude that the relative efficiency of the spot and long term markets for gas, for example, respectively in the USA and in continental Europe, need to be ascertained, while in fact it seems that both markets are still linked to oil prices, see also Pindyck (1999), Stern (2009), and Brown and Yucel (2008).

De Bock and Gijon (2011) study through VAR models the possible decoupling of oil and gas prices in the USA and in Algeria. After noticing the ongoing structural changes in market patterns, mainly driven by the shale gas supply explosion in the USA, and diversification of LNG sources, they conclude that there is some evidence of decoupling of oil and gas prices in the USA, but less evidence in Algeria. In fact this study points to a still close link, but with a shortened effect in terms of the impulse–response behaviour of the time series.

On the very complex legal issues raised by the evolution of the EC position about vertical long-term supply contracts, the differences between the upstream and downstream markets, and the relationship of this regulatory issue with Community Competition Law, see several papers in Delvaux et al. (2010). Verhelpen (2010) and respectively Talus (2010) discuss the legal trade-offs between the objectives of EU competition law, and of security of supply. See also Zafirova (2010), who presents the functioning of the gas trading hubs, that is the NPB in the UK, Zeebrugge in Belgium, TTF in the Netherlands, and E.ON Gas Trading in Germany (for information on other minor hubs, see below the reports by DG Energy). Zafirova (wisely) observes that even if regulators push for the emergence of the spot markets, and they increase their range of activities, in the end only the market players will decide on the balance between risk profiles in different types of markets. For example municipalities will be hesitant to feel comfortable with trading, particularly after the sad lesson of Enron's bankruptcy, and many upstream players are also risk averse: 'Such more risk-averse parties or parties which need longer term security in order to be able to secure large investments in exploration and production will be the main reason why the hybrid model of short term, medium term, and long term gas sales coexisting in the European market, cannot expected to continue into the future' (Zafirova, 2010, p. 227).

Breton and Kmarbach (2012) show through a theoretical model that vertical integration of gas storage, with accounting and management separation, can be a better option than ownership unbundling in welfare terms.

For updates and detailed information on gas market trends in the EU, see the Quarterly Report on European Energy Markets, available at http://ec.europa.eu/energy/observatory/gas/gas_en.htm, and the annual report by EC Directorate General for Energy, 'Europe's Energy Position', available at http://ec.europa.eu/energy/observatory/annual_reports/annual_reports_en.htm. Additional data are available with Eurostat and IEA. For a US perspective see www.naturalgas.org.

On depletion policies for oil and gas, see Kemp and Stephen (2007) and on oil outlook, other chapters included in Helm (2007). On political risk and gas market integration in Europe, see Asche (2002), and Biava (2007) for the emergence of a solidarity issue in the EU following the Russia–Ukraine dispute. On regional integration see for example Brito and Rosellon (2010) on the pricing of gas between Mexico and USA, based on the netback rule, which is seen as consistent with the Little–Mirrlees frame in cost–benefit analysis. The World Bank approach for South-East Europe is presented by Kennedy and Besant-Jones (2004). They suggest a gradual market opening strategy, with due attention to institutional and affordability issues.

Appendix 6.A.1: ECTR scores

Table 6.A.1. Sectoral indicator of regulatory reform: Gas

	Weights by theme (b_j)	Question weights (c_k)	Coding of data		
Entry regulation:	1/4				
How are the terms and conditions of third party access (TPA) to the gas transmission grid determined?		1/3	Regulated TPA 0	Negotiated TPA 3	No TPA 6
What percentage of the retail market is open to consumer choice?		1/3	(1% of market open to choice/100)*6		
Do national, state, or provincial laws or other regulations restrict the number of competitors allowed to operate a business in at least some markets in the sector: gas production/import		1/3	No, free entry in all markets 0	Yes, in some markets 3	Yes, in all markets 6
Public ownership:	1/4				
What percentage of shares in the largest firm in the gas production/import sector are owned by government?		1/3	None 0	Between 0 and 100% 3	100% 6
What percentage of shares in the largest firm in the gas transmission sector are owned by government?		1/3	0	3	6
What percentage of shares in the largest firm in the gas distribution sector are owned by government?		1/3	0	3	6
Vertical integration:	1/4				
What is the degree of vertical separation between gas production/import and the other segments of the industry?		1/2	Ownership separation 0	Legal/accounting separation 3	Integrated 6
What is the degree of vertical separation between gas supply and the other segments of the industry?		3/10	0	3	6
Is gas distribution vertically separate from gas supply?		1/5	0	3	6
Market structure:	1/4				
What is the market share of the largest company in the gas production/import industry?		1/3	< 50% 0	between 50 and 90% 3	> 90% 6
What is the market share of the largest company in the gas transmission industry?		1/3	0	3	6
What is the market share of the largest company in the gas supply industry?		1/3	0	3	6
Country scores (0–6)	$\Sigma_j b_j \; \Sigma_k c_k$ answer$_{jk}$				

Source: OECD.

Appendix 6.A.2: The EU natural gas directives

The core ingredients of directive 2003/55 are the following:

a) Market access: the right of third parties to non-discriminatory access to transmission and distribution systems and to liquefied natural gas (LNG) facilities.

b) Since 1 July 2004, industrial consumers are free to choose their supplier. Domestic consumers since 1 July 2007.

c) Access to storage facilities is covered by specific provisions, and access may be either negotiated or regulated.

d) In each MS, the system operator is appointed for the transmission, the storage, the liquefied natural gas, and the distribution system. The SO must ensure the safety, reliability, efficiency, and non-discriminatory interconnection of facilities with due regard for the environment.

e) When companies are vertically integrated, the transmission and distribution activities must be legally and functionally separated from other activities, such as production and supply; ownership unbundling is not mandatory (see, however, below, the 2009 Package).

f) Public service obligations and consumer protection: common minimum standards include the right to change supplier, transparent contract conditions, general information, dispute settlement mechanisms, etc. Adequate protection of vulnerable consumers should be provided (for example, by taking the appropriate steps to avoid disconnection of the gas supply). Member States can impose public service obligations to guarantee security of supply, economic and social cohesion objectives, regularity, quality, and price of the gas supply and protection of the environment.

g) Independent regulators, appointed in each Member State, are responsible, in particular, for monitoring non-discrimination, the level of transparency and competition, and the tariffs. These regulatory authorities also act as dispute settlement authorities. The European Regulators Group for Electricity and Gas—ERGEG) has been established through Decision 2003/796/EC.

h) Derogations are allowed in certain situations, such as isolated markets, emergent markets, geographically limited areas, lack of capacity, development of the infrastructure capacity or of new infrastructures; if access to the system prevents delivery of public service obligations; serious economic difficulties of a gas company, in particular due to its 'take or pay' commitments, sudden crisis in the market, or the physical safety or security of persons, apparatus, or installations or system integrity being threatened.

i) The European Commission draws up an annual benchmarking report that evaluates the progress in the implementation of competitive electricity and gas markets, on the basis of information supplied by national governments and regulatory authorities.

j) The EC should monitor the supply and demand balance in the different Member States, the interconnection capacity and the quality and maintenance level of the systems.

The 2009 third legislative package provides for:

- the separation of production and supply activities from transmission system operation, either based on ownership unbundling or on an independent system operator. Non-European operators should meet these separation requirements if they wish to acquire a 'significant share' in a European country;
- the increased powers and the independence of national regulators as well as their cooperation through the setting up of a cooperation agency of the energy regulators authorized to make binding decisions and impose penalties;
- formalization of the European groups of transmission system operators for better coordination and, in particular, the drawing-up of joint market and technical codes;
- improved market operation, notably greater transparency, effective access to storage facilities and LNG terminals.

Source: adapted from http://europa.eu/legislation_summaries/energy/internal_energy_market/l27077_en.htm

Part III

Perceptions, Quality, Affordability

7

Double checking: Perceived price fairness[1]

7.1 Introduction

This chapter uses data on self-declared individual attitudes to answer a simple research question: Are consumers happier about the price they pay in countries where the electricity or gas or telephone services have been privatized and opened to the market? There are two reasons to consider social attitudes in the context of the debate on utility reform. First, attitudes are important per se. Policy-makers and regulators are well aware that utility reforms are often at the forefront of public debate in the EU, and may raise vocal opposition or support to change. Second, subjective data can be a complement to objective evidence in order to evaluate the welfare impact of reforms.

The chapter uses four waves of Eurobarometer Surveys, (2000, 2002, 2004, 2006), for the EU-15 countries, and ECTR indicators again to describe the extent of reform by country and by year. It then tests the association of privatization and liberalization on users' attitudes. The rest of the chapter is organized as follows: after a presentation of the conceptual frame (Section 7.2), earlier findings and Eurobarometer survey data (Section 7.3), Section 7.4 offers some descriptive statistics, and some estimates of a set of conditional models of consumer dissatisfaction (Section 7.5). The concluding section discusses the findings.

7.2 Learning from what users say

It is apparent from the previous chapters that, while there is a common direction of reform, substantial variations exists over time and across EU Member States. Our empirical analysis exploits this variability. As for any

[1] With Carlo Fiorio. The chapter also draws in part on Fiorio and Florio (2011), Bacchiocchi et al. (2011), and Fiorio and Florio (2010).

public policy, the evaluation of welfare changes is the cornerstone of predicting long-term success. In developed economies, issues of access of the services are often (but not always and everywhere) a relatively minor concern. Hence, the key signal that influences consumers' evaluation of public utilities reforms is price (for a given quality, see next chapter). We distinguish here between the actual price paid as recorded in available statistical data, and perceived price fairness, as recorded by surveys among consumers.

Ideally one would like to consider micro-evidence of changes in appropriate welfare measures. One would need to know households' preferences about different uses of electricity, quality-adjusted price structures for type of users, and their equivalized income. This information is, however, not available in comparable form across EU Member States. For example, comparable national data on the price per KWh by domestic users of electricity broken down by ranges of yearly consumption are available, but comparable EU-wide matched data on the income and other individual characteristics of those users, or the number of individuals by each household, are not available.

One strategy to glean some individual-level information is to adopt a different empirical shortcut: as a complement to relying on revealed preference through the estimation of individual compensated demand functions (or their proxies) we can turn to stated preferences, that is subjective well-being measures. In other words, we analyse consumers' answers to direct questions about their own satisfaction about price paid for the utilities. While this may look like a dramatic change of perspective in economic welfare analysis, it is much less radical than it may appear when compared with the actual practice of cost–benefit analysis in project or policy evaluation. In fact, applied CBA, usually regarded as objective welfare evaluation and officially endorsed by government agencies and international organizations, routinely uses contingent evaluation methods (see e.g. Boardman et al., 2005 or Pearce, Atkinson, and Mourato, 2006, for a survey of the large applied literature in this field). Such evaluation methods revolve around eliciting, through surveys of users, direct information on willingness-to-pay or willingness-to-accept policy changes.

To an applied welfare economist, using revealed or stated preferences is more a matter of data availability than a fundamental methodological divide. Any empirical approach has possible biases. Consumers may misperceive prices, and statistical offices may misrepresent, through reporting national average prices, a much more complex structure.[2]

The discussion of empirical approaches to the evaluation of the welfare impact of policy reforms has some resemblance to the wider debate on the

[2] For an analysis of reforms and actual average prices see the previous Chapters 4, 5, and 6.

merits of the 'economics of happiness' (Graham, 2006; Layard, 2005). The typical focus of this recent research avenue is the study of the relationship between subjective well-being as self-assessed by individuals, and objective macroeconomic welfare indicators, such as national income, inflation, or unemployment (Frey and Stutzer, 2002). We propose to use a similar approach in the microeconomic context of the network service reforms.

Services of general economic interest are sufficiently important in households' budgets to influence perceptions of well-being. While such perceptions can be imprecise, they are of course based on the information set available to the respondent, plus an idiosyncratic bias. Thus, when a respondent says, in one particular country and in one specific year, that he or she evaluates the price of electricity as 'fair', this person is indeed revealing something about their subjective well-being. If an individual is happy with the price paid, then he is in a better (perceived) welfare position than somebody who feels compelled to pay too much for what is received. Moreover, and most importantly, we can test empirically some aspects of the possible individual bias, using (average) prices as a control variable. Thus, if we find that, after controlling for some possible individual bias, there is the expected relation between perceptions and actual (average) prices, we can be more confident about the interpretation of the results of data on social attitudes. The parallel with happiness economics here is that while the latter research typically relates overall subjective well-being to macroeconomic variables, in this case we are focusing on satisfaction with a specific, albeit important consumption item. If there are variations across time and across countries in the frequency of those who assess the price of a service as 'fair' or otherwise, we can exploit this variability to indirectly test the impact of the reform on (perceived) welfare of respondents, after controlling for other factors.

To further understand the potential use of subjective data, some insights from behavioural economics literature, as well as from other research strands on social attitudes, are helpful. The discussion below is very selective, and focuses exclusively on the question: What can we learn from reported users' happiness with prices? Issues related to the perceived quality of the service will be discussed in the next chapter.

One possible starting point for discussing the issue of price fairness is the seminal paper by Thaler (1985), which—building on prospect theory by Kahneman and Tversky (1979)—suggests that the mental accounting of consumers is different from the standard microeconomics optimization calculus (see also the essays collected in 'Quasi-Rational Economics', Thaler 1994). According to Thaler, the Khaneman–Tversky value function differs from the usual rational consumer's utility function because gains and losses are only perceived relative to a reference point. In our context, this would be the price for a specific network service. For the sake of simplicity we shall refer below to the example of the price of a telephone call. The value function is usually assumed to be concave for gains and convex for losses, and the losses are valued more than gains. When, however, you consider multiple dimensions of the telephone service and prices, for example the unit price of the (international, national, local) call and the charge for connection, or additional services through the wire, different outcomes and evaluations are possible. To simplify, however, we focus on one dimension, or attribute, of the telephone service, for instance the local call. Three different price concepts should be considered: the actual price p of the call, that is the one that is entered on the telephone bill; the value equivalent p', that is the user's willingness to pay for the call; and the reference price p^*, a price that the user considers 'fair'. Consumer utility comprises acquisition and transaction utilities.

Acquisition utility is defined by Thaler as the compound combined subjective value of the outcome 'unit of service acquired, and actual price paid', which by definition is equal to the subjective valuation of $(p'-p)$. Given the assumption that losses are valued proportionally more than gains, and the consumer wants to maximize his value function v, the combined outcome of having to pay a price and being able to make a telephone call will be integrated as $v(p'-p)$. This notion is seen by Thaler as comparable in principle to consumer surplus, the standard welfare notion of individual welfare, and applies to any voluntary trade. Thus, the subjective value of one unit of service is a function of the difference between actual price and willingness to pay for that unit.

The second component of perceived value is transaction utility, which arises from the comparison of actual price and a 'fair' subjective price p^*. Thus, total utility is the sum (perhaps weighted in some way) of acquisition utility and transaction utility:

$$W(z, p, p^*) = v(p' - p) + \beta v(p^* - p)$$

where W is total subjective value (leading to overall satisfaction), z is the telephone call, $v(z) = v(p')$ expresses the individual welfare associated with the marginal willingness to pay, and the last term is the difference between actual price and the fair price of the service, with β a weighting factor. This factor is related to different possible specifications of mental accounting. For example, we adopt here—slightly adapting the original idea by Thaler—the *difference* between fair and actual price; but one may consider a ratio of the two, or more complicated versions of the transaction utility function. According to Thaler, for routine transactions, the outcome is always perceived as integrated, hence the actual price is not separately accounted as a loss. The bottom line of 'mental accounting' in this framework is that if a consumer has a certain expectation of what the fair price of telephone service should be, and the actual price turns out to be different, he will be negatively or positively surprised. In turn, p^*, is (to a certain extent) influenced by the perception of costs of the supplier. Thus, the standard notion of consumer surplus should be augmented with transaction utility. In fact, in our simple version above, if $p^* < p$, there will be a negative perception that decreases subjective welfare.

In our context, it seems reasonable to assume that the acquisition of the services of utilities, such as telephone, electricity, gas, water, public transport, and so on, is a routine process, with infrequent decisions taken by the consumer about the selection of the supplier (in fact only access/exit under monopolistic provision). There is usually no scope to finance specific purchases from the service providers through debt, and very limited demonstration effect. In the sequencing of decisions, the purchase of essential services will feature among the top priorities in the budget allocation of the household. Thus, what is a fair price to a consumer in this context? We would expect this to be a more or less vague notion based on some comparative observation.

Users of essential services usually have very limited information about the cost of supplying the services. The cost of producing a telephone call, a KWh, or a litre of water at the tap is a combination of fixed investment costs sunk in the networks, and of operating costs that are mostly invisible to the user. Searching privately for this information would be very costly. For example, the consumer would have to read the income statements of the supplier and assess whether it enjoys extra profits. This is very unlikely to happen. Moreover, Thaler lists several ways for the seller to try to raise p^*, such as offering combined services, or obfuscating perceived costs. This obfuscating strategy is common practice in the electronic communication industry, as we have seen in Chapter 4, in relation to bundled contracts.

Perceived price fairness, as an empirical question, has received scant attention in the applied welfare economics literature, including contingent evaluation, which focuses on willingness to pay/accept, that is p' not on p^*. See also Biggar (2010) for the notion of price fairness in utility regulation, and Maxwell (2008) for a review.

Four conjectures on subjective price fairness seem particularly interesting for exploration in the field of network services:

a) past actual (relative) prices may be an ingredient in shaping the current reference price (Kahneman et al., 1986). Thus, a consumer will value negatively an increase in telephone prices, which is not perceived as justified by an increase of cost. Moreover, the asymmetry of the value function would make users more sensitive to tariff increases relative to tariff decreases; inflation, or changes in prices of other goods, particularly alternative goods, can also enter in the mental accounting process that leads to fixing a reference price;

b) individual characteristics should also play a role in locating p^*. Users who are more educated probably have some additional information on costs, relative to other people. For example, they may be more aware of news about wage negotiations, value of shares, and prices of some inputs. The very young or the very elderly are probably less informed or experienced about the current cost of new services, for example mobile phone or Internet VOIP, as compared with young adults. Political opinion also may be a factor, as this may be correlated with some prior notion of fairness. Most importantly, income or wealth will influence the perception of what a fair price is, because—perhaps differently from Thaler's original frame—budget constraints may alter the slope of perceived losses (in other words p' and p^* are both influenced by income of the consumer);

c) the supplier's characteristics, including their stated mission, reputation, and market strategies, may also contribute to shape p^*. Given the absence of or very limited information on production costs to the consumer, government ownership would perhaps act as a perceived guarantee that actual prices are not too far from costs. For a private monopoly the opposite perception may be more widespread;

d) in a liberalized network industry the reference price would probably also be influenced by the perception of a range of actual prices in the market, either advertised or known to the consumer in other ways.

The last point deserves a brief discussion. Under effective competition, prices must converge, and price discrimination would be difficult. Hence, p^* should be simply proxied by p, and mental accounting in Thaler's sense should be replaced by conventional consumer rationality (i.e. acquisition utility). After some switching, all consumers will pick up the best price in a competitive telephone or energy market, and this will fix the equilibrium price. As we discuss in Chapter 9, price divergence and discrimination is—however—a typical feature of liberalized markets of network services. As a result, different actual prices are potentially available to the consumer, and these prices are advertised by firms in the form of special offers, combined service packages, and so on, which will contribute to the marketing obfuscating strategies. The consumer must bear costs in acquiring the comparative information. He will often delay switching from one supplier to a competitor because of uncertainty about the outcome of the new decision. Hence, aggressive marketing in oligopolistic network industries could lower p^* against a more or less sticky p. This divergence will increase dissatisfaction. Shopping for luxury goods in fashionable exclusive streets may be exciting for some types of consumer, but shopping around for gas tariffs is not fun for most households. Multiple prices will prevail, and this is irritating if you are uncertain about the cost or benefit of changing your supplier. We shall come back to this issue later in Chapter 9.

To sum up this discussion, survey data in which interviewees are asked if they perceive they pay a fair price tell us a complex story. Declared (dis) satisfaction of the respondents depends upon an (unknown) reference price, which in turn is correlated to actual past and current prices, and individual and firms' characteristics. Most importantly, reference and even actual prices are typically private information, difficult to recover by aggregate statistical data collection. Microdata are necessary to capture some welfare effects that are hidden by the average price paid by the average household, as we have seen in previous chapters. Thus, any empirical analysis of satisfaction with prices should disentangle the different factors that shape consumer perceptions. The interpretation of the results, in turn, should be combined, as far as possible, with the previous evidence on actual prices. Does subjective welfare diverge from objective welfare? We further discuss other methodological aspects in the next chapter, where we turn to the even more difficult issue of satisfaction with quality of the services.

7.3 Some earlier empirical findings

The research line we are interested in is related to, but differs from, the empirical study of social attitudes on reform of utilities or of privatization and liberalization policies in general. Examples of the latter for Latin America can be found in Armijo and Faucher (2002), Bonnet et al. (2011), Checchi et al. (2009), Chong and Lopez-de-Silanes (2003), Graham and Felton (2005), Graham and Pettinato (2002) Graham and Sukhtankar (2004), Martimort and Straub (2009), and Panizza and Yanez (2005). Some of this research uses surveys such as Latinobarometro, which includes direct questions about the perception of privatization policies and other market reforms.

Here we are not interested, however, in social attitudes towards policies, but in the perception of prices by users, whatever their political orientation and their appreciation of the reforms that have restructured network utilities. Examples of papers in this line are Clifton et al. (2010a, 2010b) and Whitehead et al. (2008).

Clifton et al. (2010a) evaluate EU policies on public services—particularly public network services—from the citizens' point of view. Citizen satisfaction with public services, as revealed through Eurobarometer surveys from 1997 to 2007, is discussed, combining the evidence on prices of services of general interest with other dimensions, such as perceived quality, information, and others. Clifton et al. (2010b) focus on comparing user satisfaction for electricity and telecom services in Spain and the UK. They do not find in either country an impact of individual characteristics such as age, employment of the respondent, or his/her education on dissatisfaction with electricity prices.

They do, however, find that individual characteristics are important in determining the probability of dissatisfaction with fixed telephony or telecommunications services in general: those who are more dissatisfied are the elderly, the unemployed, the less educated, and those who do not own their home. They also compare these findings with information on revealed preferences from household budget surveys. The results point to a heterogeneity of effects of utility reforms on different social groups.

Previous results in this research area from the Milan UPP Team include Fiorio and Florio (2011) on user satisfaction with electricity prices and Bacchiocchi et al. (2011) on telephone prices, both papers based on Eurobarometer survey data (see next section). We briefly summarize their findings here, while we present new results in the rest of the chapter.

The analysis of Bacchiocchi et al. (2011) focuses on the perception of telephony prices. This is performed through a probit model, where the dependent variable is the probability of being satisfied (dissatisfaction is a complement to this, and will be used in the next sections). Since Bacchiocchi et al. (2011) dichotomize the data about the satisfaction of final users, their model estimates the probability of observing a satisfied respondent in the Eurobarometer survey, given the individual characteristics of the country of residence, the level of reform of the telephone industry in that country, and the price he effectively pays for the service. The model takes the form:

$$P(y_{ijt} = 1|x_{ijt}, z_{jt}, r_{jt}, p_{jt}) = \Phi(\gamma_0 + \gamma_{0j} + \gamma_{0t} + \gamma_1{}'x_{ijt} + \gamma_2{}'z_{jt} + \gamma_3{}'r_{jt} + \gamma_4{}'p_{jt})$$

where Φ is the standard normal cumulative distribution function, and γ_{0j} and γ_{0t} are country and year fixed-effects respectively, individual characteristics are (x_{ijt}), macroeconomic variables (z_{jt}), and regulatory indicators for entry regulation, public ownership, market structure, and vertical integration are (r_{jt}). Actual prices (p_{jt}) are entered among independent variables to verify whether subjective satisfaction depends on prices and whether the relationship between subjective satisfaction and regulatory variables is at all driven by the relationship between regulatory variables and market prices. Bacchiocchi et al. (2011) find that the OECD reform indicators (ECTR), after controlling for actual price levels do, in fact, contribute to the probability of users being satisfied, but with some unexpected results. The entry regulation variable shows a surprising positive sign (i.e. more satisfaction is associated with more restrictions to entry).[3] The other two reform indicators, market structure

[3] This last result is tentatively explained by observing that over the three Eurobarometer waves (2000, 2002, 2004), only Portugal and Greece for the two years 2000 and 2002 (i.e. only 1002 over the total number of 47,284 observations) present values different from 0, i.e. free entry in the trunk telephony market (as mentioned in previous chapters the ECTR/REGREF indicators by the OECD are coded from 0 = full privatization/liberalization to 6 = public ownership, legal monopoly).

and public ownership, enter with negative signs and are statistically significant. Actual competition is more important than formal entry legislation in explaining satisfaction.

Fiorio and Florio (2010) test whether liberalization and privatization are positively associated with the satisfaction of electricity consumers. When estimating marginal effects in a probit model and conditioning only on individual characteristics of the respondent, some country macroeconomic variables, country and time dummies, they find the following results. Females are slightly less satisfied than males, the elderly are slightly more likely to be satisfied than youngsters after controlling also for average prices. More educated people tend to be more satisfied than people who abandon study before they reach the age of 15, while there is no significant difference in terms of marital status. Managers and students tend to be consistently more satisfied than self-employed people, with the unemployed being the least satisfied. Respondents with moderate political views tend to complain less about electricity prices than people at both the extremes of the political spectrum and, controlling for respondents' cooperation it emerges that less cooperative people tend to declare themselves as less satisfied. In some models, population density is correlated with more satisfaction, and the Gini inequality index, on the contrary, with less satisfaction. However, when the consumer price index and the average electricity price are also included in the analysis, macroeconomic variables lose statistical significance. In this latter case it emerges that satisfaction is negatively correlated with electricity consumer prices (an increase of price by ten cents would reduce the probability of satisfaction by 10%) but it is positively correlated with the consumer price index (an increase of CPI by 1% would increase the probability of satisfaction with price paid for electricity prices by 2.7%), as expected as a relative price effect enters the perception. Average consumer satisfaction with electricity prices is significantly greater in 2004 compared to 2000, with the probability of satisfaction across EU-15 being nearly 2% higher.

When introducing reform indicators, Fiorio and Florio (2010) find that satisfaction increases the less reformed the electricity industry is. For instance, if the average composite indicator of reforms increases by one unit from its average of 2.2 in the sample considered, the probability of satisfaction is predicted to increase by nearly 4%, with no control for average prices and nearly 6% after controlling for prices. When the aggregate electricity reform index from OECD/ECTR is separated into a 0–6 score of public ownership and a 0–6 score of liberalization, with and without controlling for prices, respectively, the authors find that reform indexes have a strongly significant positive association with consumer satisfaction on the prices paid for electricity services, even controlling for average prices. If ownership of the incumbent in the industry switches from public to private, the model predicts that the likelihood of consumer satisfaction with prices decreases by around 13%.

Building on these earlier findings, in the rest of the chapter we offer new estimates of the probability of (dis)satisfaction with prices of the utilities.

7.4 Data

Eurobarometer public opinion surveys (henceforth, EB)[4] have been conducted on behalf of the Directorate-General for Education and Culture of the European Commission each spring and autumn since autumn 1973. An identical set of questions is posed to representative samples of the population aged 15 years or over in each Member State. In each household, the respondent is drawn at random. Questions are asked during a face-to-face interview at the respondent's home and in the appropriate national language. The regular sample in standard Eurobarometer surveys is 1000 people per country except Luxembourg (600), the United Kingdom (1000 in Great Britain and 300 in Northern Ireland) and Germany (since EB34, 1000 in East Germany and 1000 in West Germany). Each survey comes with a set of weights obtained, using marginal and intercellular weighting, carried out on the basis of the population description provided by Eurostat in the Regional Statistics Yearbooks. We use these weights throughout the analysis to follow.[5]

In the years 2000 to 2006 (Eurobarometer 2002, 2004, 2006, 2008), the Eurobarometer surveys included some questions concerning several aspects of various services of general interest (SGI), including telephone, electricity supply, gas supply, water supply, postal, transport, and rail services.

One ranking of countries by the extent of regulatory reforms can be constructed using again the indicators of regulation in energy, transport, and communications (ETCR).

In order to analyse consumer perception of prices, we use self-rated dissatisfaction with the prices paid for each service considered (electricity, gas, and telecoms) by a representative sample of EU-15 citizens, as measured in the Eurobarometer surveys (henceforth EB). The EB data also include a list of individual characteristics, allowing us to control for some individual heterogeneity. Using the EB data we had four separate cross-sections comprising over 50,000 individual opinions about electricity and telecommunication prices and around 30,000 about natural gas prices. As for the regulatory reform variables, including measures of entry regulation, public ownership, market

[4] EB surveys always come with an issue number. In our notation to follow, we denote EB34 issue 34 of the Eurobarometer surveys, and similarly for other issues.

[5] A detailed analysis on the Eurobarometer data can be found on the official Eurobarometer website: http://ec.europa.eu/public_opinion/index_en.htm.

structure, vertical integration, in varying levels of detail, we use sub-indicators provided by the ECTR dataset, as in the previous chapters.

We use the aggregate regulatory index as an indicator of the overall reform but then we introduce our own coding of sub-indicators, as we question the use of cardinal numbers for categorical variables such as those describing the regulatory framework of an industry. In particular, we define dummy variables where only ordinal variables are available in the first place. For instance, for each industry, we define a dummy variable for public ownership equal to 1 if the industry is 'public' and zero otherwise. A dummy variable for entry regulation is equal to 1 if there is 'no third party access, full entry regulation' and zero otherwise. A dummy variable for vertical integration is equal to 1 if the industry is 'vertically integrated' and zero otherwise. No market structure measure is available for electricity as for ECTR, while the market structure dummy for the gas industry is equal to 1 if the market share in all segments of the industry is larger than 90%.

We use subjective dissatisfaction data to be analysed with dichotomous dependent variable models. The model for consumer price dissatisfaction rests on the definition of a variable disentangling satisfied from relatively dissatisfied consumers regarding the price paid for a particular network service. This variable is recorded as equal to 0 if individual j in country i at time t states that the price he pays for fixed telephone services is fair, and is recorded as equal to 1 otherwise. We then estimate a standard probit model for individual dissatisfaction, where we control for year and country fixed effects, a vector of individual characteristics (i.e. sex, occupation, etc.) accounting for individual observed heterogeneity, a vector of macroeconomic control variables including population density, GDP levels, consumer price index, and the total price (including all taxes) of a unit of each industry's provided service. Looking at prices rather than quality or accessibility, we concentrate on a clearly identifiable economic variable, which is easily understood by customers thus presumably reducing measurement error, and a variable that is supposed to be mostly affected by recent reforms.

Although we do not present here all the detailed coefficient estimates,[6] there are some consistent patterns across all three sectors considered for individual characteristics and likelihood of dissatisfaction. In particular, on average, females are more likely to be dissatisfied than male respondents, and, similarly unemployed people and respondents showing average or low level of cooperation with the interviewer tend to be more dissatisfied. Conversely, more educated and student respondents are less likely to be dissatisfied, as well as over 75s as opposed to respondents aged 31–45. We also find evidence of a

[6] Estimates are available from the authors.

significant positive time trend, showing that average dissatisfaction decreases over time.[7]

In what follows, we focus on the set of regulatory variables and interpret any significant coefficient in the vector of regulatory variables as a sign of the fact that regulatory reforms have had some effect on consumer satisfaction regarding prices. A priori, we have no expectation regarding the statistical significance nor the sign of the coefficients of the regulatory variables.[8]

7.4.1 *Electricity*

Table 7.1 shows results for dissatisfaction with electricity prices, focusing mainly on regulatory variables only. The first three columns show no significant correlation with the overall reform indicator (*regulatory index*) and dissatisfaction, neither controlling nor without controlling for prices paid by the average consumer in each country. As the aggregate ECTR score might hide

Table 7.1. Probit estimation of individual dissatisfaction with electricity supply prices

	(1)	(2)	(3)	(4)	(5)	(6)
Regulation index	0.021	0.019	0.024			
	(0.324)	(0.335)	(0.390)			
Public ownership[1]				−0.035	−0.057*	−0.060**
				(0.303)	(0.062)	(0.035)
Entry regulation[1]				0.129*	0.125*	0.137*
				(0.089)	(0.088)	(0.060)
Vertical integration[1]				0.026	0.032	0.042
				(0.438)	(0.332)	(0.178)
Price for average household (period *t*)		1.828	2.357		2.316**	3.616***
		(0.113)	(0.162)		(0.022)	(0.000)
Price for average household (period *t*−2)			−0.583			−1.342
			(0.666)			(0.143)
Individual characteristics	yes	yes	yes	yes	yes	yes
Macroeconomic variables	yes	yes	yes	yes	yes	yes
Year fixed-effects	yes	yes	yes	yes	yes	yes
Country fixed-effects	yes	yes	yes	yes	yes	yes
Individuals	57,153	57,153	57,153	57,153	57,153	57,153
Log-likelihood	−35,333	−35,306	−35,304	−35,268	−35,227	−35,217
Pseudo R-squared	0.0578	0.0585	0.0586	0.0595	0.0606	0.0609

Sources: own elaboration on ECTR for regulatory indices (see Table 5.A.1), Eurostat for price variables and macroeconomic variables, Eurobarometer for individual variables.
P-values in parentheses.
*** $p<0.01$, ** $p<0.05$, * $p<0.1$.
[1] dummy.

[7] The estimates are available from the authors.
[8] See our discussion in Chapter 3 about the possible directions of price change following reform.

important information, we estimated the probit model replacing the ECTR score with regulatory dummies and found that the likelihood of dissatisfaction is roughly 10% lower in countries where there is public ownership of the electricity industry, even after controlling for country and year fixed effects, for some macroeconomic characteristics including per capita GDP and individual characteristics. The other regulatory variables considered show no statistical significance, while the contemporaneous price is, as expected, significantly positive, confirming that the higher the price the higher the rates of dissatisfaction. We also included the average price in each country lagged by two periods, aiming at capturing the effect of reference pricing for determining satisfaction, however its coefficient is not statistically different from zero, providing no conclusive evidence about the role of past prices as a reference point for electricity supply.

7.4.2 *Gas*

As for the analysis of gas supply dissatisfaction, Table 7.2 shows that public ownership is again negatively correlated with individual dissatisfaction.

Table 7.2. Probit estimation of individual dissatisfaction with gas supply prices

	(1)	(2)	(3)	(4)	(5)	(6)
Regulation index	−0.024	−0.041***	−0.041***			
	(0.246)	(0.002)	(0.001)			
Public ownership[1]				−0.109**	−0.096*	−0.107**
				(0.041)	(0.052)	(0.015)
Entry regulation[1]				0.036	−0.017	−0.012
				(0.180)	(0.689)	(0.745)
Vertical integration[1]				−0.112***	−0.065	−0.062
				(0.000)	(0.167)	(0.133)
Price for average household (period *t*)		0.032***	0.025		0.024**	0.015
		(0.004)	(0.148)		(0.033)	(0.254)
Price for average household (period *t*−2)			0.012			0.015
			(0.621)			(0.516)
Individual characteristics	yes	yes	yes	yes	yes	yes
Macroeconomic variables	yes	yes	yes	yes	yes	yes
Year fixed-effects	yes	yes	yes	yes	yes	yes
Country fixed-effects	yes	yes	yes	yes	yes	yes
Individuals	30,757	29,585	28,386	30,757	29,585	28,386
Log-likelihood	−18,829	−18,028	−17,274	−18,775	−18,009	−17,255
Pseudo R-squared	0.0585	0.0612	0.0618	0.0612	0.0622	0.0629

Sources: own elaboration on ECTR for regulatory indices (see Table 6.A.1), Eurostat for price variables and macroeconomic variables, Eurobarometer for individual variables.
P-values in parentheses
*** p<0.01,** p<0.05,* p<0.1.
[1] dummy.

While the other liberalization variables are not statistically significant, the overall reform indicator is estimated to have a significant and negative sign when average prices are also controlled for, meaning that the more advanced the reform process (i.e. the lower the ECTR score) the higher the rate of dissatisfaction. However, when average consumer prices lagged by two periods are also included in the analysis, the contemporaneous price variable loses statistical significance, most likely due to multicollinearity because of the high persistence of gas prices over time.

7.4.3 *Telephone*

Dissatisfaction with telephone prices, similarly to what was found for the electricity and gas sectors, is lower in countries where services are provided by public operators. In this case, the ECTR score is significant and negative, suggesting that the further reform process has gone (i.e. the lower the ECTR score), the more likely individuals will express dissatisfaction. This result also remains when prices are controlled for, see Table 7.3.

Table 7.3. Probit estimation of individual dissatisfaction of telephony services

	(1)	(2)	(3)	(4)	(5)	(6)
Regulation index	−0.109***	−0.108***	−0.109***			
	(0.002)	(0.002)	(0.000)			
Public ownership[1]				−0.205***	−0.240	−0.281**
				(0.010)	(0.298)	(0.049)
Market structure[1]				−0.038	−0.036	−0.032
				(0.198)	(0.222)	(0.170)
Price for average household (period *t*)		0.019**	0.021**		0.023**	0.024**
		(0.045)	(0.030)		(0.030)	(0.024)
Price for average household (period *t*−2)			0.012*			0.010
			(0.090)			(0.198)
Individual characteristics	yes	yes	yes	yes	yes	yes
Macroeconomic variables	yes	yes	yes	yes	yes	yes
Year fixed-effects	yes	yes	yes	yes	yes	yes
Country fixed-effects	yes	yes	yes	yes	yes	yes
Individuals	53,090	53,090	53,090	53,090	53,090	53,090
Log-likelihood	−30,432	−30,423	−30,414	−30,485	−30,463	−30,457
Pseudo R-squared	0.130	0.130	0.130	0.128	0.129	0.129

Sources: own elaboration on ECTR for regulatory indices (see Table 4.A.1), Eurostat for price variables and macroeconomic variables, Eurobarometer for individual variables.
P-values in parentheses. *** p<0.01, ** p<0.05, * p<0.1.
[1] dummy.

7.5 Summing up

A traditional approach in applied welfare economics is to consider changes in consumer surplus or in other appropriate welfare measures. While we think that this objective measurement is still the core test for utility reforms, it is also interesting to look—with due caution—to subjective perceptions. After all, if in countries where the reforms have been implemented prices go down and consumer satisfaction consistently goes up, compared to countries lagging behind, then there is a double check of the success of the reform.

We find evidence that consumer satisfaction about prices is higher in countries where the network service provider is government owned. Market opening reforms in the period of time and in the countries we considered do not seem to be systematically associated with increased satisfaction.

How can we interpret these results and do they have policy implications? One may suspect that survey respondents are biased in their perceptions, and tend to see public ownership as more protective of their interests. This interpretation implies, however, that when a consumer is asked the question 'In general, would you say that the price you pay for electricity supply service is fair or unfair?' (we cite from the English version of the Eurobarometer questionnaire) he or she knows whether in his country the industry is under public ownership or not. Thus, one may conjecture that the answer is driven by this information and by a sentiment of protection given by public ownership. An alternative and more straightforward interpretation is, however, that in responding to the question, whatever the knowledge of the respondent about ownership, they just focus on their perception of price fairness. The same applies to any liberalization variables, as here we are not testing social attitudes on those variables, but simply answers to a direct question on prices.

We cannot directly test what information is available to respondents on our variables of interests, and whether they have a bias pro-public ownership and pro-liberalization, but we control for the political views of the respondents. What we could do was to test whether price satisfaction and average actual prices, which we use as controls, are correlated. As we find that individual perceived price fairness is strongly negatively correlated to actual average price paid, as expected, we can be reasonably confident that this quite large sample of respondents is mostly driven by their perceptions of the difference between the price they pay and what they would consider a fair price. From this perspective, we conclude that, after controlling for possible individual and country confounding factors, there is no evidence that consumers are happier about prices they pay in the EU countries which have adopted the reform paradigm, as compared to those which have delayed its adoption.

7.6 Further reading

Bertrand and Mullainathan (2001) discuss the economist's mistrust of survey data and suggest a measurement error framework on subjective evidence. We further consider this issue in the next chapter. On data sources for the research presented in this chapter, Eurobarometer related publications are available at http://ec.europa.eu/public_opinion/index_en.htm. While in this book, and particularly in this chapter, we have focused on the EU-15, there are Eurobarometer surveys for the New Member States of the EU too. European Commission (2004) is about consumer opinions on services of general interest in acceding countries, with fieldwork completed in 2003, that is before actual accession. Different aspects and sectors are covered, including price. For example, price dissatisfaction for mobile telephone prices was highest in Lithuania, the Czech Republic, and Slovakia, with around 50% of users dissatisfied; for fixed telephone services in Lithuania, Estonia, and Slovakia more than 70% of respondents were dissatisfied. For electricity, supply dissatisfaction with prices was very high in Slovakia (77%) and lowest in Latvia (37%). Similar data are available for gas, water, postal services, and different transport services. Occasionally Eurobarometer also publishes qualitative research based on focus groups: EC (2004) is based on 6 discussion groups of consumers in 25 countries, plus individual interviews. This qualitative evidence, when available, can contribute to the understanding of what the determinants of consumer satisfaction with network services are.

Eurobarometer (2007) surveyed data on perceptions of quality of life in 75 European cities, including satisfaction about public transport. Fiorio, Florio, and Perrucca (2011) use these data for testing correlation with local public transport integration.

On the potential role of behavioural economics in the regulatory economics, Chang and Wilson (2006) summarize the results of a workshop organized by the Centre for Competition Policy (Norwich). Related materials are available at http: http://else.econ.ucl.ac.uk/conferences/consumer-behaviour/. Chang, Waddams Price, and Wilson (2006) show that in different service industries, including electricity, mobile telephony, fixed line telephony, calls, broadband, car insurance, mortgage, and current bank account, while on average 86% of respondents to a survey were aware of different offers, only 23% searched for alternatives to their current supplier, and only 17% actually switched. Also surprising is the finding that only 8–11% of consumers switched to firms offering the highest surplus, and on aggregate only 26–39% of the potential gain was appropriated by switchers (meaning a large majority of sub-optimal decisions to change supplier, for the minority who actually changed it). These issues are further discussed in the next chapter.

On consumers and electricity markets, see Defeuilley (2009), focusing on the distinction between active and inactive customers.

Appendix 7.A.1: Empirical models for consumer dissatisfaction

The models of dissatisfaction conditional on individual characteristics, country, and year fixed effects and other macroeconomic variables are estimated using maximum likelihood and can be described as follows.

As we do not know the exact level of individual satisfaction, S_i^*, for each service, we assume that satisfaction is generated by a latent variable model:

$$S_i^* = \beta_0 + x'_i\beta + e_i \qquad (7.A.1)$$

where $i = 1,\ldots,N$ for a sample of N individuals, $x'_i\beta = \beta_1 x_{i1} + \ldots + \beta_k x_{ik}$ includes individual characteristics (i.e. sex, occupation, etc.) accounting for individual observed heterogeneity, time-varying country macroeconomic variables (e.g. per capita GDP) accounting for time-varying heterogeneity, a time fixed-effect to capture any linear time trend, and some time-invariant country fixed-effect to capture any country-specific effects. Finally, e_i is a continuously distributed variable independent of x_i, and accounts for unobserved heterogeneity. As S_i^* is latent, for each individual i, one can only observe

$$S_i = 1[S_i^* > 0]$$

where 1[.] is equal to 1 if the argument is true and equal to zero otherwise. Assuming that e_i is distributed as a standard normal we obtain the probit model:

$$Pr(S=1|x) = \Pr(S^* > 0|x) = \Pr(e > \beta_0 - x\beta|x) = 1 - \Phi(\beta_0 - x\beta) = \Phi(x\beta - \beta_0) \equiv p(x)$$

Where ϕ is the standard normal cumulative density function. The partial effect of $\Pr(Sx_j, j = 2,\ldots,k,$ on $p(x)$ depends on x through the standard normal density function, $\phi(x\beta)$, as $\partial p(x)/\partial x_j = \phi(x\beta)\beta_j$.

As we have four repeated cross-sections of the EB data set, over 15 EU countries, we pool them obtaining the following model:

$$S_{it}^* = \sum_{s=2002,2006} \gamma_s d_{s,it} + \sum_{j=l}^{15} \alpha_j d_{j,it} + x'_{it}\beta + e_{it} \qquad (7.A.2)$$

where the year dummy γ_s is equal to one if $s = t$ and zero otherwise and the country dummy α_j is equal to one if individual i lives in country j and zero otherwise. The intercept is omitted as all country dummies are included. Model (7.A.2) is estimated as a random effect model as errors are assumed independent over both i and t, with heteroscedasticity correction.

8

Quality of service

8.1 Introduction[1]

After having considered the evidence on consumers' perception of price fairness for some services, we turn to the issue of perceived quality. As in the previous chapter we explore subjective data on consumer satisfaction. Thus we do not directly study whether utility reforms explain variations in welfare of consumers across countries and time periods, but instead whether they are correlated to the perceptions of users. Ideally, one would use both detailed microdata on quality-adjusted price paid and expenditure by households. This would require data on objectively measured service quality, along with individual attitudes on this dimension of services, to test economic welfare change and compare them with 'happiness' measures. If the two measures do not correlate, this fact would need further research to understand whether the cognitive process by the consumer is biased, or whether the objective statistical evidence does not capture some detail in services provision, possibly best known to the user, particularly aspects of quality not captured by prices. For example, for the same price you can get either a connection to the electricity grid that is frequently broken, or that instead is nearly always well functioning. You can buy a stable voltage, or large variations of it that may damage your computer or your fridge. As a telephone user the fixed line can be noisy or disturbed. Repairs can be easily obtained or not. Testing for gas leaks can be frequent and fast, or only occasional. Contract and information to the customers by the providers can be transparent or obscure. Thus, the price you pay does not tell the full story of the service. Moreover, customer satisfaction for the quality of the service is important information per se, as any commercial provider and regulator knows.

After a general discussion about service quality, we again use different waves of Eurobarometer Surveys, for the EU-15 countries, and try to test the impact

[1] With Silvia Salini. The chapter also draws on Fiorio, Florio, Salini, and Ferrari (2007), Section 8.6 and Appendix 8.A.2 draws on Grassi and Puglisi (2008).

of privatization and regulatory reforms on attitudes of users of electricity, gas, and telephone. The structure of the chapter is the following: Section 8.2 discusses the concept of service quality and of customer satisfaction; Section 8.3 presents some methodological issues; Section 8.4 presents some descriptive statistics; Section 8.5 presents a conditional analysis of consumer satisfaction with an ordered logit model; Section 8.6 turns to an individual fixed-effects approach; other empirical approaches are mentioned in Section 8.7; Section 8.8 compares objective and subjective measures of quality for electricity supply; Section 8.9 presents our main conclusions; further reading is suggested in Section 8.10.

8.2 Quality of services and user satisfaction

Quality of network services is often an elusive concept. Perceived quality depends on the amount of information available to users. This information is often relatively limited, but some final effects of technical problems in the network are apparent, particularly service disruptions, even if their origins and transmission mechanisms are unknown to the users. Before turning to quality perceptions, which we use in the empirical analysis in this chapter, we briefly mention some objective measures of quality, which are often monitored by the regulators. The services we consider are those that will be analysed later on: telephony, electricity, and gas. In the Further Reading Section, we cite additional references and also mention quality issues in other network services.

8.2.1 *Quality of telephony service*

Quality of service in telephony, as defined by ITU (1994, 2008) is related to service response time, loss of signal, signal-to-noise ratio, cross-talk, echo, interruptions, loudness level, and so on. A more general concept is related to coverage and capacity of the network, including outage probability. Most of these aspects can be directly perceived by users, even if they do not know the technical determinants of service quality. More complex issues are related to the use of the telephone line as a vehicle for data, connection to the Internet and to computer networks. Traffic engineering should be able to support a required bit rate, a minimum delay, packet dropping probability, and other technical aspects that are much more difficult to directly perceive by users. Commercial quality refers to any aspect related to billing, connection timing, management of customers' complaints, and so on.

Different metrics have been used for 'objective' quality of service (QoS) and 'subjective' appreciation, or quality of experience (QoE). In the rest of the chapter we focus on the latter, while here we briefly mention the former.

As most flows over the fixed line are now supported by packet-switched networks, the technical problems that may be encountered are more complex than with the old telephony technologies. For example, a low throughput can be determined by an insufficient bit rate when different customers share the network with the same priority. Routers may fail to deliver some data packets if some information is corrupt or electronic buffers are insufficient to accommodate the stream of data. Over long copper wires and wireless connections there may be bit errors caused by 'noise'. Latency is the technical term used to describe the time it takes one packet to be delivered to its destination. In contrast to the old point-to-point technology, the system holds up signals in long queues or diverts them to long routes. Jitter is the effect of different delays, and may disrupt the quality of streaming audio or video. For Internet telephony, disruption may occur when packets take different routes and are no longer delivered in an approximate order, and the rearranging protocols are inadequate to manage the problem. This issue is not important for voice transmission with non-IP protocols, such as ATM or GSM, which have built-in QoS, and do not need additional protocols. With digital convergence, high quality is often provided through over-capacity of the network, which is costly. In practice, the network is scaled to peak-load capacity, but with heavy traffic, VOIP for instance would require a substantial over-provisioning to manage an acceptable quality.

A broader view of telephony service quality (Milne, 2006) would include the following items:

- successful calls (call set-up or conversation quality)
- accuracy, transparency, suitability of billing and payment systems
- public phones
- response of the provider to faults, complaints, service requests
- directories
- unwanted calls.

Milne (2006, p. 8) discusses the options for regulators to monitor the performance quality of all providers, or more intrusively to set compulsory quality standards. He observes that:

> Competition can improve quality. However, cutting prices is often the preferred way of competing, partly because it can be done much faster than improving quality. When cutting prices involves cutting costs, competition can even reduce quality... A dominant operator... might try to maintain margins by reducing the quality of retail service, for example, public payphone faults and directory enquiry call failures might rise because of cuts in staff. The operator might even introduce a service at a higher quality level and lower the quality level of the "standard" service to encourage customers to move to the better service.

The monitoring strategy of a telecom regulator can then take different routes, either based on voluntary schemes or on compulsory ones. In the UK, between 1995 and 2004 OFTEL published every three months five 'objective' quality measurements and four customer satisfaction indicators. The scheme started with 13 voluntary participants, eventually only 5 were left. Then, OFTEL turned to a compulsory approach, and comparable data are presented on service supply timeliness, fault repair timeliness, fault report rate, complaint resolution, and bill errors.

The European Union has regulated the monitoring of quality of service in telephony since 1998 (Directive 98/10/EC), having selected nine measurements, reduced to seven in the 2002 regulatory framework (Directive 2002/22/EC). Directive 2009/136/EC, states that the national regulatory authorities must set performance targets for undertakings with universal service obligations and monitor compliance with these targets by designated undertakings. In fact USO is defined by the Directive as the 'minimum set of services of specified quality to which all end-users have access, at an affordable price in the light of the specific national conditions, without distorting competition'. We shall come back to USO and particularly to affordability in the next chapter.

The new EU frame for electronic telecommunications states that users are entitled to some rights, which include the right to a contract including a certain set of information; transparency of tariffs; minimum quality requirements, particularly to prevent slow down of traffic over the networks; publication of comparable quality information by providers; guarantee of handling of catastrophic network breakdown; provision of operator assistance; and a free of charge emergency number. The EC 'Progress report on the single European Electronic communications markets' every year monitors some of these aspects for each of the 27 Member States. The OECD (2003) has selected six measurements, and these are discussed biannually at regular meetings.

While monitoring and publishing comparable information can be seen as soft regulation, which leaves to competition the role to adjust prices to quality, setting explicit targets or standards is a different, more intrusive approach. This may be justified by the intrinsic information asymmetries in the telephony market, at least for services to households.

8.2.2 *Quality of electricity service*

For electricity supply, according to a detailed report by Energy Community Regulatory Board (2008):[2] 'Service quality is an important issue in the

[2] An organization based on a treaty including Italy, Romania, Ukraine, Croatia, Serbia, Albania, Bosnia, and Kosovo.

electricity distribution and retail sectors. Experience so far shows when a regulator chooses to regulate prices using price or revenue caps, a service level tends to drop. On the other hand, rate of return method could lead to over investments in the network. For this reasons, all types of regulation must be supplemented by some kind of service quality regulation' (ECRB, 2008, p. 5).

This implies establishing transparent rules on data measurement, dealing with the multidimensional nature of quality, and monitoring mechanisms. There are two main aspects of quality, which are of interest of regulators: commercial and technical quality. The first concept refers to issues such as connection times, metering, billing, and handling of customers' requests. These issues are common to most network services. Technical quality is instead technology specific. In the case of electricity the two core aspects are continuity of supply, which is related to the number of interruptions over a period of time and their duration; and voltage quality.

The continuity dimensions can be summarized by simple indicators, such as the System Average Interruption Frequency Index, SAIFI, or the System Average Interruption Duration, SAIDI:

$$SAIFI = \left(\sum_k N_i\right)/N_{tot}$$

$$SAIDI = \left(\sum_k N_i D_i\right)/N_{tot}$$

where k is the number of outages, N_{tot} is the number of customers of the utility, N_i is the number of users affected, i is the interruption index, and Di is the duration of interruption in minutes.

While continuity of supply is easily observable, voltage quality is a complex concept, usually not directly measurable by users.[3]

The consequences for users of interruptions are obvious and their welfare effects, in principle, are measurable through the willingness to pay to avoid the outage (this approach has been used, for example, in cost–benefit analyses of electricity projects by the World Bank), or to accept compensation for it (an average of WTP and WTA is found in some contingent valuation studies). The consequences of voltage instability are less easily captured by a single indicator. For example, above a certain range of over-voltage, households may experience immediate damage to appliances, but other effects may be less immediate, and depend upon the engineering of computers, televisions, and the other electrical appliances, with effects which to a certain extent are unpredictable.

[3] Voltage quality includes variations in supply voltage (within a certain range of values), rapid voltage change (i.e. within a given time span); voltage swells; flickers; voltage unbalance; harmonic voltage distortions; transient over-voltage; mains signalling voltage.

Regulators can intervene by establishing quality indicators, performance standards, and penalties for the electricity supplier that underperforms. The latter can in some cases be used to compensate users. Similarly, quality standards can also be regulated for commercial quality.

The Council of European Energy Regulators (CEER), an association of European regulators, regularly publishes reports on the performance of electricity providers in the EU, see particularly CEER (2008). This detailed report shows the evidence available on planned and unplanned interruptions (excluding exceptional events), in terms of minutes per year. In 2006 or 2007, for example, there were more than 100 minutes of interruption in Portugal and Spain; nearly 90 minutes in the UK; 70 minutes in France, and so on. In terms of number of interruptions, there were more than two significant interruptions in Italy, Portugal, and Sweden, around one in France, and so on. The EU-27 plus Iceland and Norway average of unplanned interruptions in 2007 was around 68 minutes. When exceptional events (mostly atmospheric) are also included, unplanned interruptions are much more substantial: 409 minutes in Poland, 300 in Sweden, and so on. Planned interruptions are also significant in some countries, for example in Poland 121 minutes were lost in 2007, and more than 200 in Latvia and Estonia. In some countries there is also important geographic dispersion of these data across regions.[4] In Section 8.8 the number of interruptions and the minutes lost per year by country will be considered, in particular its relationship with the users' satisfaction.

There are also several surveys on the quality of electricity service produced by ERGEG, the European Regulators Group for Electricity and Gas, which was established by the EU between 2003–11, now supplanted by the ACER. CEER (2010) offers a review of good practice on estimation of costs due to electricity interruptions and voltage disturbances. This report breaks down total socio-economic costs of disservice into different categories: private customer costs and net costs to the rest of the society, monetary and non monetary, direct and indirect, and suggests estimation approaches, see also CEER (2011).

8.2.3 *Quality of gas service*

Gas quality regulation focuses on four aspects: security of supply, technical and safety standards, commercial quality, reliability of supply, see Kema (2010). We briefly discuss each of these items in turn.

Security of gas supply has already been briefly discussed in Chapter 6. It is particularly addressed by Directives 2003/55/EC and 2004/67/EC. Security of

[4] For a detailed overview see Fumagalli et al. (2007).

supply is basically an issue of the reliability of partnerships with exporting countries, the nature of the contracts, and the cross-border agreements in case of an emergency. Storage services can be also considered part of security of supply. These issues are more often dealt with by national legislation, aimed at protecting the consumer from risks related to the uncertainty of physical deliveries of natural gas.

The technical standards, as set by national legislation or by the regulators, include such themes as the maximum pressure of pipelines, and some engineering aspects of the network. Technical standards in general have been increasingly harmonized by CEN, the European Committee for Standardisation. Safety standards are obviously very important, as natural gas is both inflammable and poisonous. These standards always include gas odourization, safety devices for gas pressure regulation stations, and installation with the end user, particularly regulation of shut-off devices, connections to appliances, the qualifications required of personnel involved in the installation and testing of appliances, and so on. Article 6, Directive 2003/55/EC states that 'Member States shall ensure technical safety criteria are defined and that technical rules establishing the minimum technical design and operational requirements for the connection to the system of LNG facilities, storage facilities, and direct lines, are developed and made public.' The same article states that technical standards should not prevent interoperability of different components or transport modes of the gas system. Quality indicators often considered by regulators of gas safety include total number of incidents per year, amount of third party damage, number of ruptures, spills, leaks, number of fatalities, frequency of pipeline inspections, and information to the public on the safety standards.

Regarding reliability, the typical quality indicators are the number of planned/unplanned interruptions per customer/year, the duration of these interruptions, the average number of customer minutes of gas service per interruption, and information on these issues.

Quality regulation of the natural gas itself has not yet been fully developed at EU level. When this is the case, it includes as quality indicators the Wobbe Index,[5] heating or caloric value per cubic metre, specific gravity/relative density, CARI[6] odourization levels, chemical properties, and information to customers about the technical quality. The issue is, however, very complex, as natural gas is in fact a variable mixture of many components with some consequences for cross-border trade. For example OFGEM has stated that:

[5] The Wobbe Index (WI) or Wobbe number is an indicator of the interchangeability of fuel gases such as natural gas, liquified petroleum gas (LPG), and town gas and is frequently defined in the specifications of gas supply and transport utilities. $Wobbe\ Index = \frac{Heating\ Value}{\sqrt{Specific\ gravity}}$

[6] Combustion Air Requirement Index, the required amount of dry air to burn 1 Nm3 of fuel gas compensated for the specific gravity of the gas. $CARI = \frac{Air\ Demand}{\sqrt{Specific\ gravity}}$

> Natural gas is made up of a mixture of hydrocarbon gases and this mixture can vary widely between sources. The "quality" of a given amount of natural gas is determined by the relative quantities of these hydrocarbon gases, and therefore also varies considerably. In Great Britain (GB), domestic and industrial appliances are designed to operate within a certain gas quality specification range. The current gas quality standards are based on the quality of gas sourced from the UK Continental Shelf (UKCS) as this has traditionally been the primary source of supply for the GB market. However, as a result of the natural decline in domestic gas production, GB has become increasingly reliant on imports of gas from the Continent and other world markets, and this trend is set to continue in the future. Over the next decade, gas quality specifications in continental Europe are set to become broader than those applying in GB. This could mean that in future years, any gas flowing from continental Europe to GB will not comply with GB's quality standards, and therefore not be able to be accepted onto the GB network. This could have consequences for security of gas supply in GB.[7]

Commercial quality regulation is also less developed for gas than for other sectors, as only a few EU members have passed a specific regulation of commercial quality of natural gas. The existing regulations usually consider items such as generic customer satisfaction, customer complaint rates, average time of companies to answer emergency calls, to answer customer enquiries, advance notification of planned interruptions, accuracy of bills, availability of customer service centre, and information that is publicly available. Some regulators have also considered environmental sustainability as part of the quality of service. OFGEM monitors emissions of methane from networks, and has a specific incentive for their reduction. The British regulator is probably one of the most careful in Europe in terms of collecting reliability data, focusing on planned and non-planned interruptions. It also regularly requires providers to perform quarterly surveys of customer satisfaction with the gas service. It has published data which show that planned and non-planned interruptions per 100 customers in the UK have been increasing from 2003 to 2007, going from 0.92 to 2.40 (OFGEM, 2008). Customers who experienced planned or unplanned interruptions are then asked (by a third party agency) to rate their satisfaction level about repairs on a five points scale, and penalties can be levied with the provider if certain standards are not met. We show in Table 8.A.1 (see Appendix) some of the detailed regulations about quality of service as established by OFGEM.

In the rest of the chapter we present our own empirical analysis of customer satisfaction, with telephony, electricity, and natural gas services in the EU-15.

[7] http://www.ofgem.gov.uk/Markets/WhlMkts/CompandEff/GasQual/Pages/GasQual.aspx, [accessed October 20, 2011]. For a detailed discussion in the US context see for example Foss (2004) http://www.beg.utexas.edu/energyecon/lng/documents/CEE_Interstate_Natural_Gas_Quality_Specifications_and_Interchangeability.pdf.

This is only a proxy for a more objective empirical analysis of the correlation between quality of service and the reforms, an issue for which comparable data are not available across EU countries.

8.3 Some methodological issues

The concept of customer satisfaction is closely related to the concept of quality. Quality, however, is different in the case of a product and in the case of a service. Evaluations of service quality are not exclusively linked to the observed output, as for products, but, and to a greater extent, to the whole process through which the service is provided. Consumer evaluation of the service received is determined by factors affected by psychological interactions that are established during the exchange transaction, and by factors connected to technical-specific characteristics of the service. The former factors concern behaviour, sensations, and psychological benefits, which are difficult to measure, whilst the latter factors can be evaluated by objective indicators similar to those utilized for product quality. For this reasons, the models that are commonly used for the evaluation of the service quality are often models typical of the psychometric literature.

An important paradigm for evaluating service quality is the gap model developed by Parasuraman et al. (1985). Following executive interviews and focus groups in four different service business areas, the authors proposed a conceptual model of service quality where consumers' perception on service quality depends on gaps between the service provider organization and the consumer environment. Later, they developed in-depth measurement scales for service quality.

Zanella (2001) provides an overview of measures and models of customer satisfaction. He classifies the main features of models and related techniques used for describing customer satisfaction. Following Zanella's work we could consider *Composition and Formative Models*[8] in which customer satisfaction is considered a 'multidimensional attribute', where each component corresponds to a dimension of the conceptual construct, that is to an aspect of a service considered essential in determining customer satisfaction.

Another approach, for which the regression models are the best known example, are the *Explanatory or Decomposition Models*. A self-declared questionnaire often provides an overall assessment of customer satisfaction from a

[8] Starting from the fundamental work by Parasuraman et al. (1988, 1991), the well-known composition model SERVQUAL was developed. The most serious criticism of the original gap model approach was expressed by Cronin and Taylor (1992), who raised doubts about the SERVQUAL indicator being appropriate to describe service quality. This criticism gave rise to another improved model, SERVPERF.

specific service. The response variables of the underlying customer satisfaction model are typically expressed on a semantic differential scale, with corresponding conventional scores such as a 5-point or a 7-point scale. Explanatory variables describe dimensions related to specific aspects of a service. The usual statistical analysis techniques for such data apply the least squares criteria for deriving estimates of the unknown parameters and for checking the goodness of fit. If one can assume a probability distribution for the response portraying overall satisfaction, the expected value of the response can be presented, with conditioning on the different situations described by the values of the explanatory variables. The logistic regression approach is used in this case. Rasch models are a particular case of logistic models, see the Further Reading Section.

Well known models, often used in psychometry and customer satisfaction, are the *Linear Structural Models with Latent Variables (LISREL)*.These models allow us to establish links between latent variables, which are related to dimensions describing customer satisfaction (Bollen, 1989). This is at the basis of the conceptual construct. The model is composed of two equations systems: structural equations and a measurement model.

A more recent alternative analysis of customer satisfaction questionnaires to measure the quality of services consists of Bayesian networks. They are based on graphic schemes, as is the case with structural models, which allows one to detect and study the net of relationships among all variables of interest. Moreover they use only manifest variables, thus avoiding the controversial notion of latent variables. However, a problematic aspect, which is common to Bayesian models, comes from the assignment of a prior distribution to the model parameters. These parameters express the probability that a chosen 'node' variable can assume one of its possible modalities, conditional on a given path connecting the node variable to a sequence of modalities describing the state of each element of an ordered set of 'descending' variables any of which causes the next variable. The choice of a path as an alternative to another is made by comparing the 'marginal likelihood' of the observed frequencies, once dependence on the parameters has been eliminated by integration with respect to a prior distribution, and by preferring the path which shows the larger likelihood.

In the following sections we provide some examples and some further developments of the techniques briefly mentioned in this section.

8.4 Descriptive statistics

For each service we analyse Eurobarometer data on customer satisfaction with respect to quality. The possible answers are: 'Very bad, Fairly bad, Fairly good, and Very good.' We consider that a user is to be considered (fully) satisfied if her/his answer is 'Very good' for quality. In this section, we examine the rate of satisfied people as resulting from the 2000 EB survey, as our starting point, and changes from 2000 to 2002, from 2002 to 2004, and the overall change from 2000 to 2004. Given the sample nature of data, the statistical significance of changes is also tested by the statistics $z = \hat{p}_1 - \hat{p}_2 / \sqrt{\hat{p}(1-\hat{p})\left(\frac{1}{n_1} + \frac{1}{n_2}\right)}$ where $\hat{p}_1$ and $\hat{p}_2$, n_1 and n_2 are the rates and the sizes of the two compared samples and $\hat{p}$ is the weighted mean of $\hat{p}_1$ and $\hat{p}_2$.

In the following sections we discuss the main results for each service, sketching the trend for the EU-15 on average and for a selection of relevant countries.

8.4.1 *Customer satisfaction: The fixed telephone service*

Quality of fixed telephony is judged by Eurobarometer interviewed users as 'very good' less often in Italy, Greece, and Portugal and more often in countries like Denmark, Ireland, and Luxembourg (Table 8.1). Quality level is, however, generally considered good: in 2000 93.83% of EU-15 citizens considered the telephone service quality either very good or fairly good. Figure 8.1 shows the trends for the overall EU-15 countries, for Belgium and Sweden (with the largest increases), and for Finland and France (with the largest decreases). No significant variation is observed for Austria, Denmark, Luxembourg, or Portugal.

Table 8.1. Valid answers on fixed telephone service quality

	Percentages year 2000				Percentage difference in very good response		
	Very bad	Fairly bad	Fairly good	Very good	2002–2000	2004–2002	2004–2000
Austria	1.03	4.11	43.15	51.71	−3.38	0.14	−3.24*
Belgium	0.75	2.01	58.65	38.60	8.27***	−0.30	7.97***
Denmark	0.00	2.79	38.60	58.60	−2.74	4.67**	1.94
Finland	0.49	1.47	50.98	47.06	−8.60***	−0.96	−9.56***
France	0.55	3.74	56.94	38.77	−7.16***	−0.41	−7.57***
Germany	1.02	4.57	62.61	31.80	−4.53***	10.18***	5.64***
Great Britain	0.61	3.61	48.28	47.50	−0.31	−3.14	−3.45**
Greece	1.82	7.06	62.19	28.93	−3.76**	−2.44	−6.20***
Ireland	0.00	2.21	37.50	60.29	−0.43	−3.95**	−4.39**
Italy	2.08	8.50	72.26	17.17	−4.51***	0.71	−3.80***
Luxembourg	0.00	0.00	44.44	55.56	0.00	−2.61	−2.61
Netherlands	0.15	2.77	44.84	52.23	−11.79***	17.83***	6.05***
Portugal	0.30	7.90	79.03	12.77	−5.18***	1.92*	−3.26
Spain	0.82	9.45	73.03	16.69	6.27***	−0.52	5.76***
Sweden	0.28	1.40	43.14	55.18	5.32***	2.82*	8.13***
TOTAL	0.93	5.23	59.89	33.94	−2.98***	3.26***	0.28

Notes: *** $p<0.01$, ** $p<0.05$, * $p<0.1$. For details see Fiorio et al. (2007).
Source: own elaboration on Eurobarometer data.

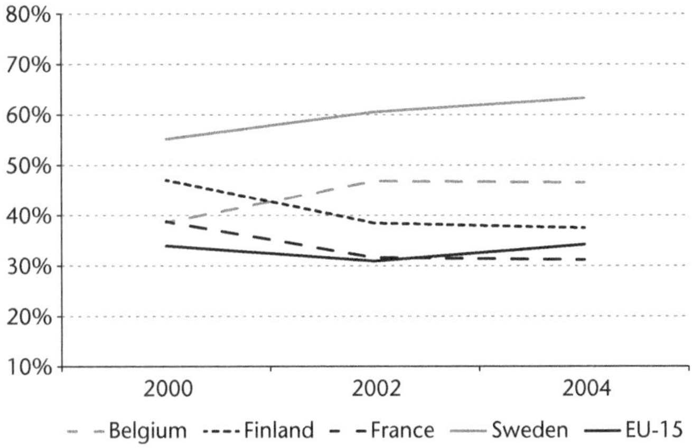

Figure 8.1. Fixed telephone service quality: Very good
Source: own elaboration on Eurobarometer data.

8.4.2 *Customer satisfaction: The electricity service*

The quality of the electricity service has been judged more often 'very good' in Austria, Denmark, Finland Ireland, Luxembourg, and Sweden (Table 8.2). On average quality standards are generally regarded similarly to price levels: overall 95.34% of EU-15 citizens consider the electricity service quality 'very' or 'fairly good' in 2000. This percentage does not significantly change in 2002

Table 8.2. Valid answers on electricity service quality

	Percentages year 2000				Percentage difference in very good response		
	Very bad	Fairly bad	Fairly good	Very good	2002–2000	2004–2002	2004–2000
Austria	0.00	1.86	29.72	68.42	−7.95***	−5.89***	−13.84***
Belgium	0.24	1.91	56.70	41.15	6.61***	−0.02	6.59***
Denmark	0.00	0.45	30.49	69.06	−1.02	2.64*	1.61
Finland	0.00	0.48	50.00	49.52	−7.21***	−3.02*	−10.23***
France	0.08	2.26	55.89	41.76	−6.52***	3.23*	−3.28*
Germany	0.90	5.83	59.79	33.48	0.32	8.02***	8.34***
Great Britain	0.63	2.40	43.85	53.12	−0.78	−7.26***	−8.05***
Greece	0.92	7.57	57.11	34.40	−6.06***	−6.66***	−12.71***
Ireland	0.00	1.34	26.85	71.81	−3.60**	−4.76**	−8.36***
Italy	0.77	6.89	72.09	20.25	−6.19***	−0.34	−6.53***
Luxembourg	0.00	0.00	38.89	61.11	−2.29	−5.88**	−8.17***
Netherlands	0.16	0.95	40.03	58.86	−17.21***	16.21***	−1.00
Portugal	0.24	7.18	79.43	13.16	−5.24***	1.45	−3.79***
Spain	0.96	3.78	74.59	20.66	5.03***	−4.41***	0.62
Sweden	0.00	2.49	37.40	60.11	4.71**	−4.44**	0.27
TOTAL	0.57	4.09	57.77	37.57	−2.44***	1.11**	−1.33***

Notes: *** $p<0.01$, ** $p<0.05$, * $p<0.1$. For details see Fiorio et al. (2007).
Source: own elaboration on Eurobarometer data.

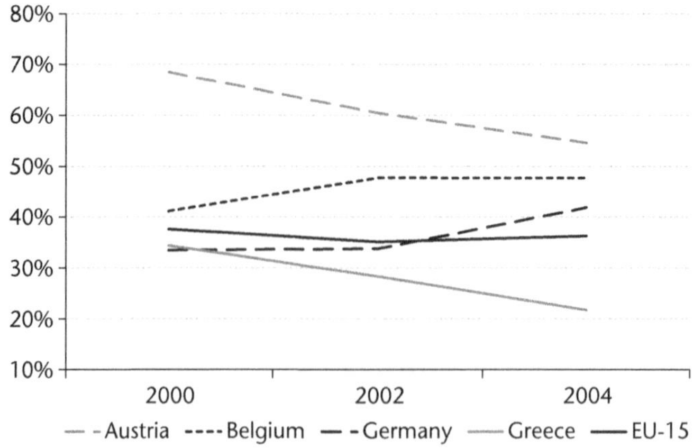

Figure 8.2. Electricity service quality: Very good
Source: own elaboration on Eurobarometer data.

and 2004. Germany and Belgium recorded the largest increases whereas Austria and Greece recorded the lowest (see Figure 8.2).

8.4.3 *Customer satisfaction: Gas supply services*

The analysis on gas supply service is deeply influenced by the large rates of no accessibility recorded in many countries, and in particular for Greece, Finland, and Sweden in 2000, see Chapter 4.

Service quality has been judged very good more often in countries such as Austria, Denmark, Ireland, and the Netherlands (Table 8.3). Overall, 94.32% of EU-15 citizens consider the gas supply service quality very or fairly good in 2000. This percentage does not significantly change in 2002 and 2004. With regard to single countries, most of the citizens do not significantly change opinion during the period. However, Greece and Sweden registered an increase of favourable opinions, whereas opinions in Ireland and Italy decreased during the period (see Figure 8.3).

From the previous analysis, we can observe that no significant variation in consumer opinion about quality of service has emerged for fixed telephone services, while a decreasing satisfaction is observed for electricity and gas supply services.

We put together these results in Figure 8.4 that shows the correlation between the percentage of consumers answering that the service quality is 'very good' and the ECTR/OECD score, in year 2004, for the three sectors. Visual inspection of Figure 8.4 suggests low correlation for each of the three services.

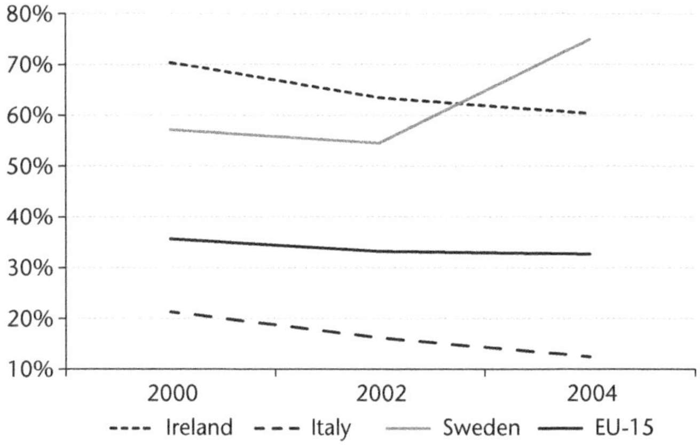

Figure 8.3. Gas supply service quality: Very good
Source: own elaboration on Eurobarometer data.

Table 8.3. Valid answers on gas supply service quality

	Percentages year 2000				Percentage differences in very good response		
	Very bad	Fairly bad	Fairly good	Very good	2002–2000	2004–2002	2004–2000
Austria	2.56	2.56	36.92	57.95	−5.97***	4.78**	−1.19
Belgium	0.85	2.56	55.27	41.31	7.65***	−0.64	7.01***
Denmark	0.00	0.00	31.33	68.67	−4.29**	3.01*	−1.28
Finland	2.38	2.38	69.05	26.19	−2.66	19.33***	16.67***
France	0.00	2.31	56.41	41.27	−4.48**	2.48	−2.00
Germany	1.13	7.93	59.21	31.73	−1.55	7.17***	5.62***
Great Britain	1.05	2.75	42.90	53.30	−1.42	−6.10***	−7.52***
Greece	7.69	0.00	38.46	53.85	−44.76***	65.91***	21.15***
Ireland	0.00	2.82	26.76	70.42	−6.93***	−3.15*	−10.08***
Italy	1.14	6.38	71.14	21.34	−5.09***	−3.83***	−8.93***
Luxembourg	0.00	0.00	46.15	53.85	6.15**	−10.00***	−3.85
Netherlands	0.16	0.65	39.35	59.84	−15.83***	16.07***	0.24
Portugal	0.30	9.97	76.44	13.29	−5.49***	−0.79	−6.28***
Spain	1.81	4.92	75.00	18.26	6.27***	−7.02***	−0.75
Sweden	0.00	4.76	38.10	57.14	−2.60	20.45***	17.86***
TOTAL	0.96	4.72	58.66	35.66	−2.44***	−0.56	−3.00***

Notes: *** $p<0.01$, ** $p<0.05$, * $p<0.1$.For details see Fiorio et al. (2007).
Source: own elaboration on Eurobarometer data.

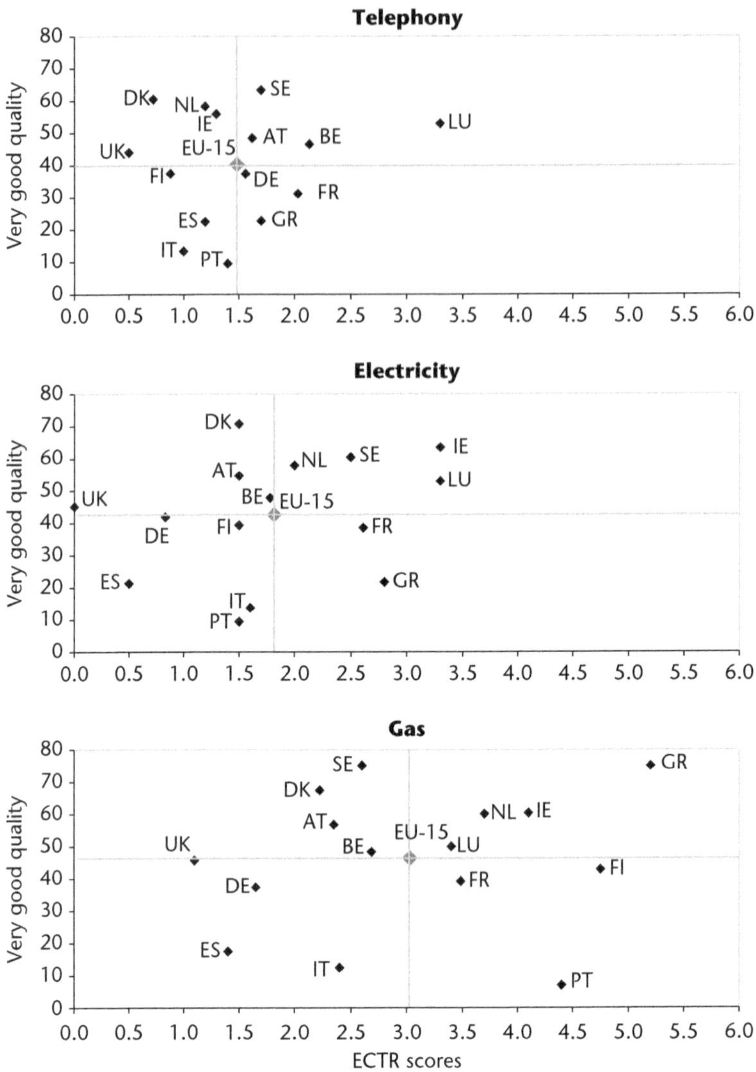

Figure 8.4. Percentage of respondents on service quality
Source: own elaboration of ECTR and Eurobarometer data.

8.5 A conditional analysis of consumer satisfaction

Although informative, the descriptive statistics presented in the previous section are unconditional to individual and country-specific characteristics of the sampled respondents to the Eurobarometer survey and do not allow one to see whether there is any pattern in satisfaction across groups of consumers and across countries. This section tries to shed some light on this issue. We now analyse consumer satisfaction with fixed telephone, gas, and electricity supply across the quality dimension, conditional on a set of information about each respondent and the country he/she lives in.

As satisfaction with different services is coded by ordinal variables, we use an ordered logit model for each of them.[9] By using this model we assume that the true level of satisfaction for each service and each dimension, S^*, is unknown and is determined by

$$S^* = \mathbf{x}\boldsymbol{\beta} + e, \quad e|\mathbf{x} \sim \Lambda(0, \pi/\sqrt{3}) \tag{1}$$

where $\boldsymbol{\beta}$ is $K \times 1$, $\mathbf{x}$ does not contain a constant, and $\Lambda(0, \pi/\sqrt{3})$ stands for the standard logistic distribution. Hence, we define our stated level of satisfaction S as:

$$\begin{array}{lll} S = 0 & \text{if} & S^* \leq \alpha_1 \\ S = 1 & \text{if} & \alpha_1 < S^* \leq \alpha_2 \\ & & \vdots \\ S = J & \text{if} & S^* > \alpha_J \end{array}$$

where $\alpha_1 < \alpha_2 < \ldots < \alpha_J$ are unknown cut points. As satisfaction on SGI quality takes four values (very bad, fairly bad, fairly good, very good), $J = 3$ for quality satisfaction.

As controls, $\mathbf{x}$, we used first: a set of individual characteristics (including sex, age, marital status, age when finished education, occupation, political views, contribution to household income, and household income, respondent's cooperation as assessed by interviewer); second: country fixed-effects, year dummies, and some country-level macroeconomic variables (GDP per capita, Gini index, population density, public procurement values, social benefits, subsidies to producers, total government expenditures and revenues); and finally the OECD regulatory indicators of market entry, public ownership, market structure, and vertical integration.

All information about individual characteristics is provided by Eurobarometer waves EB53 (for year 2000), EB58 (for year 2002), and EB61.2 (for year 2004). Although the structure of the questionnaire has remained substantially unchanged across these three years, in the 2004 issue there is no information about economic variables (respondent's contribution to household income and household income): when these variables are included in the model, the whole EB62.1 is left out of the analysis and comparisons of remaining coefficients across models are exploited for assessing the robustness of the results.

[9] In the previous chapter we used probit models. Here we use ordered logit, in order to capture satisfaction levels with the multidimensional quality indicators.

Macroeconomic variables are obtained by Eurostat. As usual, we use the OECD variable 'entry regulation', which is a weighted average of legal conditions of entry in a market and is coded from 0 (free entry) to 6 (franchised to one firm), available for telephone, gas, and electricity; the variable 'vertical integration', which is an indicator of network separation in different industries and is coded from 0 (ownership separation) to 6 (integration), available for electricity and gas; the variable 'market structure', which is an indicator of the market share of the incumbent and is coded from 0 (less than 50%) to 6 (more than 90%), available for telephone and gas; the variable 'public ownership', which measures the public ownership of each service and is coded from 0 (private ownership) to 6 (public ownership), available for telephone, gas, and electricity. In the present analysis we considered these variables lagged by one period for each year (i.e. for years 1999, 2001, and 2003).

All models were estimated by maximum likelihood using pooled datasets and a year fixed-effect dummy was introduced to capture any trend.

Table 8.4 presents the results, where coefficients reported are the coefficients of model (1): although they cannot be interpreted as marginal effects and their magnitude has no economic meaning, a positive sign shows that the J outcome is more likely and a negative sign shows that the outcome $S = 0$ ('very bad') is more likely.

Taking these results, and looking first at personal characteristics of the respondents, we notice that:

- There is in general no significant difference of satisfaction according to gender, age of the respondent, or education.
- Looking at occupation variables, with respondents holding self-employment as the reference category, unemployed people across SGI are consistently more dissatisfied, and students more satisfied than the self-employed. Managers are more likely to be very satisfied than others with the quality of telephone service, and again managers and other white collars are more likely to be very satisfied with quality of gas supply.
- As the Eurobarometer datasets also contain a question asking whether respondents have political views closer to the right or to the left, we also introduced this variable as control, finding that those who are closer to the right tend to complain the least.
- With respect to those giving excellent or fair collaboration to the interviewer, the lower the collaboration the more likely tends to be the individual dissatisfaction for different services.
- Finally, looking at household income, which is recorded only for the first two surveys, results show that the level of satisfaction is higher the income level of the respondent.

Overall, we interpret these controls as going in the expected direction. While the assessment of price is more influenced by individual characteristics (see

Table 8.4. Ordered logit analysis of consumer satisfactions about fixed telephone, electricity, and gas supply

	Fixed telephone		Electricity supply		Gas supply	
	A	B	A	B	A	B
Public ownership	−0.088**	0.073	0.133***	0.090***	0.048**	−0.283***
Entry regulation	0.014	−0.006	−0.032	−0.103***	0.413***	0.145
Market structure	0.078	0.012	0.098***	0.121***	0.050	−0.050
Vertical integration					−0.089*	0.466***
female	0.000	0.023	0.006	0.046	−0.005	0.019
age	–0.005	–0.011	–0.004	–0.008	0.002	0.000
age squared	0.000	0.000*	0.000	0.000	0.000	0.000
single	–0.087*	–0.084	–0.072*	–0.093*	–0.107*	–0.139**
separated/divorced/ widowed	–0.109**	–0.053	–0.053	–0.040	–0.061	–0.027
age when finished education	–0.008	–0.003	0.010	0.009	0.016*	0.046
age when finished education squared	0.000	0.000	0.000	0.000	–0.000**	–0.001
manager	0.060	0.137	0.068	0.179**	0.111	0.155
other white collar	0.035	0.004	–0.063	–0.034	–0.008	–0.017
manual worker	–0.051	–0.034	–0.032	–0.005	–0.032	–0.021
house person	0.060	0.110	0.018	0.101	0.104	0.163
unemployed	–0.140*	–0.100	–0.058	–0.048	–0.113	–0.097
retired	–0.018	0.011	0.038	0.092	0.111	0.158
student	–0.009	–0.025	0.227*	0.174	0.320*	0.590*
political views: right	0.099***	0.073	0.087**	0.063	0.074	0.049
respondent's cooperation: average/bad	–0.099**	–0.070	–0.112**	–0.075	–0.122**	–0.121*
resp. contrib. to hh income: least		–0.021		–0.074		–0.028
resp. contrib. to hh income: equal		–0.074		0.038		0.104
Household income: II quartile		0.122*		0.077		0.054
Household income: III quartile		0.098*		–0.002		–0.029
Household income: IV quartile		0.270***		0.161**		0.204**
Country fixed-effects	Yes	Yes	Yes	Yes	Yes	Yes
Year fixed-effects	Yes	Yes	Yes	Yes	Yes	Yes
Population density	0.012	–0.010	0.028*	0.023	–0.032**	–0.062**
GDP, per capita	–0.003	0.004	–0.031***	0.001	–0.007	–0.044**
GDP growth rate	0.084**	0.226***	0.050	0.243***	0.254***	0.535***
Employment growth rate	–0.015	–0.117***	–0.064**	–0.174***	–0.135***	–0.149***
Gini	–0.037*	–0.007	0.022	0.092***	0.020	0.013
Individuals	39,788	27,915	42,548	29,344	25,030	18,364

Notes: *** $p<0.01$, ** $p<0.05$, * $p<0.1$. For regulatory variables see Tables 4.A.1, 5.A.1, 6.A.1. For details see Fiorio et al. (2007).

Source: own elaboration on Eurobarometer data.

the previous chapter), particularly income and occupation, quality perception in contrast is less dependent on the socio-economic conditions of the respondent, but there is still some evidence of an influence of social status (and of political views) to satisfaction.

Country fixed-effect coefficients (see Fiorio et al., 2007 for the details) show that there is large variability in consumer satisfaction looking at different countries across SGI. Several positive coefficients for country fixed-effects are significantly different from zero. However, the main message is that the variability across countries is a lot more complex than it seemed using an unconditional analysis as in the previous section. The UK, for example, seemed to score among the best without conditioning for individual and country characteristics. This country is a benchmark in terms of privatization, liberalization, and vertical disintegration policies. Taking this country as the reference point, and conditioning the analysis for the sample structure, British consumers are less satisfied in the quality of telephone and electricity than an unconditional analysis would show.

Looking at macroeconomic controls, results show that the higher the population density and the rate of growth of GDP, the more likely consumer satisfaction seems to be, although GDP levels, employment growth rate, and Gini index coefficients do not provide the same message for all services.

We turn now to our main research question: do privatization and regulatory reforms influence the perception of service quality?

As for telephone services, quality seems to be uncorrelated with any of the reform items, except public ownership, an effect that vanishes, however, when we consider the income variables. As for gas supply, when the income of respondents is included, entry regulation and vertical disintegration have opposite signs, while public ownership plays no role. Vertical integration seems to be correlated to decreasing quality satisfaction.

Finally, with respect to electricity, the results are more robust: the larger the public ownership, the more likely is consumer satisfaction; the more regulated the entry, the higher is satisfaction with quality, while the vertical disintegration coefficient has the opposite sign. These findings, taken together, do not support the view that the reforms are associated with higher consumer satisfaction with service quality.

8.6 An individual fixed-effects approach

A possible objection to the previous analysis is that using subjective data is based on the assumption that responses to surveys are acceptable proxies of the unobserved true level of satisfaction of respondents. Experimental psychologists have shown that this assumption may be false, as the responses tend to be biased for a number of reasons. From an econometric perspective this issue was raised by Bertrand and Mullainathan (2001), who frame the discussion in terms of measurement error, that is the difference between self-reported satisfaction and unobserved true satisfaction. In this section we report the findings in the UPP paper by Grassi and Puglisi (2008), who approach this problem by taking advantage of the fact that in surveys such as Eurobarometer there are different questions about different services, asked to different individuals at different times. They stack the observations about reported satisfaction over these different dimensions and run a regression with individual-specific fixed effects.

Grassi and Puglisi (2008) assume that many of the biases affecting the respondent are specific to the individual but are common across similar questions being asked, and then can be captured by an individual fixed-effect. They suggest a panel-like estimation, which can be implemented if the same individual i at time t is asked about their satisfaction regarding more than one issue (in their case study individuals are asked to report their level of satisfaction from consuming public utilities services). Since the Eurobarometer is a repeated cross-section, they focus on a situation where individual i at time t is asked to report their level of satisfaction of different services $s \in S$, for $s \in \{1, \ldots S\}$, but is not interviewed again at a different time. See Appendix 8.A.2 for details.

The analysis again uses Eurobarometer 2000–2, 2002–4 and the ECTR/OECD regulatory indicators.

We report here only some of the results by Grassi and Puglisi (2008) when specific regulatory variables are included, as they are easier to compare with the previous empirical analysis. The services considered in some of their models include again electricity, gas, and telephony. Post and railways are also considered to enlarge the dimensions of the panel.

As Table 8.5 shows, there are some interesting results in this approach that allow us to distinguish the impact of reforms, if any, on perceptions of quality of the five services considered. The analysis is based both on an ordered and a binary version of reported satisfaction.

The results can be summarized as follows:

a) there is no impact of entry on satisfaction for the two energy sectors (quality perception is associated with restrictions to entry in the binary model); the only service where pro-entry legislation is more clearly correlated to satisfaction is telephony;

b) public ownership shows no association with perceived quality in energy and is associated with a higher satisfaction in telephony;

c) market dominance by the incumbent is associated with higher perception of quality in telephony;

d) vertical integration increases perception of quality for gas in one model.

Overall, this analysis, which tries to correct possible estimation biases in the previous more traditional approach, rejects the view that higher user

Table 8.5. Consumer satisfaction on quality and regulatory variables: Sector-specific effects

Dependent variable	Ordered satisfaction variable			Binary satisfaction variable		
	EGTPR	GTR	EGR	EGTPR	GTR	EGR
Entry regulation						
Quality in electricity	0.005	–	0.011	0.014	–	0.020*
	[0.32]		[0.53]	[1.62]		[1.69]
Quality in gas	−0.007	−0.007	−0.021	0.015**	0.022***	0.013*
	[0.55]	[0.44]	[1.45]	[2.36]	[2.81]	[1.67]
Quality in telephony	−0.052***	−0.098***	–	−0.013	−0.041**	–
	[2.61]	[3.68]	–	[0.89]	[2.04]	
Public ownership						
Quality in electricity	0.012	–	0.009	0.002	–	−0.001
	[1.01]		[0.53]	[0.35]		[0.17]
Quality in gas	0.016*	−0.003	0.01	−0.003	−0.009	−0.006
	[1.73]	[0.21]	[1.01]	[0.63]	[1.43]	[0.96]
Quality in telephony	0.036***	0.035***	–	0.016**	0.018**	–
	[4.88]	[3.61]		[2.60]	[2.14]	
Market structure						
Quality in gas	–	0.027	–	–	−0.007	–
		[1.50]			[0.63]	
Quality in telephony	–	0.032*	–	–	0.027**	–
		[1.85]			[2.13]	
Vertical integration						
Quality in electricity	–	–	0.016	–	–	0.009
			[0.96]			[0.77]
Quality in gas	–	–	0.060***	–	–	0.007
			[3.07]			[0.57]
Individual-specific fixed-effects	yes	yes	yes	yes	yes	yes
Sector-specific fixed-effects	yes	yes	yes	yes	yes	yes
Sector-specific quality dummy	yes	yes	yes	yes	yes	yes
Individuals	46,829	45,146	45,611	46,829	45,146	45,611
Clusters	217	129	130	217	129	130
Observations	419,743	245,508	251,125	419,743	245,508	251,125

For regulatory variables see Tables 4.A.1, 5.A.1, 6.A.1. E: Electricity; G: Gas; T: Telephony; P: Post; R: Railways. Notes: The first group of columns displays findings with the ordered measure of satisfaction as dependent variable, while in the second subgroup the dependent variable is a dummy which equals one when the individual declares to be 'very satisfied'. Regressions are in both cases OLS, with individual-specific and sector-specific fixed-effects, plus a sector-specific dummy for responses about quality. The partial correlations with the regulatory reform measures are allowed to vary across sectors, and—within sectors— between quality and price evaluations. Standard errors are clustered at the (country x year x SGI) level. Robust t statistics in brackets.

*** $p<0.01$, ** $p<0.05$, * $p<0.1$.

Source: adapted from Grassi and Puglisi (2008).

satisfaction for quality is correlated with the typical ingredients of the reform paradigm in the two energy sectors.

8.7 Other empirical approaches

In this section we briefly report other results by the UPP team, which complement the analysis of user satisfaction. Ferrari, Annoni, and Manzi (2010a; 2010b) focus on the 2002 Eurobarometer survey and consider four services in the EU-15: telephone, electricity, railways, and the post. They build a combined satisfaction indicator based on a non-linear principal component analysis, and test for robustness through a Monte Carlo approach. Differently from the previous analyses they combine together the responses on price, quality, information, and contract fairness (see Table 8.6 below). Their findings are as follows: countries where citizens are least satisfied are Italy, Portugal, Spain, and Greece, followed by Germany and to a certain extent France. Countries where citizens are most satisfied are Denmark, Ireland, Luxembourg, and the UK.

Table 8.6. Items on SGI in the EBS8 survey questionnaire

Questions	Criteria	Category codes
In general, would you say that access to XXXX service is easy or difficult for you?	Access	1 Easy access 2 Difficult access 3 Excessive (spontaneous) 4 Don't know
In general, would you say that price you pay for the XXXX service you use is fair or unfair?	Price	1 Fair 2 Unfair 3 No access (spontaneous) 4 Don't know 5 Not applicable
Overall, what do you think of the quality of XXXX services that you use? Would you say it is very good, fairly good, fairly bad, or very bad?	Quality	1 Very good 2 Fairly good 3 Fairly bad 4 Very bad 5 Don't know 6 Not applicable
In general, would you say the information (bills, contracts, advertising, tickets, leaflets, etc.) you receive from XXXX service provider is clear or unclear?	Information	1 Clear 2 Unclear 3 Don't know 4 Not applicable
In general, would you say the terms and conditions of your contract with your XXXX service provider are fair or unfair?	Contract type	1 Fair 2 Unfair 3 Don't know 4 Not applicable

Source: Eurobarometer and Ferrari et al. (2009).

The findings of this analysis, which is a different way to aggregate consumer perceptions, do not suggest any relation between the country pattern as emerges for the data, and the scoring of reforms in these countries. In fact, the aggregate OECD regulatory score in 2002 would place Luxembourg and Ireland among the least 'reformed' countries, while indeed the UK is the network industry's reform champion and Denmark has a similar score. Thus two reformed and two relatively unreformed countries perform well. On the bottom of the satisfaction scale, while Greece and Portugal, for example, are less reformed, Germany is similar to Denmark, but gets a much lower satisfaction score than expected. France performs better than Germany, and is much less reformed.

We will briefly mention other approaches that broadly confirm the previous results. Ferrari and Salini (2011) compare non-linear PCA (NLPCA) with the Rash Model (RM). They consider three Eurobarometer waves 2000, 2002, 2004, EU-15, and satisfaction for accessibility, price, and quality of fixed telephone, electricity, gas, and water supply (not considered in the other models that we have mentioned above). They intend a study that evidences which items (services or aspects of service) are more important for the consumer as indicated by the component loadings (NLPCA) and which of these are perceived by the consumer to be of higher or lower quality as indicated by item parameters (RM).

A joint reading of the findings of RM and NLPCA is derivable from a scatter plot (Ferrari and Salini, 2011—Figure 3) where each item is represented by its component loading (horizontal axis) and item parameter (vertical axis). This type of representation might be a helpful tool for decision-makers who want to increase the level of satisfaction of users. In particular they identify items with low quality and high weight, for example quality of electricity supply, for which an action could be welcome, since it would lead to a substantial increase in the general level of satisfaction.

They again find that countries with the highest reported satisfaction are Ireland and Denmark, followed by Luxembourg, Netherlands, and the UK, but also Sweden and Austria. The lowest performers are, as usual, Greece, Spain, Portugal, and Italy.

A different approach is tried by Salini and Kennett (2009), who use Bayesian networks, a technique that is based on probabilistic graphs in order show stochastic dependency between variables (represented as nodes, with arcs as conditional dependency). They considered the same data mentioned before (Ferrari and Salini, 2011). Taking advantage of 47,000 respondents, 12 dimensions (4 sectors x 3 aspects of satisfaction), and 15 countries, for three waves, they discover a complex pattern of probabilistic interrelationships among the variables. For example good quality and easy access positively influence the opinion about price, as should be expected, given our discussion at the beginning of the chapter. Moreover, attitudes are also interrelated across sectors: for example when prices are considered unfair in other sectors, the satisfaction will decrease for the remaining sector. Overall this Bayesian Network analysis offers some support to the view that respondent's answers show a certain logical structure. This approach also offers some support to the results by Grassi and Puglisi (2008).

8.8 Objective quality vs. subjective quality: The case of electricity supply

An interesting analysis could involve the study of the link between the objective quality of services and perceived satisfaction. Unfortunately there are not enough data (particularly long time series) to answer this question by testing causality.

We, however, study correlations using the CEER data, already mentioned in Section 8.2.2, in particular the minutes lost per year for both unplanned and planned interruptions of electricity supply. Table 8.7 reports the data available from the 4th Benchmarking Report on Quality of Electricity Supply (2008).

Considering all years and all countries, the Pearson Correlation Index between the percentage of 'very good' responses and the minutes lost by year for unplanned and planned interruptions are −0.580 and −0.625, both significant at the 0.01 level. As expected, the percentage of users who are 'very satisfied' is lower when the number of minutes lost due to interruptions is higher.

Figure 8.5 shows the negative correlations and represents the path of different countries by year. It is not possible to show clear increasing and decreasing trends, which would suggest a cause–effect link between interruptions and satisfaction. It would take more years to be able to apply more complex models

Table 8.7. Minutes lost per year of unplanned (MUI) and planned interruptions (MPI) of electricity supply and percentage of 'very good' responses for electricity service quality (year 2000, 2002, 2004)

Country	MUI			MPI			Very good		
	2000	2002	2004	2000	2002	2004	2000	2002	2004
Austria		83.08	30.33		7.04	20.07	0.63	0.57	0.58
Finland	115.60	139.80	104.70	38.00	32.00		0.37	0.28	0.50
France	40.00	52.00	60.30	6.00		6.60	0.43	0.41	0.41
Great Britain		101.33	87.33		9.04	6.95	0.54	0.52	0.47
Italy	187.40	114.74	0.53	82.62	77.97	62.62	0.21	0.16	0.13
Netherlands	27.00	28.00	24.00				0.58	0.43	0.59
Portugal		467.98	217.79		52.21	49.16	0.13	0.06	0.10
Spain	145.41	142.56	123.60	37.05	30.66	21.60	0.21	0.25	0.20
Sweden	89.17	101.84	78.08	34.53	37.12	24.83	0.67	0.50	0.66

Source: own elaboration on CEER (2008) and Eurobarometer data.

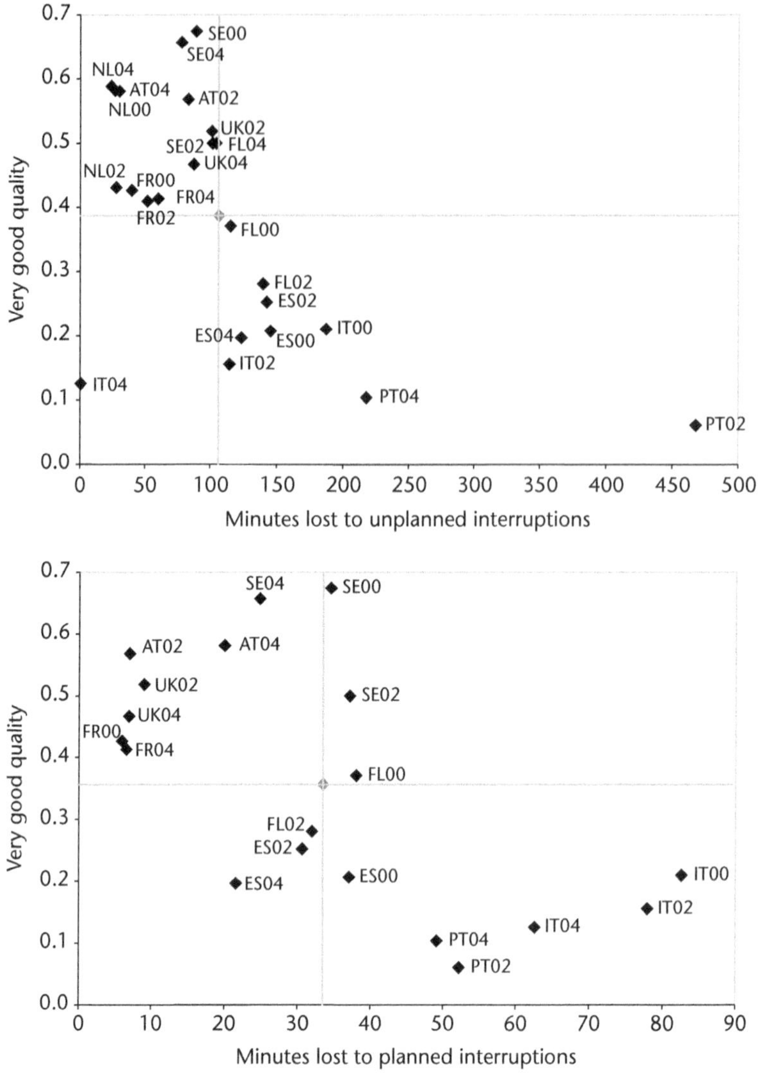

Figure 8.5. Correlation between interruption of electricity supply and quality perceptions

Source: own elaboration Eurobarometer and CEER (2008).

that also take into consideration the effect of time present and delayed. The negative correlation is, however, consistent with the view that in countries where there are more interruptions, consumers are less satisfied, particularly for unplanned interruptions.

We also examined whether the quality of service is correlated with the level of privatization (ECTR score). We consider, again from the 4th Benchmarking Report on Quality of Electricity Supply 2008, both the number and minutes of

Table 8.8. Pearson Correlation Index between ECTR scores and number and minutes of unplanned and planned interruptions of electricity supply

Pearson Correlation	Unplanned Interruptions		Planned Interruptions	
	Number[1]	Minutes[2]	Number[3]	Minutes[4]
Entry	0.016	0.155	−0.053	0.189
Public ownership	−0.301**	−0.090	−0.137	−0.055
Vertical integration	−0.023	0.015	−0.051	0.088

Source: own elaboration on CEER 2008 data, various years. For regulatory variables see Tables 4.A.1, 5.A.1, 6.A.1.
[1] Number of sample units 96.
[2] Number of sample units 98.
[3] Number of sample units 79.
[4] Number of sample units 85.
** Correlation is significant at the 0.01 level.

unplanned and planned interruptions for various year (from 1999 to 2007). Table 8.8 shows the Pearson Correlation Index between different form of interruptions and the main ECTR regulatory variables.

It is interesting to notice the negative correlation between public ownership and the number and minutes of interruptions, significant at 0.01 for the number of unplanned interruptions. The result allows us to formulate the conjecture that the objective quality improves, that is the number of interruptions decreases, when public ownership is higher. This would be consistent with the results that satisfaction increases if the public ownership increases, as suggested by Table 8.4.

Correlation analysis was also done country by country. In most cases the correlations are not significant. In some countries, we notice significant negative correlations that confirm the previous result: in Austria between vertical integration and planned interruptions, in France the three ECTR variables are negatively correlated with the planned interruptions. In other countries, Italy and Spain in particular, significant positive correlations are found instead.

8.9 Summing up

After a discussion of concepts, measurements methods, and regulation, this chapter has presented evidence of consumer satisfaction on the quality of the service provided by some network industries in the EU-15 Member States. Although an unconditional analysis across countries points towards some countries, notably Italy, Greece, and Portugal showing a more significant extent of dissatisfaction, once one controls for individual and country effects the picture is significantly more complex. While some individual characteristics in the samples and some macroeconomic controls may contribute to explain the degree of satisfaction in the Eurobarometer surveys 2002 to

2004, in these concluding remarks we focus on the correlation with regulatory variables.

a) As for privatization, the ECTR variable we use is the share of public ownership in these industries. This variable in most models is uncorrelated to satisfaction with quality, with some findings suggesting a positive correlation between public ownership and perceived quality of electricity supply.
b) One obvious measure of liberalization is the market share of the incumbent, and one would expect that the smaller such a share, the more competitive the market, the lower the price for a given quality, and the higher the access, hence the higher consumer satisfaction. This expectation in general is not supported by data.
c) A second liberalization variable is 'free entry': this variable does not work as expected, particularly for perceived electricity and gas service, while there is some evidence in the opposite direction for the telephone service (reversed with individual fixed-effects).
d) Vertical integration has a negative impact if any for gas service satisfaction, but a positive one for electricity services (which disappears with individual fixed-effects).
e) For electricity supply, the objective quality, that is the number and minutes of interruptions are negatively correlated with public ownership. Moreover, the percentage of users who are very satisfied is lower when the number of minutes lost due to interruptions is higher. Public ownership and satisfaction are positively correlated.

Taken together, these results show that the economic, institutional, and social environment that shapes attitudes towards services of general interest is a complex one, and the interpretation should be very prudent.

In fact, our empirical analysis suggests that any expectation of a simple EU-wide positive correlation between consumer satisfaction about service quality and the extent of privatization–liberalization–vertical integration reforms—to say the least—is not supported by the available evidence. To put it bluntly, reforms are uncorrelated to perceived quality, and particularly for electricity the evidence runs in the opposite direction.

8.10 Further reading

A non-exhaustive list of books dealing with customer satisfaction data analysis includes Salini and Kenett (2011), Grigoroudis and Siskos (2010), Allen and Rao (2000), Johnson and Gustafsson (2000), and Vavra (1997).

The major part of them cover topics like survey questionnaire design, data collection approaches, validation of questionnaires, sampling problems, descriptive statistics, and classical statistical inference techniques. Salini and Kenett (2011) present such topics, but also provide a large range of non-standard techniques for customer satisfaction data analysis. It addresses modern topics such as web surveys and state-of-the-art statistical models such as the CUB model, Fuzzy methods, and Bayesian Networks. The book chapters, written by leading researchers in the field, use practical examples in order to make their content accessible to non-statisticians too. Moreover, various techniques are compared by application to a common benchmark dataset.

Grigoroudis and Siskos (2010) describe service quality models and the Multicriteria Satisfaction Analysis (MUSA), with examples of satisfaction barometers. The book by Allen and Rao (2000) is the most comprehensive in terms of statistical methods. It provides a useful and well-written description of techniques of descriptive analysis of univariate, bivariate, and multivariate data; it also describes dependent models (linear and logistic regression), explanatory techniques (factor analysis, principal component analysis), causal models (path analysis), and structural equation models. Appendix C of Johnson and Gustafsson (2000) presents an interesting comparison of alternative data analysis methods, in particular considering (1) gap analysis, (2) multiple regression, (3) correlation, (4) principal component regression, and (5) partial least squares (PLS). Vavra (1997) covers theories of customer satisfaction and loyalty with several examples of scales, analytic procedures, and best practices.

Besides these books there is a large technical literature on the measurement of service quality through customer satisfaction surveys.

On the non-linear principal component see Ferrari et al. (2010a; 2010b). On the application of Bayesian networks to the customer satisfaction context, see Kenett and Salini (2009), Salini and Kenett (2009) and Renzi et al. (2009). On the application of Rasch see Nicolini and Salini (2006), Bacci (2008), De Battisti et al. (2010), Bartolucci et al. (2011). On the CUB model, see Iannario and Piccolo (2010). On Tree-Based Methods, see Nicolini and Salini (2006) and Tutore et al. (2007). On service quality using conjoint analysis see Barone and Lombardo (2004). For comparison of models and integrated analysis, see Kenett and Salini (2011).

We cite below just some examples of more applied and policy-oriented literature. Baker and Tremolet (2000), in a NERA report, deal with regulation of quality in developing economies, and suggest that the poor should be offered lower quality for a lower price. We mention this issue in the next chapter.

Sappington (2003) reviews and discusses the evidence of the impact of incentive regulation on retail telephone service quality in the USA, as compared with rate-of-return regulation. He does not find clear evidence of a better performance. Currier (2007) suggests a practical approach to quality-adjusted price-cap regulation, in order to solve the price quality trade-off faced by regulators. Sutherland (2007) reviews the regulation of quality for mobile telephony. Clements (2001) offers some empirical evidence of local telephone QoS in the USA.

Appendix 8.A.1: Quality regulation of gas service in the UK

Table 8.A.1. Natural gas guaranteed standards of performance/penalty payments UK

Regulation	Guaranteed standard	Compensation if not met
Regulation 7 Supply restoration	GT* must restore customer's gas supply within 24 hours following an unplanned interruption. Further compensation must be paid for each additional period of 24 hours until supply is restored, subject to a cap. If the interruption is caused by another GT, the other GT is either required to make the payment to the GT to whose network the customer is connected or to the customer directly.	£30 (domestic) £50 (small non-domestic). Cap of £1000
Regulation 8 Reinstatement of customer's premises	GT must reinstate customer's premises within 5 working days. Further compensation must be paid for each additional period of 5 working days.	£50 (domestic) £100 (non-domestic).
Regulation 9 Priority domestic customers	GT must provide alternative cooking and heating facilities to priority domestic customers when supply to the customers' premises is discontinued. GT must provide these facilities within 4 hours for planned and unplanned interruptions affecting less than 250 consumers, or within 8 hours for an unplanned interruption affecting 250 or more customers.	£24 (domestic) if claimed by the customer within 3 months
Regulation 10 Connections: Provision of quotation	GTs must provide a quotation for providing a new or altering an existing connection within: 6 working days for standard connections 11 working days for non-standard ≤275KWh connections 21 working days for non-standard >275KWh connections Further compensation must be paid for each additional day that failure continues subject to cap. If the quotation is inaccurate it is treated as if it was not provided on time.	£10, or £20 for non-standard >275 KWh connections. Cap is lesser of £250 (or £500 for non-standard >275 KWh connections) or contract sum

Regulation 10 Connections: Accuracy of quotations	Where a customer challenges a quotation under the GT's published accuracy scheme and the quotation is found to be inaccurate the GT shall refund the amount of any overcharge.	Amount of any overcharge
Regulation 10 Connections: Response to land enquiries	GT must respond to land enquiries in respect of a new connection or alteration of an existing connection within 5 working days. Further compensation must be paid for each additional day that failure continues subject to a cap.	£40 Cap is £250 (or £500 for >275 KWh customers)
Regulation 10 Connections: Timing of work	GT must offer a date for commencement of the work and substantial completion within 20 working days of the customer accepting the quotation. Further compensation must be paid for each additional day that failure continues subject to a cap.	£20 (≤275 KWh) £40 (>275 KWh) Cap is lesser of £250 (or £500 for >275 kWh customers) or contract sum
Regulation 10 Connections: Completing the work	GT must substantially complete a connection on the date agreed with the customer. Further compensation must be paid for each additional day that failure continues subject to a cap.	Initial payment between £20–£150 (depending on quotation amount). Cap varies depending on quotation amount, up to £9000 for quotes between £50k–£100k
Regulation 12 Payments	GTs must make payment required under the guaranteed standards to the customer within 20 working days from when the payment became due. Payments to other GTs under Reg. 7 (Supply restoration) must be made within 10 working days of receiving notification of the interruption	£20
Regulation 10A Notification of planned interruption	GTs must notify consumers at least 5 working days in advance of a planned supply interruption	£20 (domestic) £50 (non-domestic) if claimed by the customer
Regulation 10B Responding to complaints	GTs must respond to a complaint within 10 or 20 working days depending on whether a site visit or making enquiries of third parties if required. Further compensation must be paid for each additional period of 5 working days until the substantive response is provided, subject to a cap.	£20 (domestic and non-domestic). Cap of £100

[*] Gas Transporter.
Source: Ofgem (2007), Gas Distribution Price Control Review Final Proposals. 3 December 2007, Ref 285/7, pp. 54/55.

Appendix 8.A.2

According to Grassi and Puglisi (2008), let the reported satisfaction for service s of individual i at time t be

$$A_{it}^{s} = A_{it}^{s^*} + \alpha_{it} + \varepsilon_{it}^{s} \quad s \in \{1, \dots S\} \tag{8.A.1}$$

where α_{it} is an individual-time specific component which is constant across SGIs. The true satisfaction depends on two vectors of observable variables: X_{it} and W_{ti}^{s}. The vector of observable variables X_{it} also affects the satisfaction derived from the consumption of other services, but they allow the coefficients of the variables contained in it to be specific to the public service s. On the other hand, the variables in W_{ti}^{s} only affect the satisfaction derived from the service s, that is they do not appear in the equation for the satisfaction of individual i regarding service $\tilde{s} \neq s$.

Hence the equation for the true satisfaction can be written as follows.

$$A_{it}^{s^*} = \beta^s X_{it} + \gamma^s W_{it}^{s} + \eta_{it}^{s} \tag{8.A.2}$$

Combining equations (8.A.1) and (8.A.2), the following holds:

$$A_{it}^{s^*} = \beta^s X_{it} + \gamma^s W_{it}^{s} + \alpha_{it} + \eta_{it}^{s} + \varepsilon_{it}^{s} \tag{8.A.3}$$

Stacking the observations about the reported satisfaction A_{it}^{s} across different public services s, it is possible to construct a panel-like model, and estimate the individual-time specific effect α_{it} across these different public services. To the extent that α_{it} is correlated with the explanatory variables, explicitly controlling for α_{it} would help to circumvent the measurement error problem, or at least the portion of it that depends on α_{it}.

Depending on data availability, it is also possible to control for a SGI-specific fixed effect μ^s. The estimated model would look as follows:

$$A_{it}^{s^*} = \beta^s X_{it} + \gamma^s W_{it}^{s} + \alpha_{it} + \mu^s + \eta_{it}^{s} + \varepsilon_{it}^{s} \tag{8.A.4}$$

This specification allows us to account for response biases that are specific to the service, but are common across individuals. Moreover, if individuals are asked to report their level of satisfaction about feature f of each public utility s (in their case study, price and quality for each public utility), one can estimate the following model:

$$A_{it}^{s^*} = \beta^s X_{it} + \gamma^s W_{it}^{s} + \alpha_{it} + \mu^s + \zeta^f + \eta_{it}^{s} + \varepsilon_{it}^{s} \tag{8.A.5}$$

where ζ^f is specific to a characteristic (price or quality) of the different SGIs.

9

Affordability of network services, social and territorial cohesion

9.1 Introduction[1]

A correlation between utility reforms and poverty may arise because of four trends, well identified by the literature, see for example Estache et al. (2002).[2]

First, privatization may lead to higher average prices for households, either because of the need to cover the costs of previously subsidized public enterprises or because of the higher profit margins of the privatized provider and of the entrants. Second, regulation may impose higher quality standards, which in turn imply higher service costs passed to consumers. Third, cross-subsidies among types of users are decreased by tariff rebalancing. Eventually, connection rates may increase, as metering is more adequate, illegal connections are no longer ignored by providers, and so on. The burden for the poor resulting from these changes may be considerable.

While richer on average than other regions, the European Union is far from having removed poverty. Actions against poverty and social exclusion are in fact firmly on the European agenda. The Commission Recommendation (2008/867/EC), and several EU Parliament resolutions advocate common principles of active inclusion policy, which focus on local and regional circumstances and emphasize the goal of ensuring access to high quality services for all European citizens. Figure 9.1 shows the share of the EU population at risk of poverty or social exclusion across the 27 Member States in 2009, according to three criteria (relative poverty risk; or severe material deprivation; or low work intensity).

[1] The Appendix of this chapter is from Lidia Ceriani. Section 9.7 draws on Poggi and Florio (2010). Carlo Fiorio has contributed to the analysis of price data.

[2] Indirect connections are mainly linked to the effect on substitute goods and to public sector budgetary constraints. The distributive impacts of such indirect effects are complex, see e.g. Chisari, Estache, and Romero (1999).

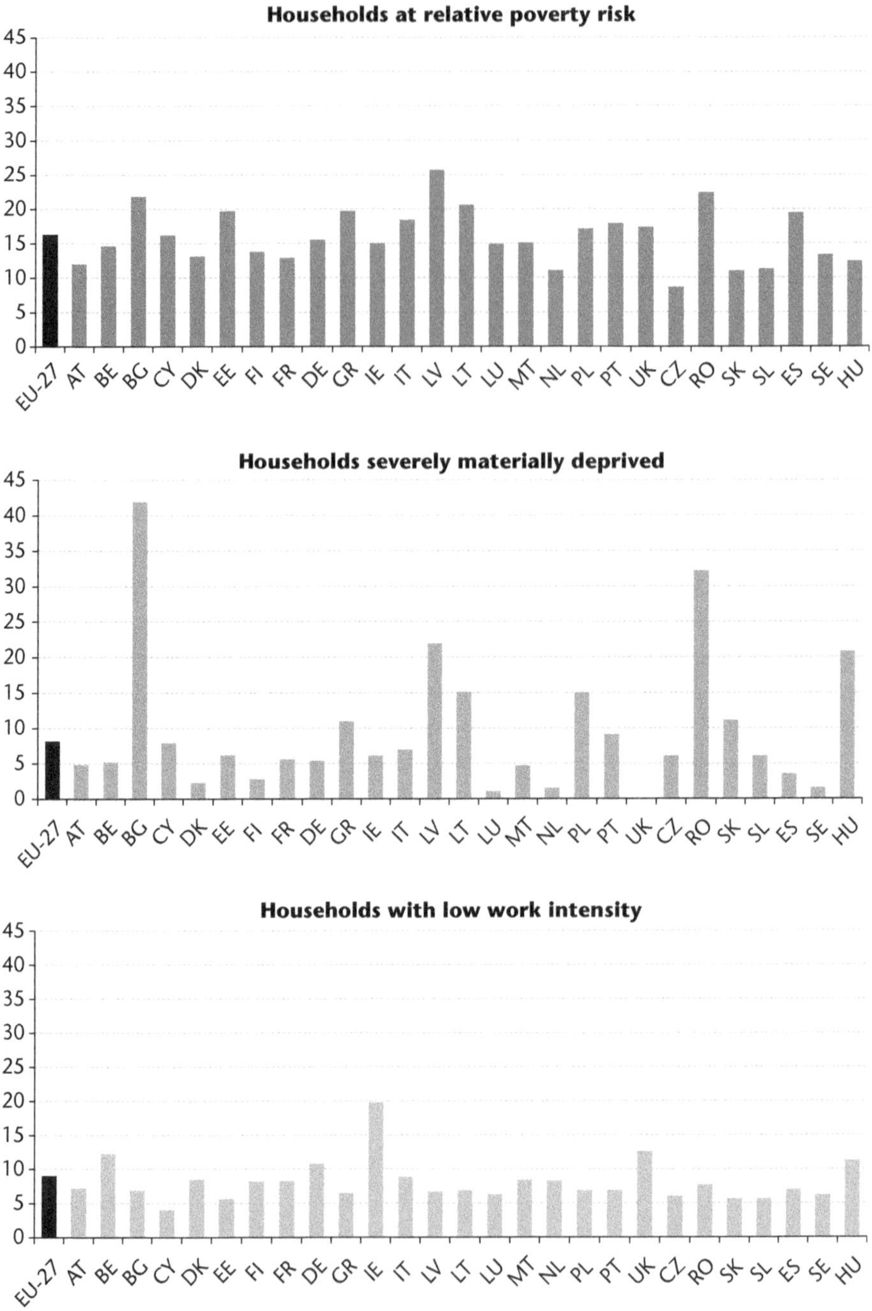

Figure 9.1. Share of population (%) living in households that are severely materially deprived, at risk of poverty, or with low work intensity, 2009

Source: Eurostat, EU-SILC (2009).

According to these combined criteria, around one in four EU residents are at risk of poverty, with, however, significant variations across countries, from 15% in the Netherlands, nearly 30% in the UK, and over 40% in Romania. There are also substantial variations between regions within countries. For example, in Italy, only 6% of the population is at risk of relative poverty in the northern province of Trento, while the share is over 35% in the southern region of Calabria.

Looking at relative poverty, in 2008, around 17% of the EU population had an income after transfers below 60% of equivalized median income in their country—a commonly used indicator of poverty risk (EC, 2010). In the context of the wide disparities in per-capita income across Member States, an absolute indicator of poverty offers perhaps a clearer view of the social conditions. Material deprivation of a common set of goods allows for direct comparisons. Conventionally (EC, 2010, p. 107) a household is considered to be in absolute poverty if it is unable to secure at least three out of the following nine items: ability to face unexpected expenses; ability to pay for one week of annual holiday away from home; *existence of arrears on bills* (mortgage or rent payments, utility bills, or hire purchase instalments or other loan payments); capacity to have an adequate meal with meat, chicken, or fish every second day; *capacity to keep the home adequately warm*; ability to afford a *washing machine, telephone, TV set*, or car.[3] As it happens, the absolute poverty data, based on EU-SILC—a wide household survey—give exactly the same aggregate result of the relative poverty indicator: 17% of Europeans suffered from absolute poverty in 2008. This amounts to around 86 million citizens, and the EU has set a target to bring at least 20 million of them out of poverty by the year 2020, as part of the Europe 2020 strategy. Around 8% of Europeans were severely deprived (on four out of the nine indicators) in 2009.

Several items from the absolute poverty indicator are directly or indirectly related to network services. These include particularly the ability to pay utility bills, to keep the home warm, and to have a telephone. In fact, the services we have considered in the previous chapters—electricity, gas, and fixed telephony—are key from this perspective, with water and district heating (heating different buildings from the same plant) (in some countries) also needing to be considered.

Thus, in this chapter we focus on the way that reforms of network industries may affect social cohesion. By the latter term we understand a broad concept of equity, which includes different aspects that will be discussed below.

Our main concern is the affordability of tariffs for vulnerable citizens. The traditional government policy for public utilities has been often (but not

[3] Our italics. In this chapter we are going to add water and public transport to energy and telecoms, to gain a broader picture of affordability issues.

always and everywhere) to keep tariffs low enough for some disadvantaged users. In many circumstances the poor are also low users of the service, and their connection to the network may be unprofitable for the service provider. Even when the access problem for the poor has been resolved, their share of the expenditure on utilities relative to total expenditure or income can be too high to ensure a minimum standard of consumption. The solution to preserve the balanced budget of the provider, and to offer affordable prices to everybody, has traditionally been a cross-subsidy among users under public ownership or tariff regulation.[4] This has increasingly been seen as inefficient and cost-reflective prices have been advocated as a part of the reform package. Tariff rebalancing, however, has created a tension between efficiency and equity objectives, and has required some remedial actions, for example 'lifeline' tariffs.

An increase in tariffs for the poor can create an undesirable combination of high payment arrears, reduction of consumption, and shifts to lower priced but also to less environmentally sustainable substitutes (e.g. coal for heating the home). These changes are not well captured by statistical aggregate data, and it is important to know who the vulnerable users are in order to achieve better designed reforms. A second point raised by the introduction of market opening, is the possibility that information and transaction costs are unequally shared by participants in the markets. For example non-switchers and mistaken-switchers between competing providers may be particularly concentrated among the less educated, or the elderly.

An additional concern for social cohesion is with the territorial dimension of access, which may be related to the spatial distribution of the welfare level of the users. Typically there is an urban–rural divide in the profitability of supplying the services, but also between regions in different economic conditions. As we shall see, the equity issues posed by network industries are multifaceted and this chapter reviews the evidence on the impact of the reforms on different social dimensions. It also enlarges the discussion to include some other utilities beyond those considered in the previous chapter, as is frequently done by the empirical literature on this topic.

The chapter is structured as follows. Section 9.2 presents a simple conceptual frame for the analysis of equity issues related to the provision of network industries. Section 9.3 discusses some of the earlier empirical research on utility affordability. Section 9.4 looks at data on price structure broken down by type of user, in order to show whether the reforms have been neutral from the point of view of welfare distribution across users. Section 9.5 mentions

[4] The regulator can impose cross-subsidies to private providers as well as in the UK. I am grateful to a reviewer for this comment.

some research on switching and its possible equity implications. Section 9.6 reports some findings on the effects of energy reforms on the probability of experiencing deprivation and on the probability of persisting in the state of deprivation. Section 9.7 adds a territorial dimension of cohesion to the picture, in terms of variations between countries and regions of access to some services of general interest in the EU. Section 9.8 discusses the policy issues, particularly the remedies to some regressive aspects of the reforms and concludes. A simple example of redistributive welfare effects of reforms is shown in the appendix.

9.2 Conceptual framework

To evaluate the social affordability impact of the reforms, a simple example may be helpful. We start from a standard microeconomic setting and we show the equity implications of the peculiar nature of network industries, and of different regulatory arrangements. Then we consider some information and behavioural issues that are outside the standard setting, but may add some realism to the picture. The discussion may be seen as a continuation of the simple framework of Chapter 3, as we shall use some of the intuition from it. For a diagrammatic exposition see Appendix 9.A.1.

Let us start with a competitive industry, where government has no specific policy objectives related to the good provided. To fix ideas, let us suppose that the production of a good is competitive. The government does not consider the good as a merit nor as a bad. Thus the government is neutral, and it is happy to just leave the market undisturbed. As equilibrium prices are parametric consumers to the firms, and they must be equal to marginal costs, the market leads to efficiency. Equilibrium price, however, rations demand differently from that of the perspective of different consumers. The latter may be differentiated in terms of income (or welfare) and preferences. There are four classes of consumers. Consumer 1 is poor and has low preferences (demand is more elastic to own price) for the good, Consumer 2 is also poor, but he has high preferences (more rigid demand). Consumer 3 is rich and has elastic demand. Consumer 4 is also rich, but has a more rigid demand. These features may be represented by a parallel outward shift of linear demand for the rich compared to the poor, and by a change of the slope for those who like the good less than the alternatives.

For the sake of simplicity, we consider consumers' surplus on the Marshallian demand as our welfare indicator, as in Chapter 3. The total surplus is actually the sum of the different contributions to social welfare from the four types of consumers. We turn to this point later on.

The traditional public economics approach to equity issues is that markets should be left undisturbed to achieve efficiency by balancing supply and demand (First Fundamental Theorem of Welfare Economics) and, if socially undesired welfare outcomes are reached, these should be corrected by lump-sum transfers (Second Theorem). Let us look at the implications of this view of network services. Instead of a competitive industry, consider a natural monopoly, possibly combined with some competitive segments.

Let us suppose that a public provider under monopoly offers the service at a price equal to a constant marginal cost that, as in Chapter 3, is higher than a first-best benchmark and is such that all the types of consumers are served, including the poor with elastic demand. The aggregate social welfare is a weighted sum of the four consumers' surplus, with the weights given by a social welfare function. The inefficiency here is simply that there is a departure from cost minimization. This can be large or small, depending upon the country and industry.

If privatization is introduced, but a regulator bars price discrimination, there are two effects. Marginal costs may decrease because the management seeks profit maximization. But price will be fixed by the private monopoly on the aggregate demand curve, i.e. the sum of the four individual demand curves, by setting marginal revenues equal to marginal costs. The unique price may well cut off the poor with elastic demand, as the price is not tailored to exploit all consumer profiles. Moreover, the aggregate consumer surplus will shrink compared with public ownership, if the latter is not 'sufficiently' inefficient.

An alternative reform path would be to allow price discrimination under privatization, but the regulator fixes a price-cap. The implications are interesting. The private provider will be considered to have four markets, one for each type of consumer, and will determine four monopoly prices. In some cases the cap may be binding, in others not at all. The regulator who wants to protect all the poor, including those with elastic demand, should fix the cap in such a way as to include them, at least to a certain extent, in the market. In fact, it may be the case that the cap is not binding for the elastic-demand poor, and the benefit of the regulation is simply that price discrimination is allowed.

An alternative strategy for the regulator would be to avoid using a price cap, but to introduce a quantity constraint, that is a service obligation that forces the provider to offer in any case a minimum quantity of service. In general this would be the more beneficial regulation for the poor with elastic demand, but not the best for the poor with more rigid demand, or for the rich, as they will face uncapped monopolistic prices.

The planner can then rank the social welfare implications of the four schemes: public provision; unregulated private provision without price discrimination; private provision with price-cap and price discrimination; and private provision and minimum service obligation. As the Appendix shows with a diagrammatic example, some surprising rankings may emerge, as different inefficiencies should be compared.

Oligopolistic arrangements would further complicate the picture, but not the basic intuition that either public or private provision may include or exclude the poor, and the social cost and benefits of inclusion should be assessed under different pricing mechanisms. In other words, there is no way to predict the welfare dominance of one type of socially inclusive policy against another, as different inefficiencies arise in any case and must be evaluated empirically.

Eventually, to add information constraints to the picture, suppose that the four types of consumers are constrained in their understanding of the prices they are offered, because the contracts are complex. You may also consider that some consumers do not actually understand their own preferences very well. If this cognitive dissonance is related to education, for example, and education is correlated to income, there will be a higher probability of the poor to making mistakes and switching to a contract which is welfare inferior for them as compared to other alternatives. This uneven distribution of mistakes will also have some consequences for the final aggregate welfare, further reducing the slice of the poor.

It is apparent that, even in this extremely simplified framework, three issues arise: a) the fact that markets are not competitive creates a potential wedge between marginal costs and prices, which has different implications for different types of users; b) regulation that focuses on average prices (or on an access price to the network only) allows for (usually regressive) price discrimination; c) the government policy aiming at all potential users receiving a minimum quantity of the service may not be incentive compatible with profit-seeking providers and needs to be implemented by a subsidy to the provider under the universal service obligation.

Interestingly, advocating lump-sum subsidies to those not receiving their basic need supply would not solve the equity issues in practice. Individual lump-sum taxes and subsidies are usually not available to governments. The information requirements and ethical issues raised are important, as shown by the fact that in this context the government should know not just the income but also the preferences of each consumer, and income per se would not be the right welfare indicator in the first place. But, and perhaps more importantly, the policy objective may be such that the government considers that the supply and consumption of a 'basic need' quantity of the good is a merit good per se.

There is a wide literature in public economics about subsidies 'in kind', and it has usually concluded that the alternative to a lump-sum subsidy would need to be too high to achieve the government's target for consumers with elastic demand. Part of the subsidy will be spent on the less socially deserving of the good, see for example Brent (2007). Combined with the distorting effect of taxation, the welfare loss of trying to combine the provision of a merit good with generic subsidies in cash is often inefficient. This has motivated, for instance, food stamps in the USA, the public provision of primary education in most countries, mandatory health programmes in Europe, and so on.

In network services, while the legislation on universal service obligation is sometimes more vague, in several countries it exactly prescribes individual entitlements to essential services, for example the right be connected to the networks at an affordable price. Regulators establish in practice what is affordable, and this decision has implications for the fixed connection rates and also for the variable part of the tariffs.

While not worded in this way, the service obligations imply that governments or regulators treat some services as merit goods, at least for a certain level of consumption. If this is the case, a 'Second Theorem' approach could not work, and other solutions are needed. One of these solutions (among the many explored in the literature on optimal tariffs for public enterprises) is a public enterprise or a regulated private enterprise, which is allowed to incur losses on some consumers and to recoup the loss by higher margins on other consumers. This was the logic of the traditional cross-subsidy for budget constrained public enterprises, which had to balance average revenues and costs, and had to offer affordable prices to some categories of users. We come back to this and other policy issues related to the distributive impact of network service in the concluding section.

9.3 Earlier empirical analyses of affordability

The empirical literature on the social affordability of public utility services is well established, see Boardman (2010b) and Buzar (2007), but it often suffers from uncertainty regarding the conceptual framework, including the definition of utility poverty,[5] and from a lack of adequate data, particularly those allowing international comparisons. For specific countries and sectors, there is, however, some evidence on the burden of utility bills for households in the lower income brackets, or in specific disadvantaged groups. Earlier examples of this in the literature include Wolak (1996) on telecommunications in the USA, Gomèz-Lobo (1996) on energy services in the UK, Waddams Price and Hancock (1998) for different services also in the UK; see also Waddams (2005), and Florio (2004) for a review. There is a specific literature on the distributive

[5] Utility poverty differs conceptually from general poverty because the former is related to access to or affordability of specific services, which are crucial for individual functioning, as assessed by governments, see discussion in Chapter 10. General poverty is a broader concept, usually based on some consumption or income standard.

effects of utility reforms in Latin America, see particularly McKenzie and Mookherjee (2003). Checchi, Florio, and Carrera (2009) review some of this literature and use Latinobarometer data (a large-scale survey similar to Eurobarometer) to establish a correlation between indicators of social well-being (including for example having certain durable goods) and perceptions of privatization. They find a clear inverse relation between the well-being of the respondents and their evaluation of the privatization of some utilities. Clifton and Ahedo (2010) use Eurobarometer and household budget data on fixed and mobile telephony and electricity in the UK and Spain, and show how the perceptions of the respondents are related to their social vulnerability. See also Clifton et al. (2010) for a wider discussion of affordability and perceptions in the EU regulatory context.

Ugaz and Waddams Price (2003), in a book on Latin America, show why households in the bottom deciles of the income distribution may suffer in absolute or relative (to their income) terms from tariff rebalancing particularly when, under public ownership, tariffs are cross-subsidized. There is also a close relationship between the literature on affordability of the utilities and the much wider research on consumption or needs-based research on poverty. For example, Hentschel and Lanjouw (2000) consider markets for electricity, cooking gas, and water in Ecuador, and show how poverty estimates can change when appropriate adjustments are made on consumption survey data.[6]

The most used concept of affordability in practice is a crude ratio between the household total income (or better total expenditures) and the expenditures on a basket of services, sometimes defined as the 'affordability ratio'. Fuel poverty and social affordability of energy bills are then usually defined in 'objective', albeit arbitrary, ways, as related to an 'excessive' share of income tied to the payment of energy bills. The British Department of Energy (DEFRA,

[6] Hentschel and Lanjouw (2000) show that the data from household surveys may not adequately capture welfare conditions. The conventional assumption that total expenditure represents the money metric of utility for a household holds if prices actually reflect marginal utilities of goods. Prices, however, can be affected by rationing, subsidies, increasing marginal tariffs, etc. These distortions usually vary across households for a number of reasons. Hence the comparison of welfare levels across households is less straightforward than one would hope. Focusing on adjustment of prices, i.e. using virtual prices in these situations goes back to contributions by Tobin and Houthakker (1950), see also Deaton and Muellbauer (1980) for a broader discussion. Expenditure surveys, which usually do not report either quantities or prices, would not provide the right welfare measure. In electricity the public utilities charge higher prices per KWh to intensive users, as compared to low users, in some countries (e.g. Italy). A reverse problem would appear in liberalized markets, if—for example—high users of telephone services got a discount. In the case of electricity in Ecuador the authors use the average cost of consumption per KWh, including a fixed charge across all type of consumers, and re-estimate a virtual expenditure based on this proxy of a shadow price. They then show how some poverty indicators would change after the adjustments.

2010) defined a household as being in 'fuel poverty' when they spend 10% or more of their income to buy the energy fuels they need.[7]

For water—an industry privatized in the UK, mostly private in France, but mostly public elsewhere—OECD (2003) suggests that affordability concerns are not negligible even in Europe. A conventional threshold of 3% or more of income spent on the water bill has been used to define 'water poverty'. Fankhauser and Tepic (2007) and Miniaci et al. (2008), report the following thresholds of the affordability ratio (in percentage of total income or expenditures), from different institutional sources, for electricity: 10–15% (World Bank, 2002); 10% (WHO, 2004; IPA Energy 2003). They also report thresholds for water and waste water: 3–5% (World Bank, 2002), 3% (UK government); 2.5% (US Environmental Protection Agency), 5% (Asian Development Bank). Given the substitution between electricity and other energy sources for space heating (natural gas, coal, wood, kerosene, and diesel) and the fact that heating is provided at the district level in some countries, it is also interesting to consider the thresholds for these energy inputs for this service. The affordability ratio thresholds for space heating vary from 6% (US government) to 20% (IPA Energy), and 15% (UN/ECE). As mentioned, the UK threshold of 10% for fuel poverty is often cited in the literature.

Fankhauser and Tepic (2007) consider 27 transition economies. Moving from the 2005 baseline, the authors' estimates for year 2007 are based on the hypothesis that these countries had moved to cost-reflective tariffs, while the baseline reflects actual data (mostly from household surveys). Estimates for 2010 are based on growth forecasts (before the recent global crisis), a constant income elasticity of 0.3 for all the three services. Price elasticities are known to be lower in the short run than in the long run for essential services, because of the difficulty in immediately shifting to substitutes. The authors report the findings of a number of studies, which suggest a median elasticity estimate of −0.5 for electricity; −0.4 for heating; and −0.2 for water.[8] Assuming that the incomes of the poor grow as per GDP, these simulations show that, if the combined threshold of the ratio utility expenditure to total expenditure is 25%, then only in the Slovak Republic, FYR Macedonia, and Serbia, will households in the bottom income decile be at risk. Most of this estimate is driven, however, from growth forecasts, which can be less realistic after the 2008 recession. A sensitivity analysis with lower income growth and

[7] For example: 'whilst fuel poverty can be described and presented in many ways, a commonly used definition is of a household that needs to spend more than 10% of their total expenditure on fuel. On this definition 5 million households in England alone could be considered to be in fuel poverty in 1996' (DTI, 1999, p. 53).

[8] By comparison, Lampietti et al. (2007), Annex 5, p. 209, report eight empirical papers with estimates of prices and income elasticity of residential electricity demand. Price elasticity is more often around -0.5, and income elasticity around 0.6, but with larger variability across countries.

price elasticity would suggest that other countries, such as Romania, Bosnia, Croatia, and Estonia the poverty threshold would be met or be close to being met.

A different way to consider the affordability of utilities is to focus on specific vulnerable groups, such as pensioners or beneficiaries of social assistance. While people in the bottom income deciles usually include a large proportion of pensioners, the unemployed, disabled persons, and lone mothers, the income-based statistics may be less reliable than those concerning specific disadvantaged social groups (also because of unreported income). One concern with utility affordability data is that there may be a difference, occasionally a very wide one, between bills and payments, because a person with insufficient income to fully pay for an essential service will accumulate arrears. When combined with errors in household surveys, the actual meaning of the data can be questioned.

An alternative is the minimum entitlement approach, which is related to a wider definition of poverty based on standards of living (Ravallion, 1998). This is a core concept in our discussion of network services, as it is closely related to what can be considered as 'essential' to ensure a socially inclusive provision.

Fankhauser and Tepic (2007) use the following minimum standards: 600 KWh/year for electricity, 1500 KWh for heating, and 22 mc for water. Other standards can be found in the literature, see particularly WHO (2004), OECD (2003), and Eurelectric (2003). Under this entitlement approach, the expenditure virtually required for a pensioner to buy a minimum of heating, water, and electricity would have been more than 10% of income for a pensioner in Lithuania or an unbearable 45.6% of income for a person receiving social benefits in the Czech Republic.

The importance of looking into micro-information as a complement to more aggregate data has also been stressed by research at the World Bank on poverty and energy reform in transition economies in the ECA Region (Lampietti et al., 2007). Table 9.1 shows some of their results for selected EU members and for other non-EU countries.

While most of the literature we have cited considers country data, Miniaci et al. (2008) discuss the evidence on Italy, based on family expenditure surveys, and suggest that national average data on electricity, gas, and water may conceal substantial differences across regions within a country.[9]

[9] Their analysis relies on the comparison of affordability data over time (1997–2004), based on Surveys of Family Budgets. The main contribution of Miniaci et al. (2008) is to show that a country threshold to define affordability is not very informative when there are considerable variations in climatic conditions and family size within the country. For example typically an Italian family in the north would consume 1400 mc of gas, while one in the south only 900 mc, in part because of differences in the need for heating. Thus the authors differentiate their analysis according to four climatic regions, after presenting detailed data on housing, appliances, number of members of the household, and other features to better describe the regional well-being and consumption patterns.

Table 9.1. Power, gas, heating, and water sector affordability for all households in 2002. Expenditures over income

Country	Electricity		Network gas		Central heating		Cold water	
	Bottom 20%	Total	Bottom 20%	Total	Bottom 20%	Total	Bottom 20%	Total
Bulgaria	10	9	3	2	16	16	5	4
Hungary	7	6	11	7	20	17	5	4
Poland	10	7	7	5	13	14	4	3
Romania	6	6	7	5	na	na	6	5
Serbia	8	6	7	5	15	11	na	na
Turkey	10	7	29	8	13	13	5	4

Source: selected data from Lampietti et al. (2007).

According to the European Commission (2007), the average share of income spent by low users (1200 KWh per year) was 1.23% for electricity, and 0.96% for gas (8.37 GJ per year). There are, however, large variations across countries and income groups. Between 1994 and 2004, in the EU-15 countries, the overall change in electricity expenditure in the bottom household income quintile, as a combined effect of price and volume change, was negative in eight countries and positive in the remaining seven. The direct effect was negative in just 5 countries out of 15, and among them were both the UK and France, which—as we saw in Chapter 2—adopted very different reform patterns. For the majority of the EU-15 Member States, however, 'electricity's share of expenditure increased more for low-income household than for high-income household' (European Commission, 2007, p. 58). Moreover, according to this evaluation, the coefficient of variation of electricity prices across the EU-15 between 1994 and 2004 by categories of users, while it decreased or was stable for median and high consumption respectively, slightly increased for low consumption levels.

9.4 Are tariff structures regressive?

This section discusses whether there is any empirical evidence of price discrimination among users, and whether this is regressive or progressive. The

Then, their definition of the affordability ratio, for each region, is the share of households living in the area, for which the 'virtual expenditure' for energy and water would exceed a certain threshold of total expenditure. In turn the virtual expenditure is based on a reference basket derived from the initial level of expenditure on one utility by the 25th percentile of expenditure in that utility. The threshold is related to the average of the ratios between virtual and total expenditure for that percentile. Hence, this threshold is a relative concept, as it depends on the distribution of expenditure, and from the initial levels, not from absolute quantities.

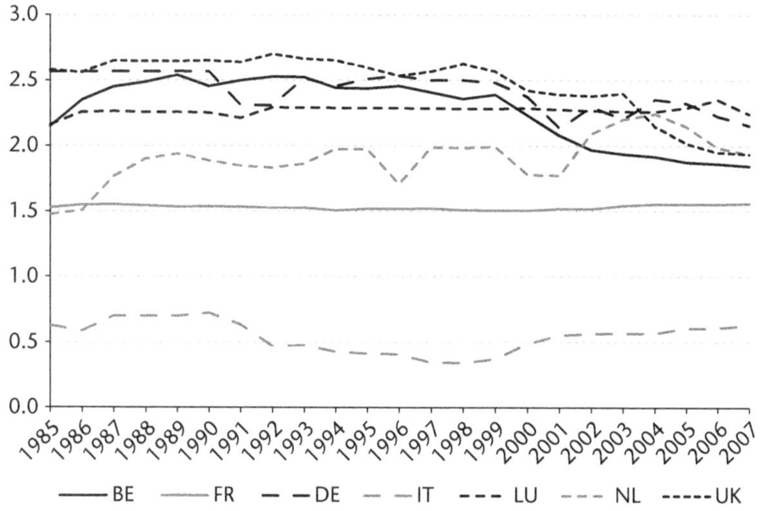

Figure 9.2. Electricity: Ratio of household prices for low and high users, selected countries (1985–2007)

Source: own elaboration of Eurostat data.

Note: data were integrated with those of the cities of Paris (1985–2003), Hamburg (1985–2002), North-East, North-West and Central Italy (1985–93), Rotterdam (1985–99), and London (1985–99).

most interesting example is that of electricity supply, as we have longer time series for a number of countries, the product is homogenous, and some data are publicly available.

Figure 9.2 shows the trend between 1985 (or the closest pre-reform year) to 2007 (or the most recent available data) of the ratio of the net-of-tax current price in Euros paid by different households in selected EU countries. The households considered are those at the extreme of the statistical consumption classes. At the bottom of the consumption pyramid are households with a yearly consumption of 600 KWh (Eurostat type Da), which indicatively have a contract for the supply of 3 KW, and on average use only 1.64 KWh per day. Energy efficiency has greatly changed over the years, but this consumption level is certainly very limited.[10]

In one of the poorest EU countries, Bulgaria, according to EBRD (2003) the lowest income decile in 2001 consumed 756 KWh per year, the top decile

[10] The power (Watt) needed for some common appliances is as follows: Light bulb 60–100 W; toaster 800–1500 W; microwave 600–1500 W; dishwasher 1200–1500 W; washing machine 500 W; vacuum cleaner 200–700 W; iron 1000 W; electric mower 1500 W; clothes dryer 4000 W; ceiling fan 10–50 W; table fan 10–25 W; air conditioner (room) 1000 W; electric blanket 200 W; hair blow dryer 1000 W; electric shaver 15 W; laptop computer 20–50 W; desktop computer 80–150 W; TV (colour) 70–150 W; clock radio 1 W; fridge/freezer 500 W; electric kettle 200 W; power shower 240 W. Source: various websites.

7644 KWh, and the average household 3636 KWh per year. According to Lampietti et al. (2007, p. 27), 123 KWh per month should be considered the basic need, that is two times greater than Fankhauser and Tepic (2007, p. 288,) or the lowest class of household consumption according to Eurostat. At the other extreme of the distribution, there are households with a consumption of 20,000 KWh, or 54.8 KWh per day.

Most consumers will not fall at these extremes, and the 3500 KWh per year (Eurostat type Dc) consumption level is often taken as the typical or 'modal' benchmark.[11] According to Eurostat the standard consumption type is also related to the surface area of the dwelling as follows:

- Da: annual consumption of 600 KWh (standard dwelling of 50 m^2)
- Db: annual consumption of 1200 KWh (standard dwelling of 70 m^2)
- Dc: annual consumption of 3500 KWh among which 1300 KWh overnight (standard dwelling of 90 m^2)
- Dd: annual consumption of 7500 KWh among which 2500 KWh overnight (standard dwelling of 100 m^2)
- De: annual consumption of 20,000 KWh among which 15,000 KWh overnight (standard dwelling of 120 m^2).

Thus, instead of taking the extremes of the price distribution, one may like to consider the ratio of prices below or above the Dc threshold. As the price pattern is, however, usually monotonic, taking the extremes simplifies the presentation (and the results are robust when instead we take the ratios between tariffs for other types).

Visual inspection of Figure 9.2 shows that the electricity tariff structure is regressive everywhere, except in Italy. The tariff paid by the end user usually includes a variable price of supply and a fixed charge, accounting for transmission, system services, meter rental, distribution, and commercial services. We briefly comment below on price ratios in some countries (in some countries and for some years, the data refer to specific cities, as reported in the notes to the table).

In the UK in 2007 the standard bottom consumption household, as defined above (Da type), had to pay a net of tax price per KWh of Euro 0.1758. The standard high-usage consumer (De) instead had to pay a price of 0.0825. This is less than one half (De/Da) of the consumer at the bottom class. One may consider that in the bottom class there are not just poor households, but also

[11] According to unpublished data from Eurostat (personal communication to the authors) the consumption volume is highest in class De in Finland; Dd in Belgium, Cyprus, Estonia, Slovenia; in class Dc in Bulgaria, France, Germany, Hungary, Ireland, Italy, Latvia, Netherlands, Poland, Portugal, Slovakia, UK; in class Db in Lithuania, Romania.

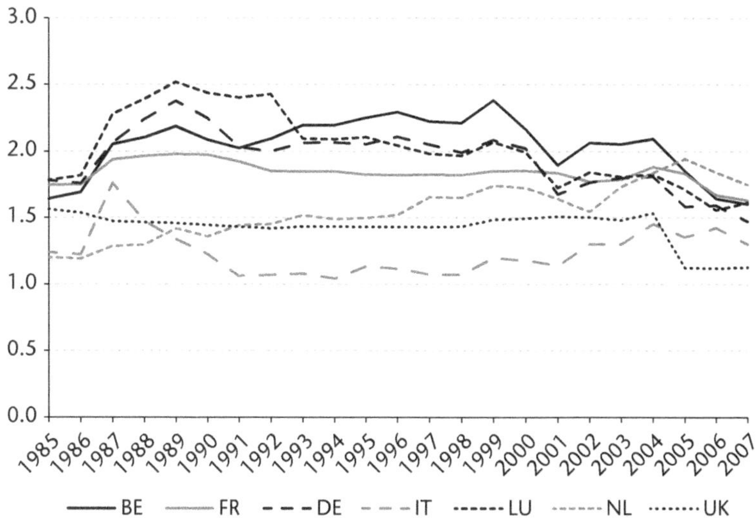

Figure 9.3. Gas: Ratio of household prices for low and high users, selected countries (1985–2007)

Source: own elaboration of Eurostat data.
Note: from 1985 to 2004 country data were integrated with those of the cities of Brussels, Rotterdam, Luxembourg, Paris, Milan, and London (1985–1990).

second dwellings, which certainly raise fewer concerns regarding social affordability. Some low users, however, are actually poor households. While less extreme, the discount for users in the 7500 KWh (Dd) as compared with those in the 1200 KWh class (Db) is still significant. The former in the UK in 2007 had to pay 0.1254 Euros, against 0.1599 Euros per KWh of the latter, that is those with typically larger flats enjoyed a 22% discount on the price. If we go back to 1985, the first year for which Eurostat publishes the data, the ratio between the extremes (De/Da) was around 0.30 (in London). In the year of privatization (1990) the ratio was even lower, 0.28, and it worsened for some years, before gradually increasing.

Figure 9.3 shows the ratio of prices for gas again between the lowest and the highest consumption brackets. Except again for Italy, the evidence of a regressive price structure is apparent, albeit decreasing over time in Belgium and Germany.

Trends over time, and comparisons across countries, do not suggest that privatization and liberalization is systematically associated with an increase in second degree price discrimination[12] in electricity. Going back to Chapter 2

[12] Second degree price discrimination arises when prices vary according to the quantity supplied. In contrast, first degree discrimination implies that the supplier knows the preferences of each consumer. Third degree price discrimination differentiates prices according to consumer groups, see Pigou (1932).

and the OECD electricity reform scores reported for these countries, there is in fact no clear association between changes in price discrimination and reforms.

This analysis of price ratios across types of users is, however, too rudimentary to show the full picture of change in social affordability, as several other dimensions should be considered.[13] The combination of regressive price discrimination and increasing income inequality particularly would explain in some countries the increasing affordability problem of energy bills, at least in some years and countries. To see how in general price discrimination may have an impact on the social affordability of utilities, we may adapt a simple formula for consumer surplus change (see Estache et al., 2002, 35):

$$\frac{\Delta CS_i}{C_i} = \frac{p_{1i}q_{1i}}{C_i}\left(\frac{\Delta p_i}{p_{1i}}\right)\left[1+\frac{\varepsilon_i\ \Delta p_i}{2p_{1i}}\right]$$

where ΔCS_i is the change of consumer surplus related to specific service from year 1 to year 2 for individual i; $\Delta p = p_1 - p_2$ is the change of price paid by individual i; ε_i is demand elasticity to own price; and C is total expenditure. Price discrimination implies such that $p_{1i} \neq p_{1j}$; $p_{2i} \neq p_{2j}$ where j is a different individual. A redistributive effect arises simply when:

$$\frac{\Delta CS_i/C_i}{\Delta CS_j/C_j} \neq 1$$

If the individual j is the rich one by assumption $C_j > C_i$, it is likely that $\varepsilon_j > \varepsilon_i$ with ε fixed over time. In general, however, ε is a function of consumption levels $p_i q_i$ and of total expenditure C_i.

It is then easy to see that even small price changes may lead to different welfare losses for the rich and the poor. This is because the share of expenditure for the poor is higher than for the rich and elasticity and total expenditure may be reduced by any price increase. This effect is greater if $p_j < p_i$.

For a more detailed discussion on the measurement of the welfare effect of price changes of the services of general interest, see Brau and Florio (2004), who study the effect of privatization in the UK in telephone, rail, bus, electricity, and gas (and coal as a substitute for energy network services). While their interest was in evaluating the 'before–after' effect of privatizations between 1984 and 1999, their methodology in principle can be applied to the evaluation of the welfare effect of any change in the price structure of a service. They consider both marginal and non-marginal price changes, and suggest some empirical shortcuts for the evaluation of second order approximations, using aggregate data only. The benchmark first-order measure that Brau and Florio (2004) use is simply Marshallian surplus, socially weighted by the distribution characteristics of each service, as in Newbery (2000). The

[13] Including connection charges and method of payment, see Section 9.5.

intuition is simply that some services have a greater share in the total expenditure of poor households than others. Given the inequality aversion parameter of the social planner (for example derived by the Atkinson 1970) in an iso-elastic utility-based social welfare function, and given the shares in expenditures, a 'double weighting' determines the aggregate Marshallian measure. For large price variations, which would imply second order approximations, Brau and Florio (2004) show that under some assumptions the following holds:

$$\Delta W/\Delta p = -d^{*}X\left(1 + (\Delta p/2p)\right)\varepsilon$$

where the welfare change per unit of price change (*lhs*) is a function (*rhs*) of the distributional characteristic of the good (d^*), the share of expenditure X, the price p change, and ε aggregate demand elasticity to own price. This skips the more usual approach involving estimation of the individual demand functions. This formula can be adapted to our context just by considering prices indexed by the type of consumer for each service, and a welfare weight for type of user (assuming that on average low-intensity users have a higher welfare weight).

The average distribution characteristic by service would usually give a higher welfare weight to water, close to 1, then to gas and electricity (with often very similar weights), then to telephone, with bus and particularly rail having a lower weight. Thus, the evidence of regressive price discrimination for these services, and particularly for water and energy, is of some concern using individual welfare weights or distributional characteristics.

9.5 Competition, switching, and affordability

The main driver of benefits to the consumer, including the poor, of the reforms of public utilities should have been the advantage of competition as a price-capping market mechanism. In a fully liberalized market, consumers are given the opportunity to pick up better choices in terms of price, and their welfare should increase. We have discussed in the previous chapters some reasons why, for the average household, the effects of competition in network industries may have been limited. The discussion so far has not, however, considered the specific mechanisms of competition and consumer choice from the perspective of the lower end of the market, where users have low incomes or are vulnerable. We review some of the scant available evidence.

Whatever the structure of competition upstream, for example in electricity generation, or in gas contracting, or in network services in telecommunications, to the end user a liberalized market usually takes a specific form. The consumer needs to pick up a contract between the offers proposed by a relatively small number of competitors. Typically one competitor is the incumbent, and the alternatives are a small number of entrants. Each of the oligopolists will offer different pricing structures, as usually regulators do not constrain the competitors to a specific type of contract. Thus, the consumer in principle has to compare the pricing of the incumbent, and the offers of one or more entrants. In network services, there are some variations in the contracts, but a typical contract would be in two parts: a fixed charge plus a variable price. In some cases there may be two marginal prices, with a threshold for consumption intensity, plus the fixed charge. In other cases there will be only the fixed charge, that is a flat rate contract. These basic structures are then complicated by time-varying tariffs in two directions. The marginal price can change according to the day of the week or the hour of the day, but also over the year if the contract is indexed to some input price, as for energy supply. There may be other aspects of the contract that can introduce further variability in the tariff structure. One of these aspects is the method of payment, which can be by prepayment, as for mobile telephone cards and some types of energy contracts; by direct debit on a bank account; or by standard credit.

The consumer can acquire information on the offers available from the competing firms in several ways, for example advertising in the media, mail, company websites, telephone calls received, contact by door-to-door sales agents, comparative websites of consumers' associations, or promotions by the regulators. Most of this information tries to convince the consumer to switch from one company to another, or in some cases from to one contract to another with the same company. In some cases the information is about an offer to bundle different services, typically electricity and gas, or these with water; and fix mobile telephony, internet, cable TV, and so on. In some more complex cases the horizontal bundling of offers may be related to a combination of public utilities with financial services and others.

To what extent can poor households benefit from these choice opportunities? There is limited evidence to answer the question. Even the answer to the more general question about the benefits for the consumers in general is limited. Davies, Waddams Price, and Wilson (2009) review the theory of non-linear pricing under oligopolistic competition, and find that in the UK energy tariffs, consistently with the theory, tend to offer non-linear schedules in order to introduce second-degree price discrimination, but—differently from the theory—the marginal price component of the tariff is not cost-based, there is no price convergence across competitors, and in fact there is evidence of increasing asymmetry of the price structures. Wilson and Waddams Price (2010) look at two household surveys in England and Wales, and conclude that many users failed to select the best offer in the retail electricity, and more than one out of six worsened their position after switching, while the remaining (i.e. the majority if switchers) improved their position. The possible determinants of the errors include search costs and information noise; generic cognitive dissonance; own consumption forecasting errors (a topic analysed in an earlier paper on telephony by Miravete, 2003, see the Further Reading Section); a bias in favour of some types of 'easier' tariffs, such as the flat-rate ones, or sensitivity to some particular aspect, such as discounts for introductory offers; or eventually mis-selling, that is the deliberate activity of sale agents who confuse the consumer. While Wilson and Waddams Price (2010) find some evidence in favour of the latter explanation, we report here another aspect of their empirical research, which is directly related to the topic of this chapter. First, in their sample, British consumers living in rented homes are less likely to gain from switching their utilities. Second, both young people and the elderly tend to pick up the wrong contracts more than the middle aged. Third, according to this study, there is little evidence that low-income households make more errors. The empirical analysis is such, however, that probably the first two variables (rented home and age) already capture most of the information about income. Moreover, it seems likely from an examination of the overall sample of the original surveys, that non-switchers are more likely to be people in the lower income brackets. Thus, there is some evidence that the poor (most of them living in rented homes in the UK, and the elderly who on average are poorer than people in their middle age) either do not switch at all, or if they dare to do so, have a not negligible probability of making mistakes, or even worsening their position.[14] We report and discuss additional evidence on the UK later on in this chapter, see Section 9.8 (point h).

Why don't people who would benefit more from differences in prices fully take advantage of this opportunity? The answer seems to be that the utility poor are usually also facing higher information barriers than other consumers. The marketing efforts of competitors are expected to be targeted first to the upper end of the market, as price discrimination based on quantity discounts should identify customers who are likely to be intensive users. While some of them may be poor households (for example with many children), most of the high intensity users are also high income. Thus, the poor are a less interesting target for advertising and offers of discounts to induce them to switch. Second, the poor are usually less educated and less able to take advantage of information on the Internet. The websites promoted by the regulators for comparisons of tariffs are visited by a small minority of potential users of network industries, and certainly not typically by the most vulnerable ones, such as the disabled, lone mothers, and so on.

Only further research will unveil the actual dimension and reasons behind the poor's missed opportunity to take advantage of competition through

[14] Obviously the poor also have a probability of improving on the previous contract.

effective switching, but the preliminary evidence points to social barriers to a successful information search.

9.6 Household survey data on deprivation

As mentioned in the introductory section, one way to look into equity issues in network services is to focus on deprivation data. According to the EU-SILC survey, the variability of deprivation data varies greatly among countries. The EU-15 average is 13% of households in absolute poverty in 2008 (EC, 2010), but in both Portugal and Greece the share is 22%. Severe material deprivation occurs when four out of the nine well-being items are reported as lacking, including as mentioned telephone connection, adequate heating, and capacity to pay utility bills. While less than 2% of the population faces this condition in rich countries such as the Scandinavian ones, the share of households in absolute poverty is over 30% in Bulgaria and Romania. The UK is, among the EU-15, the country where there is the highest percentage of people in absolute poverty, with a share of 12%.

Is there a correlation between the probability of household experiencing deprivation of public utility services in one country and the adoption of privatization, vertical disintegration, and market opening of such services, after controlling for other factors? Poggi and Florio (2010) focus on non-payment of utility bills, or those in arrears, as reported by respondents to surveys in different countries and years. As mentioned, this is one of the items considered by the EU in the definition of absolute poverty. The advantage of this approach is that no arbitrary assumption is needed about income or consumption thresholds to define the utility poor. Two cross-country sources of data are available.

The ECHP is a multi-country comparative household panel survey conducted annually by following the same sample of households and persons in the Member States of the European Union. To define whether a household suffers deprivation, Poggi and Florio (2010) take advantage of the information given by answers to the following question: 'Has the household been unable to pay scheduled utility bills, such as electricity, water and gas during the past 12 months?'[15] In principle, the ability to pay may depend on (i) a decrease in the income level, (ii) some changes in the household necessities, or (iii) more expensive bills due to an increase in price. In order to focus on the last issue,

[15] Note that, for Italy and the Netherlands, this question has been asked only if the household has declared some degree of difficulty in making ends meet. The OECD/ECTR does not report data on the water industry, a sector mostly unreformed in the years we consider. Typically, the water bill is much lower than the electricity and gas bills, hence water poverty is perhaps less important than energy poverty. Disconnection of a domestic water supply is illegal in most countries.

one needs to control for differences in the household attributes (i.e., number of working members in the household, household size, and household income) and in the social characteristics of the reference person (i.e., sex, age, and education).

As a first step, and given data availability, seven European countries are considered (other countries are excluded because of data availability): Denmark, Belgium, France, Ireland, Italy, the Netherlands, and Spain. Ireland and France are the countries with the highest deprivation rates in 1994 (about 9%). The Netherlands has, instead, the lowest proportion of population deprived in utilities (about 2%). In 2001, the percentage of people suffering deprivation was lower in all countries. The country that experienced the largest decrease in deprivation was Ireland, reflecting its sustained income growth, before the recent financial crisis.

Around 44% of the deprived households are also income poor (while only 16% of the not-deprived households are income poor).[16] In fact, deprived households are mainly households in the lower half of the income distribution; thus, the above result may be explained by increased consumption, financial constraints that may explain low or absent savings, and some degree of subjective perception of the financial situation.

Poggi and Florio (2010) also use microdata from the European Statistics on Income and Living Conditions survey (EU-SILC) 2004–2005. This allows us to extend the analysis to ten countries: Belgium, France, Ireland, Italy, Spain, Austria, Finland, Luxembourg, Norway, and Sweden. The EU-SILC question is whether the household experiences: 'Arrears on utility bills—whether the household has been in arrears on utility bills (electricity, water, and gas) in the last 12 months.'[17] Information about household attributes (i.e., number of working members of the household, household size, and household income) and social characteristics of the reference person (i.e., sex, age, and education) are also available.

OECD Reform scores are used to represent the reforms. Eurostat price data are used as additional controls. Poggi and Florio (2010) construct a binary indicator of deprivation that is equal to one if deprivation occurs and zero if it

[16] The percentage of households that are income poor is computed using as the poverty line 60% of the median equivalized household income. Note that this measure is sensitive to the poverty line—if the poverty line is 80% of the median income, then about 66% of utility deprived households are poor. In general, 80% of the deprived households have incomes lower than the median.

[17] According to EU-SILC, the proportions of households deprived in utilities in 2005 are the following: 5.27% in Belgium, 6.83% in France, 5.59% in Ireland, 9.15% in Italy, 3.9% in Spain, 1.54% in Austria, 6.35% in Finland, 2.23% in Luxembourg, 5.26% in Norway, and 4.58% in Sweden. These proportions could be only partially compared with those computed from the ECHP—in some countries the difference in the meaning of 'being unable to pay the bills'(ECHP) and 'being in arrears on utility bills' (EU-SILC) can be large.

Table 9.2. Probit estimation results

Dependent variable is deprivation	Unbalanced panel[1] (A) Coef.	Unbalanced panel[1] (B) Coef.	Balanced panel[1] (C) Coef.	Balanced panel[2] (D) Coef.
Electricity: entry regulation	0.003	0.004	0.032	0.707**
Electricity: public ownership	−0.017*	−0.022*	−0.002	−0.057*
Electricity: vertical integration	−0.036***	−0.051***	−0.043*	−0.458**
Gas: entry regulation	0.016	0.029*	0.011	−0.038
Gas: public ownership	−0.039***	−0.054***	−0.044*	−0.017
Gas: vertical integration	0.030	0.018	−0.026	0.446***
Reference person: male	−0.233***	−0.311***	−0.283***	−0.149***
Reference person: age	−0.011***	−0.017***	−0.012***	−0.015***
Reference person: medium education	0.145***	0.283***	0.157***	0.169***
Reference person: low education	0.325***	0.525***	0.382***	0.433***
No. of household members	0.085***	0.130***	0.071***	0.120***
No. of household working members	−0.204***	−0.274***	−0.199***	−0.241***
Household equivalent income	0.000**	0.000***	0.000***	0.000***
Year fixed effects	Yes***	Yes***	Yes***	Yes***
Country fixed effects	Yes***	Yes***	Yes***	Yes***
Constant	−0.541**	−1.226***	−1.191***	−2.400***

Note: *** p<0.01,** p<0.05,* p<0.1. For regulatory variables see Tables 4.A.1, 5.A.1, 6.A.1.

[1] Data source: ECHP 1994–2001. Included in the sample are households from Belgium, France, Ireland, Italy, Spain, Denmark, and the Netherlands. Deprivation is defined as the inability to pay utility bills.

[2] Data source: EU-SILC 2004–2005. Included in the sample are households from Belgium, France, Ireland, Italy, Spain, Austria, Finland, Luxembourg, Norway, and Sweden. No. of observations: 83,141. Deprivation is defined as being in arrears on utility bills.

Source: Poggi and Florio (2010).

does not.[18] The static estimates are presented in column (a) of Table 9.2.[19] Some results are consistent with those found in the literature on poverty and deprivation. The household probability of experiencing deprivation is higher if the reference person is medium or low educated, and it is lower if the reference person is male. Moreover, the probability of being deprived decreases when age rises, and when the number of household working members increases. The probability of being deprived rises when the number of household members increases. These results suggest that the country samples have the desired characteristics, as they capture the expected demographic features of deprivation. These findings can be compared with those on switching errors in the previous section, as some households' characteristics are consistent in the two types of analysis.

[18] For details on the econometric approach see their paper, and Poggi, 2007, for a previous application.

[19] The reference group is composed of Danish households with a 'highly' educated female as reference person; the omitted year dummy refers to 1994.

Of interest in this regression are the estimated coefficients on energy regulation indicators. Privatization (or private ownership in general) raises the probability of experiencing deprivation. A reduction in the degree of electricity vertical integration increases the probability of being deprived. Decreases in the entry regulation indicator do not seem to have statistically significant effects on deprivation during the period of study. Finally, the probability of experiencing deprivation decreases over time and it varies significantly across countries.

The previous conclusions may well be incorrect if there are unobserved characteristics that differ systematically between households. A panel analysis is reported in Column (b) of Table 9.2 showing coefficients using the random-effect estimator and unbalanced panel (while column (c) reports coefficients using the random-effect estimator and balanced panel). The results are similar to those of the standard probit model. The estimated coefficient of the electricity vertical integration and the gas public ownership are statistically significant.[20]

The random effect model described above uses balanced data from EU-SILC for the period 2004–5. Estimated coefficients are shown in column (d) of Table 9.2. Once again a reduction in the degree of vertical integration of the electricity sector significantly increases the probability of consumers being deprived. A decrease in the degree of public ownership of electricity or gas raises the probability of experiencing deprivation (but, this result is only slightly statistically significant for privatization in the electricity sector).[21]

This evidence suggests that market opening is not yet strong enough to counteract the deprivation effect of privatization. The dynamic analysis confirms the previous result and shows that over time there is some persistence of utility deprivation.

9.7 Adding the territorial dimension: Regional disparities in essential services

While the utility affordability literature typically focuses on heterogeneity across households, there is also an important territorial heterogeneity of access to network services in the EU.

[20] These variables result as statistically significant at 1% level if the unbalanced panel is used, while, they result as statistically significant at 10% level if the balanced panel is used. The decrease in the degree of significance of these variables may be due to small variations in the sample composition when moving from the unbalanced to the balance panel.

[21] Over the period 2004–5, a decrease in gas entry regulation (i.e., formal liberalization) seems not to have statistically significant effects on utilities deprivation during the period of study, while liberalization in the electricity sector decreases the probability of being deprived.

The existence of excessive variations in living conditions across the regions and the need for remedial actions have been acknowledged by the Lisbon EU Treaty, which states that the EU 'shall promote economic, social, and territorial cohesion'. While the first two terms refer to regional disparities in competitiveness and well-being, the latter term refers particularly to access to services, sustainable development, and other spatial dimensions of social conditions. Since 1997 the Amsterdam Treaty mentions explicitly territorial cohesion in the access to services of general economic interest. Network services, along with health, education, and other services, are at the core of this concept. Energy and water industry features are also important from the sustainable development perspective of territorial cohesion. We briefly mention below some examples of the regional disparities in access to network services, particularly in ICT, energy, environmental, and public transport, drawing from EC (2010) and on Sirtori and Vignetti (2011).

a) ICT services are considered by the EC as a powerful integration mechanism from a territorial perspective, while the lack of adequate connections is regarded as a major obstacle to overcome the digital divide across regions. In the New Member States (NMS) the number of fixed telephone lines per inhabitant in the mid 2000s was only 46% of the EU-15 average, for example only 19% in Romania. Mobile telephone availability is however more widespread in the NMS, and in some countries higher than in the EU-15. The DSL network coverage in the EU was only 87% of the population in 2005. This increased to 94% in 2009. In spite of very fast progress, the average coverage in a large country such as Poland is only 75%.

   As usual, regional differences within countries are greater than those between countries. ICT is clearly a key driver of territorial and social cohesion, and of growth (Del Bo and Florio, 2012). According to DEAS et al. (2010), the standard indicator of telephone lines for every 100 inhabitants in the EU-27 in 2006 (Eurostat data) was 40, but the range goes from over 60 in France or Denmark, to less than 30 in Romania or Slovakia.

   The broadband penetration rate, defined as the number of high speed connections per 100 inhabitants, while rapidly spreading, goes from more than 20 in Sweden or the Netherlands, to less than five in Poland or Greece. The disparities are even greater for regional data, that is looking within Member States. For example, the number of households with access to the Internet at home goes from more than 80% in all Swedish regions to less than 20% in Southern Italy, which in turn is around one half of Internet penetration in Northern Italy. Everywhere there is also evidence of substantial disparities along the urban–rural divide (Sirtori and Vignetti, 2011). Interestingly, the country and regional patterns are more complex for mobile telephony and for e-government (described as the availability of a package of 20 services for citizens and business), for example with Slovenia more advanced than Luxembourg. As discussed in Chapter 4, DSL is often not adequate, however, to ensure the kind of broadband connection needed to effectively use the range of services now available and expected in future.

The Europe 2020 Strategy (COM 2020) (EC 2010)has set the objective of achieving universal coverage of Internet broadband by 2013 and to increase the speed of connection by 2020 to 30 Mbps for all, and to 100 Mbps for 50% of users. This is a very ambitious goal, given that currently broadband coverage in thinly populated areas lags behind considerably in most countries, for example it is less than 50% in Romania and Bulgaria. Moreover the take-up rate is also limited. Taking together all the platforms (ADSL, cable, and optical fibre, etc.) only 55% of households in the EU have broadband, with the Scandinavian countries near to 80%, and Italy, Greece, and some New Member States below 40%. Sirtori and Vignetti (2011) report that the interregional variability within the NMS is also very high, but then it also is in the UK, Italy, Spain, and Portugal. According to EC (2010, p. 62) 'The lack of private investment in next generation networks outside large conurbations could lead to another digital divide emerging between more developed and less developed regions of the EU.' According to the European Commission (2002), 'giving the opportunity to everyone to participate in the global information society' is a core objective of the European policy agenda.[22]

b) Energy sustainability shows dramatic differences across regions. The most polluting sources, coal and lignite, account for around one quarter of electricity production on average, but in Poland and Estonia the share is 90%. The EU 2020 objective to reduce CO2 emissions by 2020 critically depends inter alia upon adopting less polluting technologies in electricity production, as discussed in Chapter 5. Moreover, increasing energy efficiency will require changes in transport, buildings, lighting, and appliances. In order to achieve these changes over a very spatially diverse Europe, considerable investments are needed in the 'smart energy grids', which use digital technologies to control consumption, and to accommodate renewable energy sources, such as wind and solar, which are substantially affected by discontinuity over time.[23]

[22] In the period 2007–13 the programmed ERDF grants for telecommunications (broadband, ICT-Trans-European networks, e-government, e-health, e-education, etc) are 14,604 million Euros, with Poland deciding to allocate the highest amount (3714 million).

[23] Around 16% of the EIB portfolio is in the energy sector in recent years. The planned EU funds 2007–2013 for energy efficiency and renewables will amount to around 9 billion Euros.

c) Environmental services, such as water and solid waste management, also have an important territorial cohesion dimension. For example, while around 90% of waste water is treated in the EU-15, in some New Member States the situation is much worse. For instance in some regions of Romania, including the capital city, the waste water treatment is below 30%. In some urban areas of Ireland, Poland, Bulgaria, and elsewhere the treatment is not much better. Clear water extraction in 2007 was on average around 550 mc/year per capita in the EU-27, but only 300 mc/year in Poland and 275 mc/year in Romania. In the latter country only 50% of the population is connected to the public water network (Sirtori and Vignetti, 2011). Water and other environmental services, such as waste treatment, prevention of air pollution, and so on show surprisingly high differences across the EU countries, reflecting also the diversity of its geography and social development. For example the percentage of the population connected to public water networks (EC, 2007) was less than 50% in Romania (2006 data), while it is around 95% in neighbouring Hungary. There are less than two water treatment plants for 100,000 inhabitants in Bulgaria or Greece, of which only one or less is able to undertake secondary treatment of water. Solid waste management is defective is some regions of Southern Italy and Portugal, while it is good in Estonia. In waste recovery (for energy production or otherwise) Germany, Austria, and Belgium have good records, and Poland too among the New Member States, while in most EU countries the main treatment, if any, is incineration without energy recovery, and in general landfill is still widely used. The EU offers substantial grants to close the gap between regions lagging behind and the rest of Europe.[24] In the whole of the EU between 2000 and 2006, 40 million additional people were connected to waste water treatment, 20 million to drinking water networks, and close to 1000 thousand unauthorized landfills were closed. In 2007–2013 around 50 billion of EU funds are planned for investment in environmental services, most of which have features of network industries, for which the private sector is unable to offer adequate investment, affordable prices, the required quality standards, and universal coverage.[25]

[24] The ERDF and Cohesion Fund combined invested 27.4 billion Euros in the supply of drinking water, in the management of solid and water waste, etc. This amounts to around one third of total investment in the EU-10 New Member States and in the former Cohesion Countries (those with per capita PPP income below 90% of the EU average, i.e. Spain, Portugal, Greece, and Ireland), matching a similar share from the national public sectors (the remaining invested by the private sector).

[25] More precisely, in the 2007–13 programming period the Structural Funds allocated to environmental services will be 48,861 billion Euros. There are additional funds for the

d) Transport networks are the main field of intervention of the EU structural funds, with 74,734 million Euros of community grants planned for co-funding roads (e.g. motorways in Greece, or regional or local roads in Hungary), railways (e.g. high speed lines in Spain and Italy), ports, and waterways (e.g. Austria and Estonia), and other transport infrastructure (including urban transport in Ireland and even cycle tracks in the Netherlands!). DEAS et al. (2010) review the significant disparities across countries and regions in terms of such indicators (provided by ESPON data) such as indexes of railway lengths relative to area and population, potential accessibility by road, and urban transport. However, most of the disparities between Western and Central Eastern Europe in road and rail endowments are more in terms of quality than in terms of density. In one third of the Convergence Regions (i.e. those with a per capita PPP income of less than 75% of the EU average) the railway density is one half of the European average, and just one quarter of the Competitiveness Regions (those above the 75% income threshold). Many lines in the New Member States are single track, not electrified, or in a poor state of repair. Similar differences between the EU-15 and the New Member States are also reported for accessibility of airports (in terms of time and distance from the closer airport from a given region).[26]

These examples of wide disparities in infrastructure endowment across the EU illustrate two important points: first, access to affordable services in Europe, despite the dramatic improvement over the last years, is still inadequate for certain regions and services; second, market-based arrangements usually work better in some sectors and geographical contexts, particularly densely populated areas with high income residents. This leaves behind the poorest regions, and negatively contributes to economic, social, and territorial cohesion, on a European wide-scale.

According to EC (2010, p. 162), private investment per capita in the EU (2002–6) was estimated to be 'highly correlated with the relative prosperity of regions, and therefore tends to be concentrated in the most affluent ones both at national level... It was significantly below average relative to population in virtually all Central and Eastern European regions (except some capital cities) as well in many Convergence regions in Southern Europe....' Thus,

development of renewable energy sources (solar, biomasses, etc), and other funds allocated to natural gas and electricity Trans-European Network.

[26] In 2007–13 the planned Structural Funds expenditure in transport will be over 77 billion Euros, of which 23.9 goes to railways, 40 to motorways and other roads, and the remaining going to ports, urban transport, airports, and other infrastructure. In 2000–6 around 56% of the increase in the high speed rail network was co-financed by EU Funds. Other EU finance in the form of loans at low interest rates are offered by the European Investment Bank, for which transport projects account for around 37% of its portfolio in recent years.

public investment had to try to fill the gap, and in fact the ERDF and Cohesion Fund combined in some regions and countries contributed to funding up to 90% of public investment. Thus, in any empirical analysis of the functioning of the network services in the EU from a cohesion perspective, it seems worth remembering that most of the investment in the poorest regions is provided by public funds, as private investment is unable to support convergence of the regions lagging behind, as investors prefer to focus on richer regions.

EC (2010) reports the public expenditure (according to the COFOG classification) for selected EU Member States for energy, telecommunications, transport, and environmental services. As most of these services are provided by network industries, Figure 9.8 allows one to see at a glance the still important role of governments in these sectors. The combined public expenditure on energy, communications, and transport, was over 3% of GDP (or more than €1200 per capita) in 2008 in the Cohesion Countries, that is for those lagging behind (see Figure 9.4). These—as we have seen—host the regions with the greatest infrastructure needs, and the largest share of European citizens at risk of poverty. The combination of the social affordability concerns within each country, and the disparities of access to network services across countries and regions, challenges the European integration perspective, and governments and the EU are vitally important in trying to achieve a more cohesive Europe.

9.8 Dealing with utility poverty

The previous discussion in this chapter suggests that there are two main dimensions of socially inclusive policies for network services: first, allowing *access* to disadvantaged citizens, that is offering them the opportunity to be connected to the networks even if they are resident in disadvantaged regions; second, after achieving this target, service supply should be *affordable*, in terms of price.

The first dimension is more important in less developed economies, where many households are deprived of regular access to drinkable water, sewage, electricity, public telephony, and transportation. The concern about service access is not entirely irrelevant in some EU regions, as can be gauged by the previous discussion about disparities of infrastructure endowment across Member States and regions. The core of the accessibility problem for disadvantaged households is simply that they are unable to pay to the provider the connection cost. Thus, if there is a policy objective of universal access, market equilibrium falls short of the target, and cost-effective remedial actions are needed.

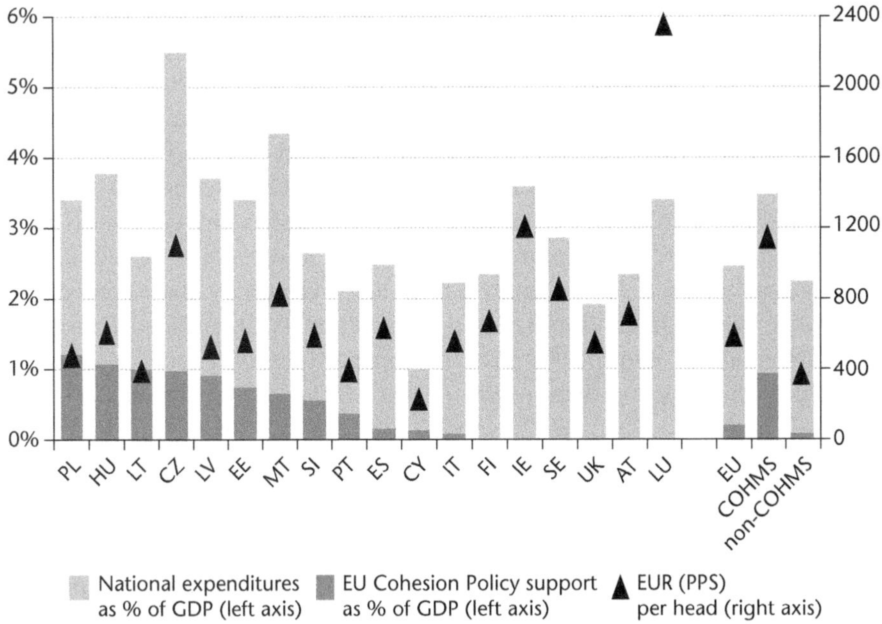

Figure 9.4. Public expenditure on transport, communications, and energy as a share of GDP and per head of population, 2008
Source: Eurostat, DG Regio.

Several instruments for promoting access have been reviewed by Estache et al. (2002), in a study on Latin America. We briefly discuss this taxonomy of instruments in the European context, having particularly in mind those regions lagging behind, or the deprived areas in the most developed EU countries.

a) Universal service obligations for the providers are perhaps the most widespread instruments for promoting access. The EC has passed some legislation in this area, particularly through the Universal Service Directives 2002/22 in telecommunications[27] and the Postal Market Directive 97/67/EC[28] as amended by Directive 2002/39/EC.[29] The Energy Market Directives 2003/54/EC and 2003/55/EC[30] use the term 'public service', with a similar meaning, see Rott (2007) for a discussion.

[27] Directive 2002/22/EC on universal service and users' rights relating to electronic communications networks and services (Universal Service Directive), O.J. 2002 L 108/51.

[28] Directive 1997/67/EC on common rules for the development of the internal market of Community postal services and the improvement of quality of service, O.J. 1998 L 15/14.

[29] O.J. 2002 L 176/21.

[30] Directive 2003/55/EC concerning common rules for the internal market in natural gas, O.J. 2003 L 176/57.

In spite of the EU directives, Member States have retained great autonomy in defining the universal service obligations, as the directives seem to follow the 'minimum harmonization' approach, that is prescribing only relatively low standards that each country can raise within its own legislation.[31] According to Rott (2007), the European legislation establishes a fundamental right to access services of general interest by all citizens as explicitly mentioned, for example, by Article 20 (2) s. 1 of the Telecommunications Universal Services Directive 2002/22/EC and by Article 3 with Annex A lit. a) of the Electricity Market Directive 2003/54/EC. Unfair termination of contracts is also mentioned as an area of regulation, with the objective of securing access over time (we discuss below the topic of disconnection). Article 3 (5) of Directive 2003/54/EC and in Article 3 (3) of Directive 2003/55/EC, provide that Member States shall ensure that there are adequate safeguards to protect vulnerable consumers, including appropriate measures to help them avoid disconnection.[32]

Policy-makers and national regulators are often quite detailed in their specification of the obligations of the incumbents. In the UK, for example, an Order of the Ministry of Trade to the telecom regulator (Ofcom) has been implemented by the latter through detailed obligations placed on BT, the incumbent, and the entrants. These include special tariff schemes for low income customers, including the connection charge to the fixed network, which includes functional Internet access; access to public call boxes with a certain geographical coverage; and a range of services for customers with disabilities, including the text relay service. In some cases, the burden of the USO is with the incumbent alone. In other countries, there are transfers from the government to the incumbent or to those companies who provide the unprofitable access points to their services. In some cases, however, the USO may be vague, for example when it prescribes 'reasonable' access. In fact 'universal' and 'reasonable' access may take different meanings in different contexts.

b) In some countries, the regulation can be more precise by stating the accessibility objective in terms of percentages of coverage of the potential users, according to a time-plan. In some cases this implies that users themselves should bear an obligation to accept the connection. While this is not the case in the targets set for broadband connection, which we mentioned in Chapter 4, it may well be for water and sewage, as in some countries it is compulsory to be connected to the network because of the related externalities.

[31] See, for example, recital (26) of Directive 2003/54/EC.
[32] See also Annex 1 lit. e) of Directive 2002/22/EC.

c) In some less developed countries, service providers have proposed that regulators use different technologies for unprofitable users. In the EU, broadband connection with DSL is offered in rural regions, while optical fibre is rolled out in the urban context. In general, however, the opportunities to differentiate the technology mix between rich and poor users is limited, as the quality standards are also defined by the legislation or by uniform regulation, as discussed in the previous chapter.

d) Cross-subsidies of connection costs is an arrangement that implies that access to new users is financed by a small (explicit or implicit) surcharge levied on existing customers, without a burden for the general tax-payer. Estache et al. (2002), after reviewing the experience in Latin America, observe that 'While in an ideal world the best subsidy is clearly the targeted lump sum cash payment, in most Latin America countries fiscal constraints impede the financing of many subsidies. In this context, cross-subsidies may be a reasonable option' (p. 44). Fiscal constraints in some of the New Member States of the EU, discussed in Florio (2011), suggest that this option should also be considered in the European context.

e) More often in the EU an explicit subsidy is given against the USO to the service provider for connecting disadvantaged users. In principle this strategy can be combined with the previous one, with the burden of unprofitable access points being shared between the tax-payers and the existing users. Historically, public enterprises in Western Europe have combined government transfers and cross-subsidies to solve accessibility problems. We have already discussed some legal issues under the EU state-aid regulations in a previous chapter. In this context the problem that can arise is when the USO is for the incumbent only, and there is a transfer from the government to cover the loss. Clearly this transfer should not be too high, or it would be a hidden subsidy to the incumbent provider, and it would be considered an infringement of EU competition law.

f) In general, a 'communal supply point' would be an alternative to private connections; less costly, but with obvious discomfort for users who would need to share the connection (e.g. water supply, Internet points). In fact it offers a different quality of service, or perhaps even a different service. While particularly important in less developed countries, in the EU this strategy seems less appropriate. For example the obligation to provide telephone boxes was often mentioned in licences released by the telecom regulators in most EU countries, but telephone boxes play only a marginal role in the current service provision.

g) The regulator can force or encourage new providers to offer access points to the networks in certain geographical areas, if the incumbent is unwilling to serve them. This policy option can be combined with allocating government subsidies to the provider who offers the least cost. Similar schemes have been used for local public transport in the UK and elsewhere, and they are rooted in an old, historical pattern of concessions associated with transfers when the service is not profitable.

h) Consortia or other forms of collaboration among the incumbent and alternative suppliers. This is a less aggressive variant of the former arrangement, and there are occasional experiences in telecommunications in some EU countries.

We turn now to the second social dimension of network services, affordability of prices, an issue that is also considered by EU legislation. An early example is Article 12 of the Postal Market Directive 97/67/EC. A more recent example is the EU Cohesion Policy regulations, where price affordability is also explicitly mentioned as a condition for assistance by the Structural Funds to major revenue-generating projects. Non-discrimination between users is required by some network service EU directives, but the details of the price control mechanisms are left to the Member States. See Rott (2007) for the perspective of consumer rights in EU legislation.

The main instruments to deal with price affordability, as identified in the literature and in regulatory practice (see again Estache et al., 2002), include the following:

a) So-called 'lifeline' tariffs. These are usually in the form of a block structure that offers low-intensity users a modest (or even zero) tariff for the first part of their consumption. Progressive tariffs, that is those inversely related to consumption, are a more extensive instrument. In fact lifeline tariffs can even be combined with a flat or regressive tariff. One way to do this is to combine in certain ways fixed charges with marginal prices. Whatever the details of the tariff structures, the advantages in terms of administrative costs of lifeline schemes through a below-cost block for everybody, are obvious, as the provider can simply discount the first part of its bill, without further enquiry into the user's characteristics. We shall discuss later the broader advantages of this approach.

b) Means-tested subsidies to tariffs are frequently advocated by some experts as a less price-distorting alternative to lifeline tariffs. The advantages, in principle, are that, in contrast to the previous arrangement, only the poor receive a subsidy. The information burden for those who manage the scheme is, however, considerable, because each deserving household should be identified by one or more indicators of social need. It is well known from the literature that poverty is a multidimensional social condition, not necessarily captured by simple data. Being on unemployment benefit, for example, may not be worse than the social condition of the working poor, particularly in developed economies. Low-income, as stated in income tax schedules, may be far from giving a good representation of the financial condition of the household (for example when capital income does not need to be included in the personal income tax base, or because of 'moonlighting' or illegal jobs). Severe disability per se does not always need to be an indicator of difficulty in paying the utility bills, but in some countries disability benefits are a substitute for income support in other forms. Taking a combination of household characteristics may work better, but finding a good fit is difficult. In practice, in several countries, means-testing implies that the household is interviewed, and a report written by a officer, with some direct administrative cost. There are also side effects, such as obvious incentives to corruption during the process, and some social stigma attached to the condition of a household needing assistance to pay for electricity, water, or heating, or other basic services.

c) An alternative to the previous mechanism is the distribution of vouchers or 'service credit' cards to the user, to be spent with a provider of their own choice. The administrative cost is greater than the previous system, even if it offers wider flexibility to the user, who is not 'captured' by the incumbent (or the current provider). The disadvantages are, however, the same.

d) Regulatory remedies can focus on changing the fixed/marginal price ratio. As mentioned, the existence of a fixed charge in the tariff structure is the typical non-linear pricing mechanism in network services. This may be intrinsically regressive, even if the marginal prices are constant (and even if they are close to marginal costs). The fixed part of the bill, if high enough, is a burden for low users, and conversely it offers an implicit discount to intensive users. Regulators can impose changes on this structure, that is 'progressive rebalancing'.

e) Service limiters are devices that simply do not allow consumption beyond a certain threshold for vulnerable users, who usually have a track record of non-payment of arrears. The consumption threshold can be combined with the lifeline tariff, with the objective to avoid further arrears or non-payment of the bills. Clearly, means-testing is implicit, but if the technology allows this (telephony or metered electricity or water) disconnection can be avoided through this form of service limitation.

Several major European utilities use this device (EDF and GDF in France developed this technology more than 20 years ago, see Wodon, 2000).

f) More frequent billing, for example monthly frequency, may decrease the burden of substantial quarterly payments for low income households, and help them to monitor their expenditure, but obviously increases the administrative costs.

g) Pre-payment arrangements have been proposed to help the poor control their own budget, and have been experimented with on a large scale particularly in the UK. A familiar example are pre-paid cards for mobile phones, but similar cards can be used for metered supplies of electricity, gas, or water (coins or tokens were used many years ago). The disadvantage is that the system is usually more costly than direct debit, or than the standard ex-post payment. Moreover, pre-payment may lead to self-disconnection, as users who cannot afford to buy the cards will simply cut off themselves from the service. Different financing schemes have also been experimented with to help the poor who have limited or no access to credit face financial shocks, which also could lead to debt and disconnection.

h) Actions that favour the switching of provider by the poor in order to avoid inertia or mistakes when better deals are available have been advocated by some regulators. A FDS International (2008) report prepared for OFGEM, the British energy regulator, showed some qualitative evidence of difficulties for vulnerable users in switching provider. This evidence offers additional insights to our previous discussion. Eleven focus groups were organized, each with one type of user within a potentially disadvantaged category, in different cities: people with low literacy or numeracy; renters; low-income single parents; low-income households: people aged 75+; Polish or Asians with limited English. An Ipsos-MORI (2008) survey in 2007, also commissioned by OFGEM, compared switching rates between vulnerable and non-vulnerable electricity and gas customers. The results of these researches show that vulnerable customers are much less able than other social groups to profit from switching their supplier. Informing them about this possibility would not be enough, however. The surveys show that in the UK 80% of vulnerable users are already aware of the possibility of changing supplier, or tariff, or payment method, but several barriers create difficulties for their choice, such as inability to compute the differences in tariffs, lack of easy access to comparable information, fear that something can go wrong during the process of changing supplier, and passivity. Some regulators have targeted specific actions to encourage switching by vulnerable customers. There is some evidence that 'digital exclusion' is often an important barrier, but it is not easily solved as in some cases it is not just a matter of access to the Internet, but also a lack of minimum competence to take advantage of it.

In the above discussion of remedial actions drawing on Estache et al. (2002) to deal with service affordability, we have mentioned more often the UK context because this is the country where historically utility poverty was associated with network service reform, and where the public debate, social research, and regulatory interest have been more developed. The fact that elsewhere the debate is less developed does not imply, however, that their problems are necessarily smaller than those in the UK.

An evaluation of the impact of the remedial activities in the EU goes beyond the scope of this book. Overall, it seems that in Europe the issues of access for the poor and affordability of network services have not been solved, and perhaps have worsened in recent years because of the economic slump and increasing inequality.

One way to look to the whole affordability issue would be to look into the data on arrears and forced disconnection because of non-payment by the users, or self-disconnection when users decide that they are unable to pay, as was done by Poggi and Florio (2010).

It is difficult to collect these data in a comparable way in the EU, as the regulators do not always require the providers to give this information or make the information available to the public. In certain cases disconnections are not related to the social vulnerability of the users, as they may simply be the consequence of billing mistakes by the provider, the user forgetting to pay, or other circumstances. In general, however, most disconnections are a symptom of affordability issues.

As mentioned, the European Commission has delegated the concrete policies in this area to the Member States, and the 2009 directives for the internal electricity and gas market (2009/72/EC and 2009/73/EC), state that:

> Member States shall define a concept of vulnerable customers which may refer to energy poverty and, inter alia, to the prohibition of disconnection of such customers in critical times.

In the UK, where recent estimates suggest that around one in four households may be in utility poverty (more than 6.2 million for fuel poverty), since 1999 it has been illegal to disconnect water supplies to domestic users. According to a report by the water industry representatives[33] five years later, 700,000

[33] See Annex 2 of *Water UK Debt, Charging and Social Impacts Think Tank Report* http://www.water.org.uk/static/files_archive/0Water_UK_Debt,_Charging_and_Social_Impacts_Think_Tank_Report.pdf [accessed 8 January 2013].

households were unable to pay their water bills. This is a disappointing record for one of the most advanced economies in the world.

According to another report,[34] again in the UK, there is evidence that in the mid 2000s around one out of four British families with an income between £7500 and £15,000 per year have accumulated utility bill arrears. In a sample of low-income families, 13% had phone arrears, 12% gas, 7% electricity, and 15% water. There is also evidence that around one quarter of pre-payment metered service users (3.5 million households for electricity and 2 million for gas) report self-disconnection, and in telephony disconnections because of non-payment were around one million per year, and 60% of them were not followed by re-connection in the same year. OFGEM reports (2008, 2009) show that disconnection rates, the most important indicator of fuel poverty, after rising in the 1990s following privatization, started to decline in the UK when social pressure not to disconnect the poor induced suppliers to introduce voluntary safety nets. The FDS (2010) report offers also qualitative insights into the motivations behind self-disconnections. These are said to have been increasing again in recent years. According to the 2008 EU-SILC survey data, arrears in utility bills (electricity, water, gas) are reported by 3.31% of the respondents in Germany, 3.09% in Spain, 10.81% in Hungary, 5.84% in Ireland, 3.81% in the UK, 5.02% in Sweden, and 11.18% in Italy. The average for the respondents is 6.53%. If we just use this average to make a guess about how many people are involved in utility arrears in the EU-27, given the average household of 2.4 persons, and a population close to 500 million in 2008, there are around 33 million people living in households who are unable to regularly pay their bills.

The overall European situation of fuel poverty, which is a wider concept than being in arrears with payments—as most poor do in fact pay their bills—is estimated by a recent EPEE study.[35] This study mainly considers five countries: UK, Italy, Spain, France, and Belgium, before the recession, which can only have worsened the figures. It does not include any of the New Member States, where fuel poverty can be higher than in Western Europe. The EPEE study suggests that in recent years between 50 to 125 million EU citizens have been in fuel poverty, with the wide range of the estimate depending upon the possible definitions and the uncertainty surrounding the data. These households are reported to share certain characteristics: inability to pay energy bills; cold, damp living conditions; disconnection from supply, including self-disconnections in some countries; debt owed to suppliers; and disease because of poor health conditions related to insufficient heating.

[34] Public Utilities Access Forum report, http://www.puaf.org.uk/docs/news/Affordability-%20a%20discussion%20paper.pdf [accessed 8 January 2013].

[35] Tackling fuel poverty in Europe.

9.9 Conclusions

We have found evidence that, around 2010, the utility poor—either defined as those who need to pay more than a certain critical share of their incomes to pay their energy, water, and so on or those who report difficulties in paying their utility bills regularly, or who are in specific vulnerable groups and cannot achieve a minimum standard of consumption—are to be found among the around 17% EU households under absolute material deprivation[36] or relative poverty, and within the 25% of the population for whom the combined criteria of absolute and relative poverty and low labour participation applies.

As discussed in Section 9.1, the variability of deprivation (according to EU-SILC data) across countries and regions according to the scant available evidence is high, from around 2% in the Netherlands to more than 10% in the UK and in some New Member States. A conservative guess estimate for the EU-27 is in the region 6.5% of households under utility deprivation on average.[37] This corresponds to more than 33 million people, with a risk area (for at least one utility) equal to the population under absolute material deprivation, that is 86 million people, or 17% of EU population, see the data in the previous section.

Fuel poverty is the main concern. The water poor are possibly a slightly lower share of the total, particularly because of the resistance to increase tariffs for this service in the former transition countries and elsewhere by municipally-owned providers. There may be larger areas of utility poverty in district heating, public transport, and telephony, while this is an under-researched subject. One possible explanation of the importance of fuel poverty is the widespread second degree price discrimination in energy prices for households, as shown in Section 9.4, and this leads to regressive structuring of electricity and gas tariffs.

There is some cross-country empirical evidence (Section 9.6) of the adverse effects of energy reforms on the probability of households experiencing deprivation in several European countries. Looking at the persistence of deprivation over time, there is also some evidence of household heterogeneity and true state dependence, even after controlling for observed household differences. Further research is needed in this area.

Moreover, there is clear evidence (Section 9.7) of considerable variation in access to high quality network services across countries and regions of the EU, with those lagging behind in terms of the per capita income needed to support

[36] As mentioned, according to EU-SILC, deprivation indicators would consider the utilities (electricity, gas, water, telephony, heating) and some appliances.

[37] A less consultative and more realistic estimate is simply to consider the share of severely deprived households (8%). This would bring the figure to around 40 million.

infrastructure investment and service provision by substantial public expenditure and assistance from the EU Cohesion policy.

The abovementioned results suggest that the opening of markets in network industries has not solved the issue of social inclusion of the most vulnerable users. We discussed in Section 9.2 why, without strong competitive pressures, the privatization and liberalization reforms are unlikely to deliver lower prices to the lower end of users and may possibly have the opposite effect if not counteracted by regulation. In fact, the removal, after privatization, of cross-subsidies that favoured low users and unbundling—seen as the first step of market opening—may have benefited other users, when combined with liberalization at the upper end of the market. Competition is not necessarily beneficial to the poor if tariff structures are regressive and if switching between suppliers is affected by errors correlated to social vulnerability, as discussed in Sections 9.5 and 9.8.

Since deprivation may increase due to deregulation and overall trends in income inequity, remedial actions should target consumers who cannot pay, with an emphasis on the rights and the needs of consumers in low-income percentiles. Recent experiences in the UK and Argentina point in the direction of social tariffs as mechanisms to promote access and affordability (Petrecolla, 2003; Baker, 2006). However, remedial actions are a matter of controversy both in mature markets (Waddams, 2009) and in developing countries (Lampietti et al., 2007). Some think that cash subsidies targeted at the poor are more beneficial than lifeline tariffs or other progressive price mechanisms, but empirical testing is often insufficient.

One may consider that some social exclusion is unavoidable, or that it would be sub-optimal to avoid disconnections of non-payers. EU legislation, however, tends to see energy, water, telephony, public transport services, and so on, as necessary for social inclusion, and favours a degree of protection of vulnerable users in this area. This is a broader policy issue that we will discuss in the next chapter.

Protection of the utility poor may be particularly necessary for the New Member States of the EU, or for less-developed regions, where poverty is more widespread, regulation weaker, and the tariff increases following the reform of water and energy utilities are substantial because of the previous pattern of cross-subsidies. Moreover, with increasing unemployment following the global crisis of 2008, the trend may be worsening.

When we consider the combined household expenditures for electricity, heating, water, telephony, and public transport (rail and bus), i.e. the typical network services, the average household expenditure is around 15–20% of income. For those in the lower income brackets, and particularly for some vulnerable groups, any price change is highly sensitive. After all, the expenditures of a poor household are mostly composed of few items: housing, food, utilities, and some social services (health, education) as far as not provided by the welfare state for free or at low price. When looked at from this perspective, the evidence of regressive price structures or public utilities has important policy implications.

While the impact of price changes might have been limited (either negative or positive) for the median consumer—according to the different services, countries or regions, and type of reforms—for poor households the welfare effect is certainly not negligible. Most of the utility poor, or those at poverty risk, are in Central East Europe and in the regions of Southern Europe lagging behind, but they are also in the UK and in the disadvantaged social groups of many more developed countries, where inequality has increased over the last two decades.

There are different policy approaches to this issue. Some would argue in favour of doing nothing or doing the minimum, as after all they would not see a fundamental right or entitlement to some services. The 'doing the minimum' would possibly take the form of considering that some utility expenditure is taken into account when calculating monetary social benefits, but without explicitly targeting subsidies to the network services. Others would make a case for specific programmes targeted only to those who deserve social support. This seems to be the orientation of many regulators and governments in Europe. Discounted local bus tariffs are offered to the elderly, lifeline tariffs for electricity to some categories, and so on.

International research on support through this form of targeted assistance is mixed. Kennedy (2005) finds little evidence that means-testing in this form actually works in less developed economies. It implies a multiplicity of schemes for different services, and it seems easily appropriated by users who are less in need than others.

If one takes the view of a universal entitlement to a minimum consumption standard, a block tariff would probably be more effective than means testing, even if it is not the most equitable solution. Under this arrangement, any household would pay a zero or low tariff for the first amount consumed, up to a threshold corresponding to the required minimum standard. Thus, if there is a consensus that—say—1200 KWh/year are a basic entitlement for electricity for a household, or 50 litres of water per day, and so on, then these amounts should be offered either for free, or at a very low price, similarly to what is done in most public health and education systems in Europe, where a certain level of free services or at very low price are given to whoever deserves them, even if they are not poor. There may be a case for explicitly reconsidering the price (and tax) structure of network services in this policy perspective.

The economic benefits of such an arrangement are as follows. First, the informative burden for social assistance to locate the fuel poor, the water poor, the digital information poor, and so on is eliminated. Second, the cost of administering a number of different support schemes, each one tailored for one type of citizen, also vanishes. Third, even if the 'basic need' quantity is priced zero, the impact of this distortion on marginal prices for the remaining supply seems reasonable. Obviously, if the threshold of the entitlement is too high, this solution would be too costly to meet the budget constraint of the providers, and government transfers to the suppliers would be increased. This is mainly a political decision, which in fact is an alternative to price regulation in other forms. Also, if the quantity of the entitlement is too finely varied with the number of family members, climatic conditions, square metres, type of dwelling and so on there is the risk of going back to targeted support, which implies costly means-testing. Some regional variations seem reasonable, however, given the territorial dimension of utility affordability. Universal service in the form of progressive block tariffs should be considered as a social compromise, as other forms of welfare state arrangements in Europe and elsewhere.

A further advantage of reconsidering the whole issue of pricing mechanisms and of the objectives of network service provision lies in the view of citizenship as founded on an equal entitlement to certain goods. This is the broader perspective that we shall develop in the next chapter, as it directly questions the current dominant paradigm.

9.10 Further reading

For a theoretical treatment of pricing by public enterprises and distributive issues see Bös (1986) and Vogelsang (1990). On price discrimination see Varian (1985, 1989), Phlips (1983) and more recently Baumol and Swanson (2003) or McAfee (2008).

The empirical literature on the distributional impact of utilities reform is growing. Several papers use the simple, but robust, approach by Waddams Price and Hancock (1998). This is based on a 'before–after' reform estimation of changes in a proxy of consumer surplus.[38] For papers using this approach see Ugaz and Waddams Price (2003) on Latin America, Bagadioglu et al. (2007) on Turkey, and Waddams Price and Pham (2009) on Albania and Bulgaria. In the latter paper the authors find that in these two countries, but possibly also

[38] The tariff at the initial year is identified and used to compute the consumption level at that tariff compatible with household expenditure as reported in survey data. Then, expenditure after reform is computed at the original consumption level. This can be based on observed data, or on simulations of the price change needed to achieve cost-reflective tariffs, which in turn can be estimated through information on the full cost of service. This approach in its crudest form implies zero elasticity to own price, and is in fact a Layspere's measure of welfare change, which is the upper bound of a Marshallian measure, with a Paasche index giving a lower bound, see Florio (2007) for a discussion.

in other countries in South East Europe, simulation of reforms leading to cost-reflective pricing would imply social affordability problems for the bottom income decile. The analysis is somewhat complicated when you consider non-zero own price elasticity, cross price elasticities, and income elasticity. For some poor consumers price elasticity can be high, but this may imply non-payment instead of no consumption. This is reported for Turkey, for example, where data on household appliances does not match data on electricity expenditure. See also Lampietti et al. (2007), which is unique as it matches households' budget data with the corresponding data on actual payments and consumption of electricity from the utilities. Short-term price elasticities tend to diverge from long-term ones and EBRD (2003) and Fankhauser and Tepic (2007) provide some empirical estimations of demand elasticities. Baker et al. (1989) discuss in general energy demand modelling with micro-data.

Lifeline tariffs as a remedial action for the utility affordability issue often take the form of a block tariff, where a certain minimum amount of the bill is priced below cost either for all users or for certain types. For example in Albania for some years a lifeline tariff was established with a 50% discount on the price of the first 300 KWh. A still useful book on subsidy mechanisms is World Bank (2000). A recent example of progressive tariff structure is given by Cape Town (South Africa), where the electricity company has a three block structure, with the lifeline charge at zero price for the first 50 KWh per month, a second bloc at 50–150 KWh, a third one at 150–450 KWh. The domestic low tariff is up to 1500 KWh per month, while the higher offers a modest discount of 10% but also a modest fixed service charge to be paid, while this is zero for lower consumption levels. This progressive tariff structure (www.capetow.gov.za) was motivated by a sudden increase of prices, in a social context where poverty is widespread. In South Africa water also is often zero-priced for the first block of consumption.

On search and switching costs, bounded rationality of consumers, and possible remedial actions by regulators Garrod et al. (2009) provide a concise but very helpful review. They frame the discussion of remedies in the context of the well known distinction between search goods, experience goods, and credence goods, the latter being those for which ex post consumers are unable to assess the quality of the offer. Internet based sites are discussed, including price comparison sites that are actually available for some network services. The expectation that this will lead to less price discrimination seems over-optimistic, as documented by the research cited by the authors. This literature shows that in some services or product markets only a tiny fraction of consumers take advantage of price comparison. Another potential remedy is regulated standardization of the firms' information on prices. This is particularly difficult with non-linear tariffs, which are typical of network industries. On switching costs in general an important reference is Klemperer (1995). Six categories are outlined: costs can arise from the opportunity cost of time in switching (transaction cost); direct costs from loyalty pricing (contractual costs); knowledge costs about the functioning of the new contract; compatibility costs (typical in the computer industry); uncertainty as a specific cost; and the pure psychological cost of switching. A specific aspect related to affordability is when a poor consumer is prevented from switching because of arrears, thus being locked-in with a higher-price supplier.

Evidence that competition in telephony may be beneficial to low income household is provided by Economides et al. (2008), in a detailed study of entry (mainly through local loop unbundling) of two competitors in New York City.[39] They conclude that in terms of compensating variations the average benefit of entry is worth USD 0.83 per month, or 2.5% of the bill, but with a very large standard deviation. The authors also state that 'because larger, lower income household have higher local telecommunication services demand, on average, entry benefits disproportionately high-usage, and thus high-valuation, households' (Economides et al., 2008, p. 724). This result should be interpreted in combination with the other results. As mentioned, these suggest that the benefits of entry were not enjoyed by many among those who did not switch. Here demographic characteristics may have played a role, particularly for the black or Hispanic incumbent's customers (marginal effect of respectively 32% and 26% of mistaken provider choice). A careful reading of this interesting paper suggests how difficult it is in fact to identify who the losers and winners are in the market liberalization game when there are so many tariff plans, even when there are just three competitors.

Miravete (2003) argues in favour of rationality of telephone users, based on an experiment in Luisville in 1986, where users were given the alternative between a flat rate and metered service.[40] Clifton et al. (2011), after reviewing

[39] With very detailed information on a sample of 7222 households, they conclude that on average AT&T users save 1.15 USD per month compared to the incumbent Verizon, and MCI customers, the other entrant, save USD 0.40 per month. In fact the saving is mostly due to buying long-distance service in addition to local service, rather than because of marginal prices (which were actually higher with the entrants in the sampling period, 1999–2003). The authors report, however, that the average value masks the heterogeneity of benefits across households. In fact a significant number of entrants' customers actually pay more than with the incumbent, by a not negligible 20% increase of their bill, and 40% of either the incumbent's or entrants' customers lose money because they do not switch to the best alternative. This share is greater than for those who select the wrong tariff plan (there are several alternatives) within their provider. Then the authors estimate the household characteristics, which correlate with overpayment through a probit model. From the perspective of this chapter, the finding that the probability of the incumbent's customers to overpay increases with them being black, Hispanic, or another minority is interesting. The opposite is likely, however, for AT&T switchers. Household size increases the chance of getting the right contract. In general the authors find that demographic characteristics have low explanatory power in the probit model, while a multinomial logit works better. The core of the paper is a complex demand model, which the authors use to perform a welfare counterfactual analysis.

[40] The sample data allow identification of expected and actual calls, and this leads to estimation of the benefit of the flat rate for more intensive users. Households with higher income tend to

insights from behavioural economics and definitions of consumer vulnerability in the field of services of general interest, consider both objective and subjective evidence in Spain and the UK.[41] Several consumer organizations and researchers are active in Europe in looking for a more comprehensive approach on utility poverty, for example the EPEE partnership (www.fuel-poverty.org).

On water poverty, OECD (2003) is a comprehensive study that concludes that in 15 out of 30 Member States affordability 'is either a significant issue now or might become in future, if appropriate policy measures are not put in place'. See also OECD (2005). For the lowest decile of the population, the share of expenditure on net disposable income for water and waste water goes from around 1.1% in Italy, the Netherlands and Sweden, to 5.3% in Slovakia and 9% in Poland. See also Hall and Lobina (2007) and Massarutto (2007).

Appendix 9.A.1: A diagrammatic example of social exclusion

As in Chapter 3, to simplify, we assume constant returns around equilibrium.[42] A monopolistic public enterprise would charge a price equal to marginal cost. The latter in turn is possibly (slightly) higher than in a first-best world, because of some inefficiency of the public enterprise. The poor with elastic demand are served a minimum quantity by construction.

Figure 9.A.1 shows before–after equilibria, with no price discrimination, but also no price regulation. The equilibrium price p is determined by the marginal costs c^* (bottom panel), which are lower than under public ownership, and by the 'horizontally'

subscribe to the metered service, while large household, or those with teen-agers),'black', or on benefit support, prefer the flat rate more. In fact, around 90% of the sampled households subscribed to a flat rate, and it is not entirely clear, given the very limited nature of the choice, why in general one would have expected more behavioural irrationality in such a simple context, with just one provider and a simple alternative. Thus, it does not seem that that the evidence provided by Miravete (2003), going back to 1986 in a special context, can reverse the evidence elsewhere that consumers make mistakes, and that poor households are less protected from obfuscation of choice in network services.

[41] On the objective evidence side they use data from family expenditure surveys, while for subjective evidence they use Eurobarometer surveys. They test three hypotheses: a) most vulnerable citizens will express greater discontent with prices they need to pay for essential services; b) vulnerable citizens have a specific consumption pattern; c) some problems of affordability should surface across sectors and countries. The vulnerability characteristics they consider include six items: age (the elderly and the young); limited education; low income; unemployment; poor health; and ethnicity. Given the lack of data they focus on some proxies. They do not find clear evidence of a link between dissatisfaction with prices and vulnerability variables in the case of electricity, while they find such evidence for telecommunications. This pattern is broadly confirmed by objective evidence on expenditure data. These results point to heterogeneity of vulnerability across sectors and countries. They also mention some of the remedial action by the regulators, and assess them as 'ad-hoc, patchy and partial solutions'.

[42] An example is a local utility which resells electricity or water, but does not own the network. Thus it faces variable costs only, while the network itself has decreasing costs.

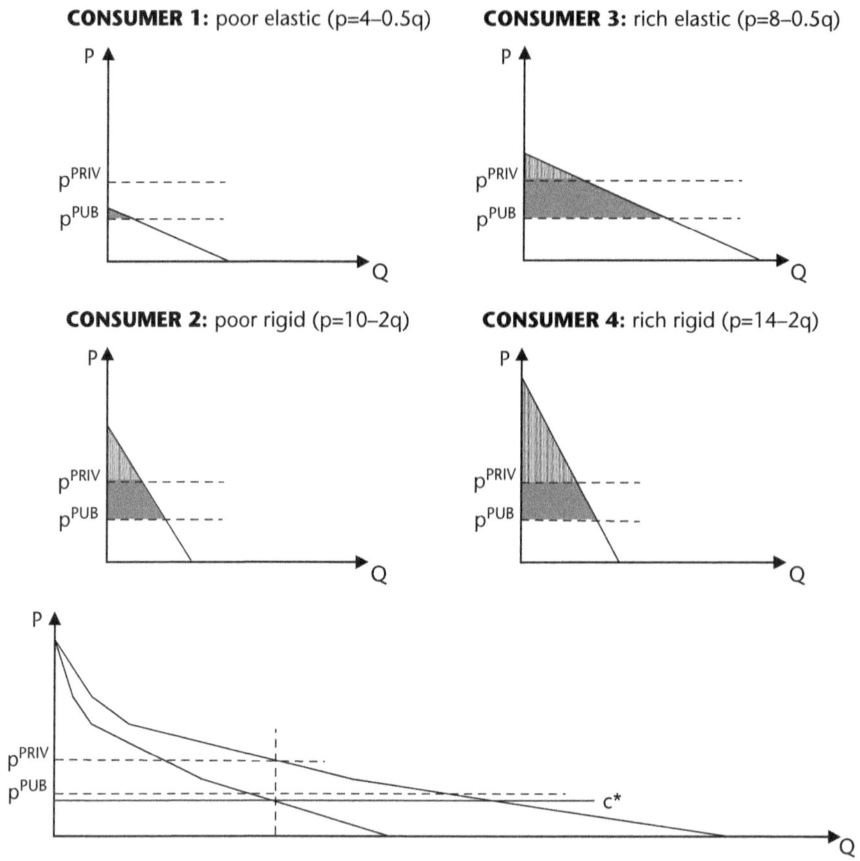

Figure 9.A.1. Public monopoly and private unregulated monopoly, without price discrimination

Note: the private monopoly equilibrium price p^{PRIV} is determined in the bottom panel, and compared with p^{PUB} (public monopoly). Consumer surplus is shown for each type. The grey area represents the loss of consumer surplus. Consumer 1 is not served under price monopoly without price discrimination.

combined demand curve. Hence, consumer 1 is not served, while by construction he was served to a certain minimum amount under public provision.

Suppose now that again the government has no equity objective, an issue that in Chapter 3 was represented by the fact that we were interested in aggregate consumer surplus only. The government, however, is concerned by the allocative inefficiency of monopoly. Figure 9.A.2 shows that if the government seeks efficiency by a common price-cap mechanism $p^\wedge$, but price discrimination is now feasible, then the poor with elastic demand (consumer 1) is only marginally supplied with electricity[43] perhaps by a too little service.

[43] We do not show the figures for duopoly or oligopoly, because the logic would be similar to the simple model of Chapter 3. If instead we have to consider duopoly, e.g. in mobile telephony, one

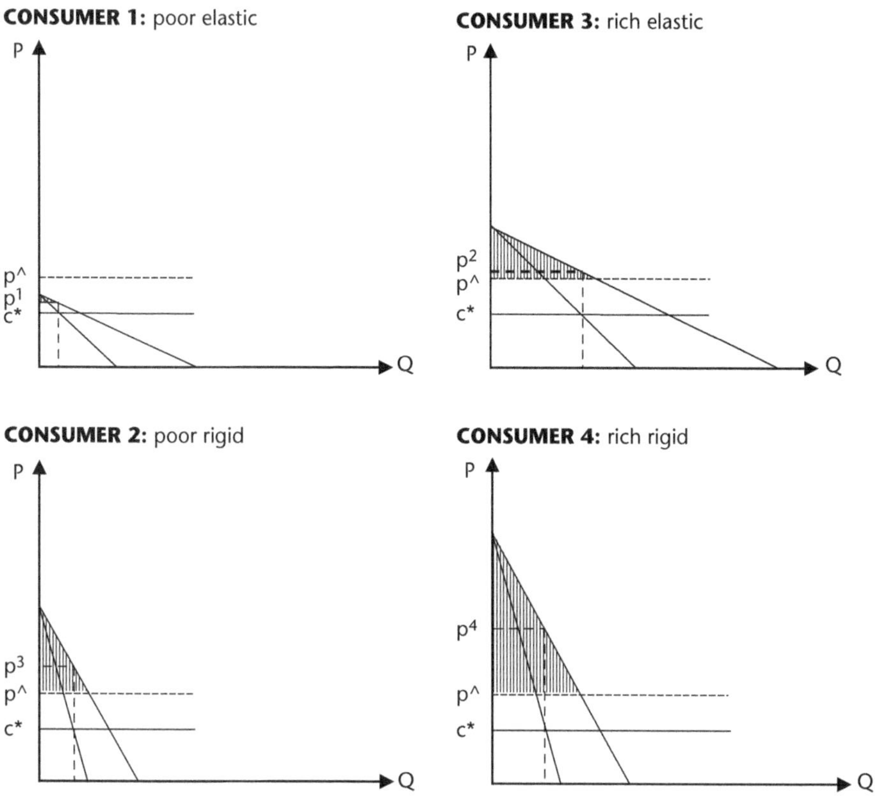

Figure 9.A.2. Private monopoly, with price discrimination and price-cap

Note: with privatization, price discrimination, and price-cap p^, consumer 1 is served, even if the cap is not binding for him.

As a final step, suppose that the government now wants every citizen to be entitled to a minimum quantity of electricity, $q^\wedge$, whatever his income or preference, and has no way of getting the information needed to set lump-sum taxation and subsidies in order to push minimum consumption at that level. Figures 9.A.3 shows that quantity regulation may be needed to include the poor in the market, with a minimum amount of service (alternatively a specific pro-poor price discrimination would be needed) (see Fig. 9.A.4).

can think back to the simple model where the entrant gets the residual market, as in Chapter 3, but this time *for each type of user*. Thus you are going to have a set of duopolies, and by extensions, a set of oligopolies, one for each market segment associated to a type of consumer. These models, however, are complex, and need assumptions on game-theoretic setting, while here we want to be as simple as possible.

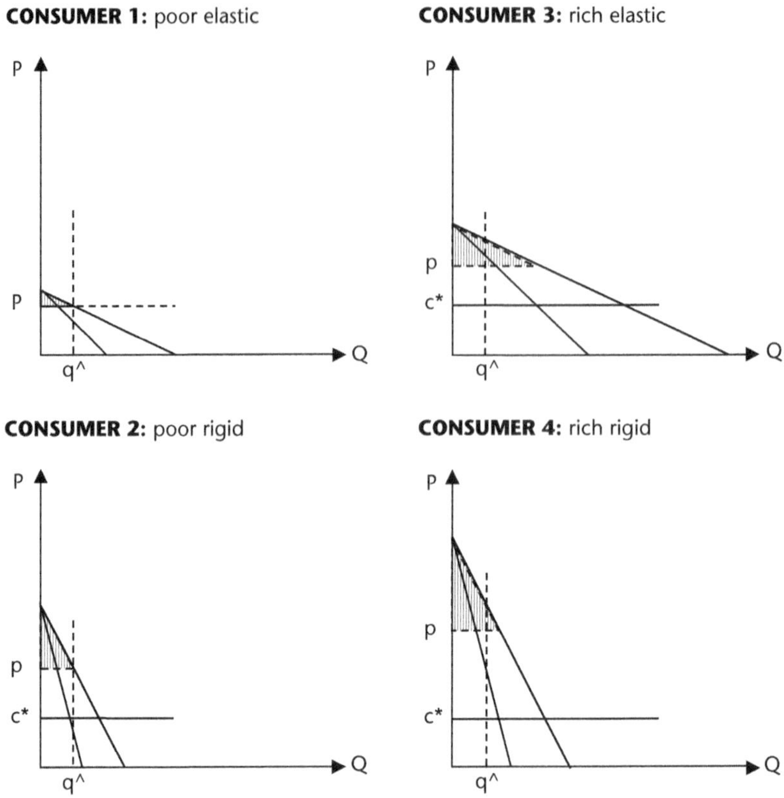

Figure 9.A.3. Private monopoly with price discrimination and quantity regulation

Note: alternatively, consumer 1 is served under privatization and price discrimination when a minimum quantity constraint q^ is introduced.

Readers can play with this very simple toy model and see that there are cases when a profit-maximizing monopoly is unable to satisfy the quantity constraint, even when its price is regulated, but still needs to allow some profit to the provider. Simply, a private firm will not serve certain unprofitable users, and would need a subsidy to allow them access at affordable prices, such that the users get the minimum quantity of electricity as required by the government. The above mentioned example shows that the same applies to competitive markets, the only difference being that the government there does not consider that citizens are entitled to a minimum supply of the good.

Price discrimination plays a crucial role in this picture. If the privatized natural monopoly is capped on its *average* price, but not otherwise, it maximizes its profits by offering different prices to the customers. The consumer welfare impact of the resulting equilibrium can be shown by 'welfare pies'. The area of a 'welfare pie' is given by the sum of consumer surpluses, each multiplied by an appropriate welfare weight. Figure 9.A.4 shows the 'welfare pies' with the individual contributions, when the government has a social welfare function such that welfare weights of the form

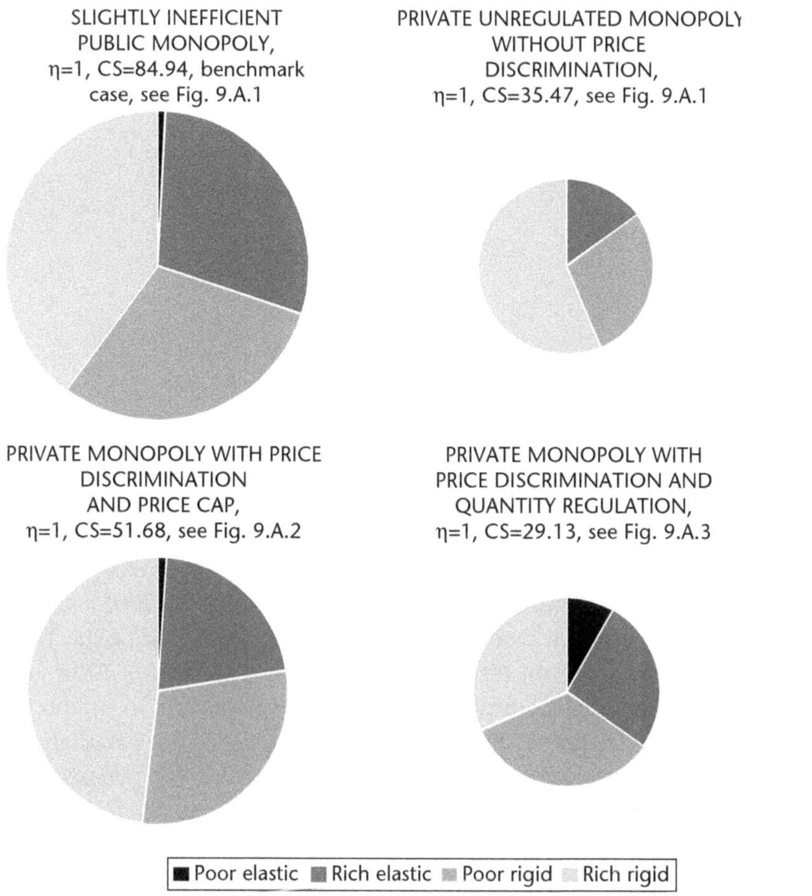

Figure 9.A.4. The welfare pies

Note: the figure shows four 'welfare pies' each corresponding to a welfare weighted sum of consumer surpluses (CS) from the previous figures, with modest government aversion to inequality ($\eta = 1$). The area of each pie is proportional to the contribution to social welfare of the consumption by the four types of users. The benchmark case is for a public monopolist that supplies at a price equal to marginal cost, but the latter is slightly higher than the minimum cost, as achieved by private providers.

$(Y^{Poor}/Y_i)^{-\eta}$ and $\eta = 1$. The function that generates this government preference for equity is well known in applied welfare economics, see for example Brent (2006, chapter 10) or Evans and Kula (2011) who discuss the extension of the welfare weights concept to the regional dimension.

The figure shows that in the illustrative case the cost inefficiency of the public monopolist does not change the welfare dominance of a price-marginal cost regulation. Interestingly, a private monopoly with price-cap and price discrimination would be better than quantity regulation.

10

Conclusions

10.1 The dominant policy paradigm re-examined

Over the last 20 years, in Europe and elsewhere, policy-makers have reversed the way they previously considered network services and their role in the society. The change has involved the regulation and governance of industries which contribute around 15% of EU-15 GDP (Martin et al., 2005). It has been a huge policy experiment. Some of the rationale, progress, and expected effects of the reforms have been discussed in Chapters 1, 2, and 3. After empirical analyses, mainly of telecommunication and energy services in Chapters 4 to 9, this concluding chapter evaluates the overall impact of the paradigm shift on the welfare of households. The discussion will broaden the perspective of the evaluation to the social and political dimensions of network industries.

A policy paradigm can be depicted as the achievement of a widespread consensus around a set of objectives and their implementation mechanisms in a broad area of government intervention.[1] Some observers may have seen the reforms of network industries in the 1990s as precipitated by new technological developments, or by a better understanding of economic principles, or by a combination of these facts.[2] Paradigm shifts, however, are never just a technical or intellectual matter. To win a consensus, and build a new policy orthodoxy, the change of paradigm needs to be widely shared by political and economic elites. In Europe this change originated in the UK under Mrs Thatcher's government at the beginning of the 1980s. There was then a gradual spread in other Western Europe countries. Once the new consensus was reached, it was promoted on a larger scale by European institutions.

The experiment in the EU has affected more than five hundred million citizens, those who use these services in their everyday life. It would seem

[1] Hall, 1993.

[2] Melody (2011) suggests that causality goes from reforms to technological change and not the other way round.

hard to deny the importance of this story of change. After around 20 years, what can we say about the welfare effects of the reforms from the perspective of households?

Three conclusions may frame the re-assessment of the reforms. *First*, in spite of technological change, network industries still retain the peculiar pattern that attracted the attention of economists as early as the nineteenth century: a natural monopoly core, combined with a very widespread and rather inelastic demand, often associated with large production and network externalities.

Second, because of these features, profit-maximizing firms had an incentive to adopt un-competitive strategies. These lead to monopolistic or oligopolistic market equilibria, associated with significant mark-ups over costs and price discrimination. To counteract these outcomes, governments have extensively intervened in the network industries. Historically, public policies in this area have taken different forms, from nationalization to intrusive regulation of private companies. In general, 'standard' competition policy has not yet supplanted these special mechanisms.

Third, the services offered by the network industries are often (albeit not always) important for the overall functioning of the economy, for the social inclusion of citizens, territorial cohesion and for the achievement of specific policy objectives. These features still make security of supply, pricing, quality, accessibility, affordability, and other aspects of service provision politically sensitive issues. As an introduction to the rest of the chapter, these three propositions are briefly discussed below.

On the persistence of a natural monopoly core, earlier views in the 1980s suggested that the separation of competitive operations from the capital-intensive network would be facilitated by the waning of economies of scale in other segments of the industry, upstream or downstream. Thus, the mandatory insulation of the non-competitive segment, regulation of access to bottleneck infrastructures, and wide liberalization would structurally transform the industries. Consumers could then enjoy the benefits of competition, thanks to an act of surgery (separation of the networks) and to the reshuffling of the building blocks of the service provision architecture. EU regulations for energy and telecommunications (and similar legislation for transport, water and other services) were increasingly based on this view.

The passage of time permits us to discern that the hope of achieving effective competitive outcomes in a reasonable time span was exaggerated in the first place. The reasons lie in those permanent features of the industries that we reviewed in Chapters 4, 5, and 6. Most of the upstream and downstream competitive potential was already available in Europe in several public utilities even with the old technologies, well before the paradigm shift in the 1990s. The concessions, franchised monopoly, or oligopoly arrangements that were experimented in Europe before nationalizations or municipalizations were often under private ownership. Oligopolistic competition, when occurred, tended to oscillate between collusion and predatory price wars and disruption. Many of these arrangements were then rejected as unsatisfactory by governments, in some cases as early as the late nineteenth century, as a part of their broad public national policy agendas (Millward, 2011).

The fundamental features of core activities under increasing returns-to-scale, and the systemic linkages between these and other activities, lead de facto to one or a limited number of viable operations on the supply side. This structure is still well reflected in the current industry organization in Europe. What the reforms have actually achieved in most countries is a transition from a franchised monopoly to regulated oligopoly, with the networks sometime separated and sometime still vertically integrated, with strict regulation of access to third parties.

In spite of this 'arms' length' supervision of the industries, governments and law-makers are far from being relieved from their more direct concerns about the functioning and outcomes of the network services. Energy, communication, transport, or environmental service policies certainly did not disappear from the agenda of ministries.

The hope that a well-functioning and mostly self-regulated market would take care of determining the optimal level of long-term investment, the most appropriate technology mix, and the best feasible protection of consumers and users at large, through a (social)[3] cost reflective pricing mechanism seems now even more unrealistic than 20 years ago.

In a nutshell: natural monopoly is still at the core of the network industries, intrusive regulation is needed, and political and social objectives related to their services have not disappeared from the agenda of governments.

10.2 Implementation: Too little or too much of it?

As far as the original ambitions of the reform paradigm, how far has it been implemented in Europe? Chapter 2 has provided a sketched history of the progress made in the EU-15 since 1975 until around 2007. A common policy trend has been discerned, with variations across countries and timing of adoption.

In the telecommunication markets, public monopoly does not exist anymore since the end of the 1990s, and private monopoly, which existed only in Spain, and as a transitory situation in some other countries, has also disappeared. In 2007 only Luxembourg had a state-owned operator, all the

[3] Including externalities.

other companies where privatized (or partly privatized) and third party access was granted to every entrant. At the beginning of the 1980s in the EU-15 there were 14 public monopolies and 1 private one, there are now 15 markets open to some competition, with a public owned operator in just one country, and the rest fully or partly under private ownership.

Networks, however, are nowhere fully unbundled. The incumbents are still the owners of the core infrastructures for fixed telephony, with special regulatory mechanisms to grant access to local loops and other bottlenecks to third parties. Mobile telephony is also highly concentrated.

In electricity, until the end of the 1980s, the industry in the EU-15 was managed as a vertically integrated monopoly, with only one country, Belgium, where the monopolist was private, 12 countries where ownership was public or mostly public, and two where there were both public and private companies in different regions (Germany and Spain). Around 2007, in all the EU-15 the industry was vertically disintegrated, with regulated third party access, but with private or mostly private ownership only in the UK, Belgium, and Spain. In Germany, generation is under private ownership, but local distribution is still often under public ownership. In Sweden, one of the most liberalized, the electricity industry is managed mostly by public enterprises.

In gas, in the 1980s, there was some limited market opening in Germany only, in the remaining 14 countries there was a statutory monopoly, with government-owned companies in 8 countries and a private monopolist in the remaining 6. Around 2007, only in Greece and Ireland there are government owned companies, in the remaining countries there are private or partly private operators. Nevertheless, in 10 out of 15 countries the market is still nearly monopolistic, and some of the partly privatized providers are under government control to a certain extent, as in France or Italy.

Thus, in the three key network industries, the paradigm, after 20 years of promotion by EU directives and national legislation, has nowhere been implemented in its full form. In telephony there is privatization and market opening, *without (full) unbundling*. In electricity, there is unbundling and market opening, *without (full) privatization*. In gas, there is privatization and unbundling, but *without (full) market opening*. The picture is similarly variable in other network industries, such as the railways, water, and postal services.

Looking beyond formal legislation, the picture is even more nuanced. Governments are minority shareholders in many of the privatized incumbents, and it is difficult in general to say whether they act as passive owners or if they have some influence in managerial decisions. Unbundling in some countries is in formal compliance with EU legislation, but is far from being the ideal outcome advocated by the paradigm. Third, and most important, competition in several countries and sectors is limited, particularly for services to the household. The latter point will be discussed later below.

Hence, in terms of speed of implementation, you can conclude that the paradigm has been an extraordinary success, as it has dramatically changed the shape of network industries in Europe. Alternatively, you can more prudently say that it is at best half way, as the ideal combination of its three pillars has not yet been achieved.

Is it then desirable to further speed up the implementation of the reform package with further privatization, ownership unbundling, and liberalization, or should policy-makers reconsider its architecture? The answer to this question lies in what we can learn from the empirical evidence on the European policy experiment, taking the viewpoint of consumer welfare.

10.3 Is the consumer dividend fair?

Empirical cross-country analysis of the correlation over time between the reforms (as scored by the OECD) and consumer prices (as recorded by Eurostat, IEA, TELIGEN, ITU) shows (Chapters 4, 5, and 6) that private ownership per se often tends to be associated with higher prices to consumers in electricity and in gas, and has virtually no effect in telephone services, for example with unit prices of local calls decreasing and fixed charges increasing with private ownership.[4] Vertical separation per se in energy has in general no statistically significant effect, and could not be tested in telephony because it has not been implemented systematically. There is some evidence that market opening, as expected, delivers lower prices to consumers, particularly in telecommunications, but the effect is often small. The privatization and liberalization effects on prices often tend to cancel out each other, or in any case are limited in either direction.

After 20 years of reforms, this nearly zero-sum game in terms of consumers' welfare does not seem to be an impressive outcome. It seems appropriate to consider possible reasons of such a modest direct[5] welfare effect of the reforms, in spite of the greater expectations it has raised.

The simple conceptual model of Chapter 3, or more detailed theoretical analyses,[6] can help to frame the discussion. The starting point of any prediction on price outcomes of the reforms is the cost differential between public

[4] In mobile telephony the ownership variable has no statistically significant impact on prices.

[5] Indirect welfare effects, through effects on other agents, have not been explored in this book. The most important of those effects may have been on the profitability of providers, but a part of profit is monopolistic rent that does not necessarily increase social welfare. The general equilibrium effects of the reforms are a difficult topic and further research on this would be welcome.

[6] Reviewed by Willner (2011).

and private ownership in the first place. A 'before–after' approach in this context is not very helpful. In a progressive economy productivity usually increases because of demand shifts and technological improvements. Input costs mostly change exogenously as well. Thus, the relevant question is not whether a privatized company is more efficient than its former government-owned counterpart, but if variations of ownership and regulation have a discernible effect on the trend of productivity and prices. The possibility of knowing this through individual company and country case studies has, however, been suppressed by history, as you cannot observe both what happened and what did not happen. You can make conjectures and simulations, but not much more than that. Cross-country and cross-firm comparison of outcomes seems a possible proxy of a counterfactual analysis.[7] This was the research avenue taken in this book.

The fact that consumer prices, after controlling for exogenous shifters, do not seem to respond in the expected way to the efficiency gains post-privatization, has two possible interpretations. Either the productivity gap between privately and publicly owned enterprises in the EU-15 was modest or, if it existed, the liberalization side of the reform did not work well enough, at least for the consumers. The two interpretations are not mutually exclusive.

Our findings suggest that most of the price variability across countries (and, implicitly, across firms) and over time is to a large extent statistically explained by exogenous demand and supply shifters, and by country fixed-effects.

To duly interpret the empirical evidence, it seems useful to restate a crucial point. The three industries we have analysed were (are) virtually never loss makers under public ownership in Western Europe.

If privatization has led to a productivity increase beyond the exogenous dynamic trend, which is uncertain, the negligible price response would imply that liberalization has failed to pass those gains on to the consumer. On a continental scale, this would imply a massive transfer of rents from those who benefited from some inefficiencies, if any, under public ownership (perhaps some categories of users, unionized employees, policy-makers, and some suppliers) to the shareholders of the privatized companies and of the entrants, with the median consumer as passive vehicle of the reshuffling of rents. Taxpayers also had little to gain from this transfer, and were possibly the losers in the game, if the privatization proceeds have been insufficient to compensate for the loss of net public income from profitable public corporations.[8]

[7] Even if it can be occasionally attempted, as e.g. in Galal et al. 1994, Newbery and Pollitt 1997, and Florio 2004. In this study we have not discussed what would have happened in one country under continued public ownership, but we observe what has happened in other countries that did not privatize the industry.

[8] As argued by Florio (2004) for the UK.

10.4 Is domestic oligopoly a second-best?

Why have efficiency gains from privatization, if they actually existed, led to so limited a benefit in terms of consumer welfare? The answer should lie in the regulatory mechanisms designed and implemented to force competition in industries, which still retain their core un-competitive features.

Price regulations and supervision of TPA were different across sectors and countries. What seems invariably similar across the reformed industries, is that the market outcome can be described as either domestic oligopoly or local monopoly, combined with the regulated monopoly of the core networks. This pattern cannot be described as full competition.

While in some theoretical frameworks (particularly under the Baumol, Panzar and Willig (1982) frame, see also Baumol and Sidak, (1994)) even a monopoly, a duopoly, or a non-collusive oligopoly can deliver market equilibria close to full competition, most economic theoretical and empirical research suggests that the conditions for this outcome are quite restrictive and unlikely. Baumol (1982) has himself stated that perfectly contestable markets do not actually exist in real world. Alleman and Rappoport (2005, p.111), in their critical discussion of telecommunication reforms in the USA observe that:

> Since the 1996 Telecommunication Act, the FCC has formalized its pricing policy by relying on notions derived from competition theory... perfect competition implies that when there are many firms—virtually an infinite number—in the industry, no single firm can affect prices in the market. From the firm's point of view, according to the theory, demand is perfectly elastic and the firm produces where prices equal marginal costs. However infinity is a great deal more than two or three or four firms in the industry.

One may suggest that leading the network industries to oligopoly from franchised monopoly (public or private) is progress. Quality-adjusted prices are the core signals for this evaluation. Some aspects of the oligopolistic arrangements of contemporary industries, such as automotive or personal computers, to cite obvious examples, may be beneficial to consumers as market dominance of one player is unstable, and can be challenged by other players. In fact, market positions in these internationally open oligopolistic markets are crucially contestable and contested because of the dynamics of innovation, that is endogenous technical progress. As a result, car or computer users get improved goods over the years for the same price or for a falling price. Oligopolistic competition in these industries, while far from perfect competition, works through innovations that range from the appearance of the product, to energy savings, quality of materials, post-sale assistance, emotional values, and so on.

This dynamic is a core tenet of an evolutionary, neo-Schumpeterian view of the relationship between oligopoly, profit, investment, and social welfare, and it was advanced many years ago (see particularly Sylos Labini, 1962, and for dynamic considerations in regulation, see e.g. Ellig, 2001). In contrast, the research and development budgets of most network service incumbents after their privatization or exposure to market opening have been cut rather considerably (Sterlacchini, 2012), and often telephony and energy incumbents have not been the front-runners of innovation.

Further empirical research is needed to ascertain whether liberalization has been beneficial to technological change through the role of entrants. This again should be a mixture of counter-factual history and international comparisons. It is often widely held that there are clear incentives for profit-maximizing competitors to adopt new technologies or to create new ones. In an oligopolistic regulated market, however, the rationale for new technologies adoption, often linked to costly new infrastructure, is less strong, because of the intrinsic regulatory risks and the more short-term perspective. Telecommunications liberalization seems to have been more successful than in the energy services. Most of the product innovation there, however, occurred before or was unrelated to the reforms, or was set in motion by research and development in other industries, for example by Internet companies or producers of electronic equipment.

Competition in network service provision reveals itself to the consumer in the form of offers of discounts for the same product quality and the bundling of different services (e.g. triple or quadruple tariffs in the electronic communication markets, electricity supply combined with gas and water, etc). When consumers have a choice among service providers, it is often not clear to them what quality, or even price, they are choosing. The extent of this lack of information is unparalleled in other oligopolistic markets, except possibly in financial and insurance services, or in some other industries plagued by asymmetric information, such as health services.[9] Thus, consumers' choice, information, and perceptions about the price and quality of the services are important issues for the realistic evaluation of the welfare impact of oligopolistic markets, as opposed to franchised monopoly.[10]

[9] Airline liberalization offers, instead, an example of a market opening that allowed consumers to experience true choice, to pick up different combinations of service and prices, and to benefit from the reform. This was related to true innovation in the management of the industry.

[10] It may be the case that business customers of utilities have better information and greater choice. This question is outside the scope of our research, but we suspect that small businesses at least are often just as information constrained as residential customers regarding the price they pay and the quality they get for energy and telecommunication services.

10.5 Perceptions, quality, and information

Consumers' perceptions are crucial for two reasons. From a political economy perspective, the support for reform may be related to the perception of their outcomes. Moreover, subjective data may reveal micro-information that is poorly reflected in aggregate price statistics or in other objective performance indicators.

The European Commission has often used surveys such as Eurobarometer to record social attitudes in different areas of policy-making. We have not considered, however, social attitudes towards privatization or liberalization as such. As reported in Chapter 7 and 8, the questions asked by Eurobarometer surveys on network services were on perception of price fairness, quality, access, and transparency of contracts. There was no hint to the interviewee about the fact that his or her supplier was a private or government-owned company, a monopolist, or a player in a fully opened market.

There are no special assumptions about the information that respondents may or may not have about the reforms implemented in their country at the time of the interview. It is likely that some interviewees, of course, know that their supplier used to be government-owned and now it has been privatized, but it is also likely that most or many users ignore the changes. This cognitive dimension cannot be directly tested, but the level of education of the respondents can act as a control variable, as it is likely to be correlated to a certain amount of generic information about policy changes, for example as far as reading the press is correlated to education.

In electricity we have found evidence that consumers in countries where the industry is under public ownership tend to more frequently consider the prices they pay as 'fair'. The effect is statistically highly significant and robust to different model specifications. Consumers are also happier in countries where they have the opportunity to choose their supplier. This effect is often statistically significant, but weaker than for the ownership variable. This result is consistent with the previously mentioned finding that average prices are higher in countries where the electricity industry is or was under private ownership. In gas the evidence points to no impact of the reforms on satisfaction. In telecom the results are also limited, and puzzling. While public ownership is associated with a very small decrease in the marginal probability of being satisfied in some models (but not in others), we find again a consistent result between the objective and subjective evidence when we focus on the market structure. It is not formal liberalization that works, but the actual market share of entrants. The higher this is the lower prices and the higher the consumers' satisfaction for telephony services.

You actually need many competitors, with fairly balanced market shares, to bring down prices, and when this happens consumers are happier. As we have mentioned in earlier chapters, this pattern did not emerge frequently in the EU countries, because utility markets were still highly concentrated, usually with the incumbent retaining his dominance.

Consumer satisfaction may also be influenced by the perception of quality of service. The results for telecom on the quality question in the Eurobarometer survey are similar to the findings on prices, and tend to stress more the role of competition as a driver of satisfaction. In electricity, the findings on quality are also consistent with those on prices, but go in the opposite direction to those of telephony. Public ownership is associated with higher satisfaction for quality, while entry regulation has no effect, or surprisingly satisfaction decreases, with the lifting of barriers to entry. A double check with data on interruptions of electricity supply, a key quality objective indicator, suggests that privatization is correlated to increased service interruptions, and this may explain the less positive quality perceptions by users.

When trying to adopt a measurement error perspective, as mentioned in Chapter 8, satisfaction over the quality of the service increases when there is public ownership in telecoms and vertical integration (in gas). Market entry and structure are confirmed to have the expected effect on satisfaction in telecoms and in gas. When using multidimensional satisfaction indicators—including again prices and quality, but also clearness of information provided to consumers, fairness of contracts, handling of complaints, level of customer service, and looking at telephone, electricity, railways and postal service together—the resulting ranking by country seems only marginally related to the reform progresses.

Italy is often the country with the least overall satisfaction, and this may fit well with the expectations of the paradigm, as in the years we have considered only telecoms were privatized (not in a very successful way), and competition was virtually absent in other sectors. But at the other extreme, countries where the satisfaction of consumers is highest include Ireland, Luxembourg, and Denmark—along with the UK. While the UK is orthogonal to Italy, with only the postal service still under public ownership, and with wide liberalizations of the above mentioned sectors, this is not the case of the other best performers in terms of happiness of consumers. Ireland in those years had no third party access to telephony, and electricity was under public ownership. Luxembourg had a government-owned public telephone operator. Denmark had a mostly-public electricity industry.

A common result of all our empirical analyses both with objective and subjective data is that country effects, when introduced to account for unobserved variations, are highly significant. There may be two possible interpretations of this finding. Either the OECD summary indicators that we have used do not capture appropriately the reforms, or there are other features, different from the reforms, that explain part of the citizens' satisfaction and the prices they pay.

As far we can see, there is no reason to dismiss the OECD indicators in general.[11] Thus, we conclude that the Eurobarometer satisfaction data, when tested with the OECD reform scores and control variables, and with a variety of statistical models, tends to reject the null hypothesis that in general consumers are happier in countries where the network industries reform paradigm has been more intensively applied. We actually find some evidence in the opposite direction, but we prefer to prudently conclude that objective (i.e. price data) and subjective evidence (i.e. satisfaction data) converge in suggesting that 'on average' consumer' welfare in the EU-15, for the years and sectors we have considered, did not improve as a result of the reforms.

The 'average' effect, however, may conceal different effects for different types of users.

10.6 Affordability and social cohesion

Social cohesion, according to a study for the European Parliament[12] 'can be understood as the capacity of a society to safeguard the welfare for all its members. This implies the need for and guarantee of universal service and comprehensive coverage. It aims to minimise and reduce/avoid polarisation while generating and strengthening bonds among citizens. A cohesive society is one that has developed satisfactory democratic ways of managing its internal diversity.'

As we have discussed in Chapter 9, the new paradigm does not specifically address the social cohesion issues related to the provision of some services. In fact, based on panel data (ECHP and SILC surveys) in some countries and years, we have found some evidence that the implementation of the reforms is correlated to an increasing probability that the respondents experience difficulties in paying their bills for utilities. This effect is not large per se but when combined with increased inequality and social hardship after the recent crisis it is a matter of concern. Moreover, the consumers' ability to effectively switch from one provider to another in liberalized markets is probably correlated

[11] Obviously, they may occasionally contain errors (i.e. for a specific year and country the reform score should be different), but this applies to any of the policy databases widely used in mainstream political economy literature (e.g. databases on governance as Kaufmann, Kraay and Mastruzzi, 2005 sponsored by the World Bank, and several others, see Rothstein, 2011). Moreover we are not aware of better comparable indicators for network industry reform in the EU-15 and for a longer time span. Some robustness testing with other sources does not show major inconsistencies.

[12] DEAS et al. (2010).

to access to information, including the use of the Internet to compare the offers, and general cognitive abilities which, in turn, are correlated with education and income levels. There may have been improvements in this area, but they are not enough, as discussed in Chapter 9.

The regulatory framework that has been built to address the issue of social affordability has typically taken the weak form of some 'universal service obligations' for the incumbent, or of special provisions for tariffs in the case of the poor and vulnerable consumer.

The traditional alternative of cross-subsidies has gained perhaps too much of a bad reputation. The actual system as implemented by nationalized industries or municipal companies—whatever its theoretical justifications—was rather crude, and never fully satisfactory. In some cases it actually created inefficiencies. Nevertheless in most European countries cross-subsidies historically ensured wide service coverage, at the same time as avoiding an increase in the general tax burden.[13] It is doubtful whether this wide service coverage, which was related to a clear political objective of social inclusion, could have been achieved in a more efficient way by cost-reflective tariffs and targeted subsidies to the poor at the time when most of the infrastructures had to be created in the first place.

The information and transaction costs of computing a range of tariffs and subsidies, that are variable according to urban density and income or other characteristics of users, would have been significant, it the policy objective was to supply electricity, water, or public transport to everybody at affordable tariffs.

A private provider with some market power is rational in offering different tariff schemes, with a discount to large users. This amounts, however, to expropriation of consumer surplus, and it is not necessarily cost-reflective either. Price discrimination implies, in practice, that low users have to pay higher tariffs. Often those users are households with low incomes. This type of price discrimination creates social access problems, because if the share of expenditure for utilities bills is too large to be affordable, the consequence will be difficulties for low-income household in paying their bills. This fact results in arrears, forced disconnections, or self-disconnections, creating problems under a universal service perspective.

This is why the post-war European welfare state moved away from the 'cost-reflective' pricing mechanism for some essential services, such as health, education, and to a large extent also for water, transport, energy, and some communication services. You cannot at the same time be willing to implement a softly regulated oligopolistic market, and avoid the possibility that supply is

[13] It was often combined with progressive taxation or public debt to finance the infrastructure.

rationed for those users who are unable to afford the price of the service (including the investment to connect them to the network).

While in advanced economies, with high average incomes, you may expect that the share of the utilities bill is a minor item in the household budget, this is not true everywhere. In the New Member States, in some regions that are lagging behind, in rich economies, and for certain social groups—such as the elderly, disabled people, unemployed, lone mothers, immigrants, and other low-income groups—the expenditure on necessary network services is substantial. Growing inequality in developed economies also has implications for those essential services, which typically absorb more than 10% of expenditure of low income households, and even more than 20% of income for the bottom decile.

If one wants to preserve both universal coverage and allow for regressive price discrimination, a peculiar form of means-testing has been advocated. Governments are urged to go back to a form of public charity, where those in a disadvantaged social position must apply for assistance, e.g. for electricity cards or vouchers, for special arrangements for water or transport and so on. Whatever the efficiency of these support schemes, administered by the incumbents or by the welfare state, this would in Europe amount to having probably several millions of people under a form of 'utility assistance', subject to scrutiny by social services. The numbers are simply too large for this. According to EC (2010) around 85 million citizens are exposed to poverty risk in the European Union. As a consequence of the 2008 recession, many of these citizens are at risk of deprivation of some essential utility service and more than 30 million suffer severe deprivation (see Chapter 9). Means-testing on such a scale would be a fundamental departure from the universalistic nature of the European welfare state and would also be inefficient. We shall discuss this issue again in the last section, in the context of a more general assessment of the political dimension of network services.

10.7 Is regulating services easier than providing them?

The prediction that privatization could have unintended negative effects on consumers because of the remaining market power of the incumbents, possible collusion among them and the entrants, and overall inefficient equilibria, motivated the establishment of regulatory offices in the first place. In Europe there are now several offices for each of the 27 countries, and for each of the industries or group thereof. Typically, in each Member State there is a regulator for electronic communications, one for electricity and gas, one or more for some transport, water, and postal services, and other public sector bodies with specific regulatory tasks. These activities are complemented by 'horizontal' competition authorities, including the European Commission itself, and the new EU networks of regulators. The judiciary is also involved, as firms in most countries can appeal to the courts against the decisions of the regulators. The European Court of Justice de facto acts as a further regulator. Authorities' competencies vary from country to country, and have widely evolved over time. There are now several thousand staff members in these offices, mostly highly qualified and well paid (relative to the other public sector staff), as legal experts, economists, and auditors. The regulated firms have in turn established special task-forces to deal with the regulators, and there is a profitable market for consultants and lawyers specializing in this area. Indirect costs are less easy to be seen, but may be much larger than the direct ones, for example when efficient investment is delayed and inefficient investment is sunk as a response to regulation. Thus, regulation is not a free lunch.

Regulatory activity in the EU is a subject dealt with by an extensive legal and economic literature, not covered in this book. The perspective here has been to look at the outcomes of regulation rather than to its output in the form of decisions, monitoring activities, tactics and strategies, legal and information constraints, and so on.

The evaluation of regulation is difficult. We discuss here just one example. Edwards and Waverman (2006) study the impact of ownership and regulation on interconnection rates in telephony services in the EU-15. They consider public versus private ownership of the network, and twelve characteristics of the regulators. They consider that the more of these characteristics are present, the stronger is the expectation that the regulator will be independent and able to push down interconnection rates, that is the price that the entrant must pay to the incumbent to access the network.

The characteristics they consider are: a) the fact that a jurisdiction is single or multisector; b) that the regulator is one person or a board; c) that funding of the office is through the government or levies; d) that the regulator reports only to ministries or to the law-makers; e) existence of specific power to fix access prices; f) whether the power is exclusive or shared with the government; g) if the appointment is by the government or involves parliament; h) if it is for a fixed-term; i) and if the term can be renewed; l) if the staffing of the office is adequate; m) if they have a sufficient budget; and n) whether they have a long experience or not.

The first finding is that state ownership of the public telephone operator tends to increase the interconnection rate. The second is that a cumulated summary regulatory independence index counteracts this effect. Interestingly, just two items more from the above list in the value of the index are enough to reverse the state ownership effect, while under private ownership the index has no effect.

These results, combined with our findings on residential telephony prices, perhaps suggest a counter-intuitive regulatory failure. In the year considered by Edwards and Waverman (2006), 2003, the countries where the PTO is under public ownership, measured by a share of more than 50% of equity capital, are just three: Belgium, France, and Luxembourg. There are another three countries where the share is more than 40%: Austria, Sweden, and Germany. And another three where governments have a minority share of around 20–30%: Finland, Greece, and the Netherlands. Eventually, in only six countries was the government absent or having a negligible share of equity: the UK, Denmark, Ireland, Portugal, Spain, and Italy. As the interconnection rates in principle are a substantial share of the overall cost of the service, if competition had been working you would expect that final prices to the consumer would be lower in countries that did not have a government-owned or partly-owned PTO, after controlling for regulatory independence.

In fact, as we showed in Chapter 4, the effect of public ownership on the price of local calls, after controlling for other factors is to very slightly raise the price, but to substantially decrease the connection charges of users, which are also inversely related to market opening. After all, a simple inspection of Teligen/Eurostat prices for local calls reveals that consumers pay less in Belgium, France, and Luxembourg, the three countries with a publicly owned operator, than say in the UK, certainly the country that privatized first and that has a longer regulatory tradition. They also pay less in Germany and Sweden, where public ownership is substantial, but not in Austria. And they also pay less than in the UK in the other three countries where there was a share of government in equity capital of the PTO. Prices for national and international calls are lower in the UK, however, than in most other countries (but higher than in Sweden and Netherlands, for example), but local calls are those most important for households. Moreover, the connection charges for households in the UK are considerably higher than in countries with a government owned PTO: around two times more than in Belgium and Luxembourg, and three times more than in France, and nearly 50% higher than the EU-15 average.

Thus, whatever the effect of private ownership of the network and of independent regulation on interconnection rates for the profitability of the entrants, the benefit for final consumers of regulated TPA and the sophisticated regulatory infrastructure to support it is not apparent. Paradoxically, if interconnection rates decrease but end prices do not follow suit, or even increase, the outcome of regulation is an increase of the wedge over marginal cost, for the entrants relative to the incumbent.

The example of fixed telephony is particularly important, because one cannot say that this was a sector where there was not enough time to implement the reform, including the regulatory design. In fact, this is the only sector where we have found (weak) support for the beneficial welfare effect of liberalization. In the energy sector the regulatory failure may have been worse, particularly if long-term investment and environmental externalities are considered.

10.8 Ownership, regulation, markets: The role of the EU

The EU tends to remain neutral about private versus public ownership. Article 295 of the EC Treaty states: 'This Treaty shall in no way prejudice the rules in Member States governing the system of property ownership.' This article was included in the 1957 Treaty to allow nationalization of certain industries, and has not been changed in over half a century. The European Court of Justice in several decisions (Hunt and Talus, 2009) has clarified the interpretation. Member States are free to establish ownership rights, as long as they do not conflict with other aspects of the Treaty, namely freedom of establishment, free movement of capital, or non-discrimination.

The belief that public ownership is necessarily associated with inefficiency, corruption, and capture from vested interests must be assessed against the reality that private ownership under oligopoly in essential services may also be socially inefficient. Private interests have an incentive to 'buy' regulators and law-makers in order to be allowed to enjoy market dominance or, under oligopoly, they may tend to collude. The balance of the public versus private inefficiencies varies country by country, and industry by industry, even looking at specific segments of the industry.

Public ownership is not a panacea, but neither is it necessarily the enemy of liberalization. An example are the national electricity transmission system operators (TSO) in the Nordic power system, perhaps one of the most advanced in the world in terms of regional integration and market opening. Around 2010 the ownership structure of the TSOs across the four participating countries ranged from Svenka Krafnat (Sweden), a state agency, to the Danish and Norwegian TSOs (Energinet.dk and Statnett) which are corporations with the state as sole owner, to the Finnish TSO jointly owned by a private generator, institutional investors, and the Finnish state. In this context, public ownership of the networks has probably been an advantage for the promotion of cross-border market integration.

Even in telecoms, the industry where privatization started earliest, public enterprises have not disappeared. For example, one of the most important mobile operators in the world is Telenor which has a strong position in Norway, Denmark, and Sweden and is 54% owned by the State of Norway. Some government-owned players in the EU have contributed to market opening through direct foreign investment. The French government, through

EDF and France Telecom, has a stake in core players in the UK, and in several other countries including Greece, Poland, Romania, Slovakia, and Spain.

The motivation for the EU legislation in electricity and gas that has provided for mandatory unbundling of the networks was to avoid restrictions to third party access to entrants, in a context where duplication of the physical facility is uneconomical, see Cameron (2007). The Third Energy Package, agreed between the EC and the European Council (May 2009), has proposed making ownership unbundling mandatory in all EU Member States. Expert opinion, however, is divided.[14] Ownership unbundling is a costly and complex move. Given the state of debate among experts, and the uncertainty surrounding the empirical estimates on this issue, the EU directives in favour of mandatory ownership unbundling in 27 Member States and in industries as diverse as gas, electricity, telephony, railways, does not appear as evidence-based.

The European Commission itself has pointed out a number of problems in the implementation of market opening. While national legislation has been passed over two decades (since the Single European Act, EC, 1987), the competitive outcomes has been less widespread than expected. The Energy Sector Inquiry launched by the European Commission (EC, 2007) concluded that: a) market concentration is still very high; b) markets are still mainly national and only loosely interconnected; c) in the 2000s a new waves of mergers led to increased oligopolistic dominance; d) in electricity, concentration in generation was mirrored by concentration in spot and forward markets; e) in gas, the market is largely determined by long-term import contracts.

In the fixed voice sector, the EC Progress report (2007) admits that 'fixed incumbents have still not lost significant market shares, and in some cases, have even strengthened their position'. According to the European Commission (2009), in the EU-27, the incumbents enjoyed in 2007 on average around 65% on retail revenues, and in 2008 more than 80% of fixed subscribers had to rely on the incumbent's infrastructure to access alternative networks (cable, unbundled lines, fibre). In no country was the market share of the incumbent less than 50%. The market share of the incumbent is now higher in the UK—that started liberalization much earlier—than in Germany.

Going back to the 1987 Single Act, the overarching goal of the European Commission was to create an internal EU market. As restated, for example, in the Energy Sector Inquiry (EC, 2007, p. 4): 'Well-functioning energy markets that ensure secure energy supplies at competitive prices are key for achieving growth and consumer welfare in the European Union. To achieve this objective the EU decided to open up Europe's gas and electricity

[14] See for example Thomas (2006) for a critical view on unbundling in Britain, Glachant and Léveque (2005) for a midway view, and Pollitt (2008) for a defence of ownership unbundling in the EU, with some reservations.

markets to competition and to create a single European energy market.' Similar wording can be read in several other directives for other SGEI including telecoms and transport. An optimistic view of the last decade could conclude that progress has been made, albeit slowly, and that reforms are on the right track.

A possible interpretation of the empirical evidence is—however—that the focus on EU legislation prescribing to each Member State similar market design and regulation of industries has been excessive. The result is an often intrusive legislation, now embodied in literally hundreds of national laws in compliance with the EU directives. National regulatory authorities are struggling to implement their own views on how to find a balance between national interests and the European perspective.

This complex legal battleground is only indirectly related to creating a unique internal market, or protecting the consumer.[15]

Given the mixed empirical support to the reform strategy up to now, after several years of legal changes, the European institutions perhaps need to refocus their role. The effort of creating an internal market for services of general economic interest should still be a core concern of the EU, but it needs to be balanced with other important policy goals (see below). A policy aimed at abolishing legal, technical, and physical cross-border barriers to trade would be more promising, in principle, than trying to achieve exactly the same reform architecture in each of the 27 Member States. EU institutions can use their unique cross-border regulatory powers, and perhaps enhanced Articles 81 (anti-competitive agreements) and 82 (abuse of dominant position) of the EC Treaty. The EU can give more voice and liberty to municipalities, consumers associations, and cooperatives, which can discover fair deals for citizens better than individual consumers. Haney and Pollitt (2013) show some interesting examples of these organizations in the electricity supply sector.

At the same time, the EC can target a larger share of the EU Structural Funds to provide incentives for interconnection infrastructure and to act as a supranational social planner. At the EU level, a combination of direct price regulation, anti-trust policy, and incentives for long-term public investments seem more promising than the current intrusive 'market design' approach.

[15] The EC has recently acknowledged, for example, that roaming tariffs for mobile telephony were still 'ridiculously' (*sic*) high. On 27 June 2008, the EC proposed, and Council agreed, to impose a EU wide price-cap of 11 cents for SMS against the average price of 29 cents. In this case the EC acted not through directives to be embodied in national legislation, but as a quasi-federal regulator. Interestingly, this bold, straightforward move by the Commission was needed in a sector that is already one of the most open to competition. The example of roaming tariffs shows two things: first, liberalization in compliance with directives at country level does not necessarily avoid substantial oligopoly rents at European level. Second, in some circumstances the EC has direct powers to address these issues.

Public provision of service or a publicly owned network or a range of different arrangements (part-privatization, mixed oligopoly, mutual ownership, etc.) should be allowed without any interference by EU Institutions, as for Article 295 of the Treaty, provided that (a) borders are open for capital investment and for trade, with only the limitation of national security; and that (b) any user is given the concrete right and opportunity to pick up the best possible deal in the European economic space. Consumer rights should be given precedence over artificially sustained competition[16] when competition is socially inefficient.

10.9 Public services and European citizenship rights

According to CEEP (2010), all services of general interest (SGI), including network utilities, education, health, and others, represent over 26% of EU gross domestic product, and employ more than 64 million workers. Until the 1980s, in the European political discourse, the welfare state was associated with the provision of core social services, such as education, health, public insurance, and often with the provision of public transport, energy, water and sewage, postal services, telecommunications, solid waste management, and other environmental services. The institutional solutions (and the legal terminology) were different across the European countries, but most policymakers of any creed would have subscribed to the view that any citizen had an entitlement to access a certain bundle of 'essential' services.

The boundaries of what is an essential service have been rarely if ever well defined, however, and they have been shifting over time. In Western Europe this close relationship between some network services, social cohesion, and citizenship, largely faded away in the last 20 years, in parallel with the expectation that competition, combined with some regulation, would enhance lower prices, and hence greater social welfare, without the need for extensive public provision. This expectation was also reflected in the semantic change of legal EU jargon from 'public services' to 'services of general interest', or 'services of general economic interest', where economic stands for 'commercial', that is potentially open to competition.

Nevertheless, several EU documents still state that European citizens are *entitled* to services of general interest. This has gradually led to a process of Europeanization of the legislation in this area.[17] Examples of the legal

[16] An example is a regional franchise monopoly, but such that a group of users, a city for example, can claim that somebody else can offer them a better deal. If the local monopolist is unable to lower prices, then the claimants should be given access to the alternative provider, by the regulator, without further market regulation.

[17] For a review see e.g., Obermann et al. (2005).

framework can be found in Protocol 26, annexed to the TEU, which explicitly mentions services of general interest as playing a 'vital role' and belonging to the 'shared values' of the Union. The Charter of Fundamental Rights in Article 36 mentions SGEI for their role in promoting social and territorial cohesion, and for their contribution to growth and development of the Union. The European legal framework does not focus on residual services for disadvantaged groups of the population, but on a more universalistic perspective. According to Oberman et al. (2005), the availability of SGEI in EU legislation can be understood as the exercise of a fundamental right to satisfy the democratic, social, and cultural needs in all sectors that contribute to quality of life and sustainable development. The multiple objectives often mentioned in the legislation on the services include universality, continuity of provision in the long term equality of treatment and of access, efficient allocation of resources, safety and security of supply, accountability and transparency, adaptability to change, affordability of prices, protection of consumer rights, territorial cohesion, and legal remedies for consumers. While these policy objectives are directly related to the welfare of users, the traditional, more strategic, concerns have not disappeared—such as issues related to the role of some services in policies of national defence, security of borders, research and development, spatial planning, and labour markets.[18] The European Commission, in its Sixth Report on Economic and Social Cohesion (EC, 2010) restates the view that 'a good quality of life, equal opportunities and access to services of general interest in all territories are crucial both for solidarity and competitiveness'[19].

As mentioned above, Protocol 26—annexed to the Treaties—states the EU goals in relation to SGEI: 'high level of quality, safety and affordability, equal treatment and the promotion of universal access and of user rights . . . '. At the same time it acknowledges 'the wide discretion of national, regional and local authorities in providing, commissioning and organising SGEIs'.

The relationships between these fundamental (perhaps 'constitutional') principles and European competition law, are complex, and show tensions. On one hand, the EC wants to promote a European market for services, on the other hand it must admit that governments have their own policy objectives

[18] This development perspective is also underlined in the Barca Report (2009), an important review on the future of EU Cohesion Policy.

[19] More recently, the EC, in preparation for further revision of the framework on SGEI and on state aid in this field, has restated that ' . . . the EU and its Member States must ensure that public services are easier to operate at the appropriate level, adhere to clear financing rules, are of the highest quality and accessible to all . . . Public Services (generally identified in the Treaties as SGEI) occupy a vital role in the shared values of the Union. They promote social and territorial cohesion, foster the well being of people across the EU, make an important contribution to economic development . . . They range from large commercial services (network industries such as postal services, energy supply, electronic communication or public transport services) to a wide range of health and social services' (European Commission, 2011).

and citizens have certain expectations or even 'rights'. As a consequence of this tension, EU institutions are struggling to define which services are subject to provisions about state aid, an issue we briefly mentioned in Chapter 1. EC (2011) restates the position that competition rules do not apply to all services of general interest, but 'only to those that are "economic" in nature, i.e. SGEI'. Examples of the latter, 'non-commercial' services, are air navigation safety or anti-pollution surveillance, police, justice, and compulsory social security schemes. The matter is somewhat blurred by the fact that Art 106(2) TFUE states that, while undertakings providing SGEIs or having the character of a 'revenue producing monopoly' are subject to competition rules, it also adds: 'in so far as the application of such rules does not obstruct the performance, in law or in fact, of the particular tasks assigned to them'. In other words, if competition damages the public service mission, it does not have priority.

In practice, a provider of electricity, of Wi-Fi connections, or a local bus company can be either public or private and receive compensation for some of its unprofitable service from the public budget. This transfer will be under the scrutiny of the Commission, to prevent undue state aid. The goal of EU state aid control is 'to ensure that Member States only implement State aid that contributes to an objective of common interest, is well designed and proportionate, and does not distort competition and trade between Member States'.

The last sentence is clearly the crucial one. The next proposed reform of state aid, due to start in 2012, points to a distinction between those local public services which have a limited potential impact on trade and competition, and large scale commercial activities. For the latter 'the Commission considers that the risk of distortions in competition in the internal market is particularly high in sectors characterised by large scale commercial activity with a clear EU-wide dimension where operators may be entrusted with public service obligations . . . such as transport, telecoms, energy supply and post . . . '.

Most of the current debate stems from the position taken by the European Court of Justice, which, has stated (Altmark Case)[20] that a public service provider is not subject to the state aid rule when it receives government funds, provided that: a) public service obligations are clearly defined; b) the compensation mechanism is based on objective and transparent parameters; c) these compensations merely cover costs and a reasonable profit; d) where the undertaking is selected by a public procurement tender based on the least cost, or on the basis of analysis of the costs of a 'well-run' undertaking in the sector concerned. If just one of these conditions does not hold, any government transfer to a public service provider should be notified to the EC, which in turn has established in what cases state aid is acceptable. Basically, state aid

[20] Case C-280/00 *Altmark Trans GmbH and Regierungspräsidium Magdeburg v Nahverkehrsgesellschaft Altmark GmbH* [2003] ECR p. I-7747.

is accepted by the EC when it is said not to distort competition. Thus, competition risk shifts from the means to achieve a policy goal, to the objective per se.

In EC (2011), interestingly the Commission, after recalling the previous arrangements, restates the obvious truth that 'Member States need in fact to guarantee certain services at affordable conditions to the general population (e.g. hospitals, education, social services, but also communications, energy or transport).' The new framework proposed for the future, after consultation with more than one hundred organizations of stakeholders, is said to broadly confirm the current approach, with improvements in terms of clarification of what the EC would consider as acceptable stated aid, and a more diversified and proportionate approach. This orientation of the EC shows one important point: the complex range of applicability of EU legislation to SGEI, including network industries, and the fact that the network industry paradigm is gradually extending to hospitals, social care, and social housing.

The earlier perception that network industries are not only special because of their economic nature, but also because of the nature of the needs they satisfy, while perhaps vague and imprecise, was, and still is, not entirely void of content. This issue is best appreciated when one considers that lack of access to some of these services was a feature of large parts of post-war Western Europe, particularly in some urban areas more affected by the conflict, in the rural areas, in the mountains, in peripheral regions, and in less developed regions. Transporting drinking water to any dwelling, as well as electricity, and ensuring road connections, local public transport, and a postal service were firmly on the policy agenda of European governments. This objective of wide coverage was part of a broader project of building a socially inclusive fabric after the disaster of World War II.

Markets are not necessarily the social institutions to provide this outcome. In some cases they may work well, but markets do not exist to address social cohesion problems nor to implement political agendas. In the UK, with the Beveridge Report (1942), the principle of universal access to some services was stated as a radical departure from the traditional view of social solidarity as 'taking' from the rich and 'giving' to the poor, that is some form of private or public charitable assistance of the former to the latter. One of the founding fathers of the European Community, Jean Monnet, was also an advocate of government planning: in fact the two ideas—of political federalism among nations and of economic planning—were closely interrelated, as shown by Lynch (1984) in his study of the Monnet Plan.

The new idea of post-war Europe, which was already rooted in some of its nineteenth-century history, was that governments should guarantee the same standard of (some) services to every citizen, just because he or she was a member of the national community. Judt (2005, 2010) has provided a superb analysis of the relationship between this evolution of government and the history of post-war Europe, and the risks of abandoning this path. He argues that the rationale for 'Universalism' was a lesson won from decades of social and military conflict in Europe:

> Instead of having their benefits keyed to income—in which cases well paid professionals or thriving shopkeepers might have complained at being taxed for social services from which *they* did not derive much advantage—the educated 'middling sort' were offered the same social assistance and public services as the working population and the poor: free education, free or cheap medical treatment, public pensions and unemployment insurance. (Judt, 2010, p. 52)

In many European countries some network services were also included in the package: electricity, water, public transport, and telephony at affordable prices. Through the variable combination across countries of taxation, bond finance, and tariffs, achievement of the welfare state paradigm in Western Europe was largely successful. Its legacy is still visible today. The high rate access to social services and public utilities in Western Europe is virtually unparalleled in any other region of the world, and this is mainly due its history. It is an unfinished job, from the perspective of the EU-27 and beyond. The post-war consensus on universalism of public services has been questioned by the paradigm shift since the 1980s. Whatever the efficiency gains related to this shift, which do not seem to be overwhelming, the implications for the role and perception of government in society are wide. Again in Judt's words:

> ... By eviscerating the state's responsibilities and capacities, we have undermined its public standing. Few in England and fewer still in America continue to believe in what was one thought as a 'public service mission': the duty to provide certain sorts of goods and services just because they are in the public interest. A government that acknowledges its reluctance to assume such responsibilities, preferring to shift them to the vagaries of the market, may or may not be contributing to efficiency. But it is abandoning core attributes of the modern state.

This reluctance to acknowledge government responsibility in service provision may also pose a challenge to European integration. The EU has evolved from a mostly custom union of six founding members, to a quasi-federal entity, or political union of 27 countries, with more candidates waiting to be admitted. It has accomplished the historic achievement of offering a common economic and legal frame to Western European former enemies in the two World Wars, and to Central-Eastern countries that emerged from the end of the Cold War with the Soviet Union. The recent financial turmoil also shows how incomplete the monetary and fiscal side of the Union is. But there is a real convergence side to the unfinished story of the European project too.

It involves several dimensions, including the availability of network services, for their unique role in linking together citizens and their economic activities over spatial dimensions. From this perspective, the EU has endeavoured to offer substantial assistance to the new Member States, mostly former transition economies, through the Structural Funds for building or modernizing their transport systems, energy self-sufficiency, water and environmental services, education and health provision, with the objective of speeding up their convergence to Western European standards.

As mentioned in Chapter 9, in the period 2007–13 the European Regional Development Fund and the Cohesion Fund will provide 240 billion Euros of capital grants for co-funding infrastructures in the Member States. Around 40% of these funds are in the transport and environmental sectors, and the remaining in ICT, energy, and some social infrastructures. Most of the funds are being targeted to the New Member States, where the need is greater. Several network services are potentially involved in the EU Cohesion Policy. The new package proposed by the EC (2011) for the period 2011–20 gives high priority to infrastructures and networks. The ERDF priorities include, for example, improving access to and quality of information technologies; climate change and moves towards a low-carbon economy; SGEI; and telecommunications, energy and transport infrastructures. The Cohesion Fund will support investments in the water waste sector, in sustainable energy, and in low-carbon transport. The trans-European networks will receive additional EU grants.

Thus, in spite of the adoption of the market paradigm in EU legislation on network services, a large part of the infrastructure in the New Member States will actually be supported by a combination of Community and national public funds in the framework of the EU Cohesion Policy. This is perhaps an acknowledgement of the difficulty in finding effective and fast market solutions to address these basic needs, and a reflection of the institutional memory of how the network infrastructure was built or re-built in post-war Western Europe, including the substantial help from the USA under the Marshall Plan (see Hogan, 1987).

10.10 Beyond the paradigm: Network services as 'political industries'

One of the leading British experts in the energy sector, after having reviewed two decades of reform in the UK—the inspirational model for the current EU approach—concluded:

> In 1980s and 1990s the pendulum swung too far the other way (from public monopoly). The market enthusiasts failed to recognize how far the electricity market deviated from the normal commodity model. To recap, supply must instantaneously match demand as there is limited scope for storage: the assets are sunk or long lived, the networks are natural monopolies. There are very great environmental externalities; and critically, electricity and gas are complementary to the rest of the economy, in that failure to supply has (extremely) large costs to all economic activity. It is hard to think of any other activity in modern developed economies with quite such coincidence of major market failures. If the issue of fuel poverty and the distributional implications of electricity and gas pricing and supply are also included, it is extraordinary that anyone could have regarded these as anything other than political industries. (Helm, 2003, p. 407)

Some of these issues have resurfaced recently in the British Energy White Paper (2011) that we discussed in Chapter 5. Our findings offer some support to this reconsideration of utilities policy, and to its extension to other network services and countries.

Relatively undisturbed market equilibria and private ownership of assets is the first choice arrangement for industries that are definitively outside the scope of specific public policies. In those sectors, which vary across countries and time, in principle, no more is needed than the enforcement of contracts and property rights, the protection of agents from information asymmetries and fraud, possibly a dose of anti-trust regulatory power, protection of consumers' and workers' rights, of minority shareholders, as far as possible non-distorting taxation, and so on. Many industries, even if they tend towards oligopolistic arrangements, should stay in the domain of market mechanisms with mild government supervision.

The situation is different for other sectors, for which governments want to take a longer and wider perspective than market players. If governments want both competition in energy markets, security of supply, social affordability, and containment of climate change and of other environmental externalities, then *independent* regulators are in a difficult position. How far should they balance the different objectives? Affordable prices to everybody may not be compatible with centralized cost-reflective prices. Security of supply implies that investment are sunk with a certain guarantee that they will be recouped in the long term, and spot markets do not necessarily offer the right signals. Green technologies may be more costly than traditional ones, and profit maximizing firms will not pick them as their preferred option. The balancing of policy objectives is the core function of government, it cannot be left to independent regulators alone.

From this perspective, we have argued that some (not all) of the services provided by network industries are more similar to health, education, or some forms of public insurance than to 'normal' industries. The key issue with 'social' and network services is not just a conventional type of market failure. The reasons for potential attention by (hopefully benevolent) policy-makers are not limited to economies of scale in production, public goods, environmental externalities, and so on. What gives a special 'political' nature to some services is the fact that governments have their own overarching goals that are not achieved by market mechanisms, and these goals overlap with industry operations. Hence, investments, inputs and outputs, prices, and quality standards, must be, directly or indirectly, adapted and planned to match those objectives.

The *planning* dimension is the right word to be used in this context, a word rediscovered by the recent electricity market reform in the UK, see Chapter 5. When the European Union adopts the objective of reaching, in 2020, a share of 20% of renewable energy sources, a 20% reduction of greenhouse gas as compared with 1990 levels, and to increase by 20% energy efficiency, then well designed policies and specific plans are needed to achieve these targets. The digital agenda of the EU is another example of a long-term policy. As long as governments have long-term objectives for transport, energy, electronic communications, water, and waste management, these services will be attracted into the sphere of public policy design and management in one way or another. The mechanisms to achieve certain targets may include taxes and incentives, rationing, special industry regulations, procurement contracts, public enterprises, public–private initiatives, and so on, in various combinations.

The core of any reconsideration of privatization and liberalization policies does not lie in establishing once and for all the appropriate ownership and market structures (everywhere and at any time), but in asking what the long-term objectives of governments are, if any, in these industries. The next step would be to see which arrangements are best suited to implement those objectives in terms of social welfare effectiveness. These arrangements are intrinsically specific to the jurisdiction in charge of the policy setting, and cannot be designed without due consideration of the local conditions.

More than an alternative generic paradigm, what we need is a reconstruction of the logic of government's responsibility in certain domains. Public ownership of assets and government appointed managers without clear goals, plans, and policies, are of no use. At their best, these public sector organizations will be 'efficient', seeking profit for their governmental principals, hence they will simply imitate their private counterparts. At their worst, public providers will be wasting precious economic resources in favour of corrupt politicians and administrators and their cronies. But there are examples in

Europe and elsewhere of successful public enterprises. These examples should be carefully examined.[21]

More generally, ownership rights and legal constraints to market arrangements must be seen as part of a broader set of collective governance mechanisms. In a fundamental sense, it is the state that assigns or constrains the rights necessary to play the economic game. In some sectors the rules should stay loose, as there is no reason to interfere with the functioning of the contractual relationships between agents. In others the 'visible hand' of government is needed.

Thus, if you think that government should care about the future of the energy balance of a country, or that transport services cannot work without the long-term planning of certain infrastructures and urban development schemes, or that a certain amount of drinking water of a specified quality should be provided to every household, or that broadband should reach the countryside, then you need to look at ownership and markets as instruments of collective welfare objectives. You would consider what is the best and most feasible way to achieve a public plan through the associated policy. From this perspective, what has possibly been inconsistent in the European paradigm of the last decades for the network industries is that it has often been proposed as a self-sufficient market reform, with scant consideration for any overarching public policy objectives about the same services. This was unrealistic.[22]

What is needed is less regulation, and more responsible and accountable government.

The delegation of essential political functions to industry regulators[23] and the crowding-in of financial players has often been the outcome of the reform paradigm, with little benefit to the citizen. Excessive regulation, rigid reform prescriptions, proliferating bureaucracies, and rent-seeking financial investors should be scaled back. We need to open a new public discourse on what the governments' responsibilities should be in setting collective goals, and in stating,

[21] The Public Enterprise/Public Services Commission of Ciriec International (http://www.ciriec.ulg.ac.be/ [accessed 8 January 2013]) has recently started a research programme in this area, see also the Special Issue of the Annals of Public and Cooperative Economics (4/2011) on the future of public enterprises. See also some contributions in the coming special issue of the *International Review of Applied Economics* and a forthcoming special issue of *Journal of Economic Policy Reform* (http://www.tandf.co.uk/journals/cfp/gprecf) 'Public Enterprises and Quality of Institutions: Alternatives to Privatization' (March, 2013). See also McDonald and Ruiters (2012), who present a survey of alternatives to privatization in service delivery (electricity, water/sanitation, health) in Latin America, Asia, and Africa. The inter-sectoral and inter-regional dimensions of the book are interesting, and reveal a vibrant (and still quite heterogeneous) search for new avenues in public service provision.

[22] A symptom of lack of realism is the fact that in the European Commission and in many governments different departments are in charge of competition and industrial policies.

[23] From this perspective, independent regulators are part of government. As the courts or central banks they cannot be construed as independent from the law that define their role. Interestingly the UK, after having abolished the Department of Energy, rebuilt it under the name of Department of Energy and Climate Change. http://www.decc.gov.uk/ [accessed 8 January 2013].

in a democratic and transparent way, the reasonable boundaries of citizens' entitlements to those services, which critically contribute to social inclusion.

10.11 Further reading

Some of the arguments in this chapter are related to different lines of rethinking the rise and decline of the welfare state in Europe and elsewhere, with the network services being considered part of a social contract of a universalistic nature. Some of these broad themes were discussed in the informal seminar organized by Tony Judt in April 2010 at the Remarque Institute, New York University. Some works of the participants,[24] while not related to network industries, deal with the tension between the market reform agenda in the last decades and the principles of the welfare state. The Oxford economic historian Avner Offer is currently working on the broad theme 'From Social-Democracy to Market Liberalism c. 1970–2015'. According to Offer, the post-war 'golden age' of economic growth offered the building materials of welfare states both in Europe and the USA. The former equilibrium was successfully challenged in the 1970s by what he describes as a coalition of business, taxpayers, consumers, ideologists, and social scientists. The old, and widely dismissed, ideology of market liberalism then 'retrieved the intellectual and political hegemony it had previously lost, although the recent financial crisis has cast doubt on many of its premises'[25]. In no sector better than in health services can the clash of views between market liberalism and principles of universal access be grasped. The work by Allyson Pollock, Queen Mary College, London and her colleagues is highly critical of the corporatization and privatization of what used to be a typical example of the Anglo-Saxon version of universalism, the British National Health Service, see the NHS plc, Pollock (2005). Some contributions by other participants in the Remarque event, which are recommended to economists who are willing to think beyond the boundaries of market mechanisms as ways to solve social problems, would include particularly Rothstein (1998), Rosanvallon (1981, 2011), and Teichman (1995). The work by Judith Teichman (University of Toronto) offers a political perspective on market reforms,

[24] The symposium participants included: Sheri Berman (Barnard College, Columbia University), Neil Gilbert (UC Berkeley), Dipankar Gupta (University of New Delhi), Jingjing Huo (University of Waterloo, Canada),Thomas Meyer (University of Dortmund), Avner Offer (All Souls' College, Oxford), Allyson Pollock (Queen Mary, University of London), Bo Rothstein (University of Gothenberg), Felix Rottenberg (Rotterdam, Netherlands), Pierre Rosanvallon (Collège de France), Toshimitsu Shinkawa (Kyoto University), Judith Teichman (University of Toronto), Joerg Haentzschel (Sueddeutsche Zeitung), and Robert Silvers (*The New York Review of Books*).

[25] See http://www.avneroffer.net [accessed 8 January 2013] and particularly Offer (2003).

when transplanted from contexts such as Europe, or from the USA, to other contexts such as Mexico and other Latin America countries. Economists often tend to think that the market paradigm is so general that it can work in a similar way everywhere in the world. Teichman shows why a standard set of reforms, such as that discussed in this book, may evolve in very different ways when the social and political context is only superficially analogous to the one where the reform package was originally conceived.

Glossary

This annex offers newcomers to the subject a very personal selection and a sketchy synopsis of some concepts and ideas often encountered in network economics literature. For a systematic treatment see the references in the Further Reading Section 1.6. Words in italics in each entry are cross-referenced in other entries.

Compatibility and standards. The *complementarity* of nodes and lines requires technical compatibility, for example the same gauge of rail tracks and so on. This is a 'standard', a common rule that—if the ownership of the network is fragmented—must be either negotiated or regulated. Only in special cases is no standard needed, and compatibility intrinsic. In some cases, the lack of compulsory standards leads to foreclosure or other forms of discrimination in interconnection.

Competition law. In most countries generic competition law is enforced but doesn't substitute for sector-specific regulation. This is because the former is often ex-post, that is, is acted when a potential infringement of competition is observed, while the latter is typically ex-ante, as it dictates detailed rules of the game. The distinction is, however, often blurred and variable across countries and industries.

Complementarity. The network components are complementary (Economides, 2008) in a precise sense, as the utility (broadly defined) of a node to a user is increased by the existence of the line (or of more lines). Often, the utility of the node without the line is in fact nil, and vice versa.

Cross-subsidies. Any firm that systematically covers losses in some market segment or products with profits made elsewhere is actually managing a cross-subsidy scheme. State-owned incumbents were often credited to use this mechanism for social reasons. AT&T, the telephone monopolist in the USA until 1984, claimed that its uniform pricing policy subsidized the basic local and long-distance public services. Entry would have skimmed the cream off the market. However, 'Independent studies of prices and costs for local exchange service and the basic residential handset showed that they were not subsidized, but rather producing high benefits being used to subsidize terminals and service for industrial users . . . The conventional wisdom was wrong on every point' (Melody, 2011, p. 110).

Economies of integration. According to Lafontaine and Slade (2007): '. . . the evidence . . . says that in most circumstances profit maximizing vertical integration decisions are efficient not just from the firm's but also from the consumer's point of view. The vast majority of studies support this claim.' The deep rationale of vertical integration should be traced back to transaction costs and lies at the core of the nature of the firm (Coase, 1937).

Economies of scale and scope in networks. The former are typical of any infrastructure based industry, where fixed costs tend to dominate, and returns to scale are increasing (i.e. the production function is such that $F(aK,aL) > aF(K,L)$, where K is capital, L is labour and other inputs, a is a coefficient, and F is the production function. Economies of scope (Panzar and Willig, 1982) are also often encountered in network industries as the *infrastructure* lends itself to convey a variety of products, for example underground facilities for water, gas, optical fibre, and so on. Retail services, such as invoicing, often also show economies of scope. These are related to sub-additivity (see *natural monopoly*) of cost functions when the firm produces more than one output.

Essential facility. A node or line may be critical for interconnection, that is is a bottleneck, as many or all lines must go through them to reach other nodes. This is often a *natural monopoly*. The owner of such component can either foreclose another player, or extract a high interconnection price. This is the rationale of *regulation* of such essential facilities, typically some component of the fixed network. There is a very wide literature about what should be the socially desirable pricing of the bottleneck infrastructure, for example Baumal and Sidak (1994). Typically network industries are analysed as consisting of different layers (e.g. resources, network infrastructure, network management, transportation, final service supply) and bottlenecks located in each layer (Jaag and Trinkner, 2011). In fact this corresponds to locating an essential facility within the supply chain.

Free entry. According to Economides (2008, p. 485): 'Although eliminating barriers to entry can encourage competition, the resulting competition may not significantly affect market structures... Antitrust and competition law have placed a tremendous amount of hope on the ability of free entry to spur competition, reduce prices, and ultimately eliminate profits. In network industries free entry... fails miserably to reduce inequality in market shares, prices, and profits.' This fact is related again to *network effects* that tend to give an advantage to the incumbent.

Incentive compatibility. According to Ledyard (2008) this is a 'characteristic of mechanisms whereby each agent knows that his best strategy is to follow the rules, no matter what the other agents will do'. In a more general setting, incentive compatibility of a regulation implies that the regulated firm is incentivized to behave as desired by the regulator. A participation constraint is also to be considered, as below a certain profitability threshold a private firm would exit, Laffont and Martimort (2002).

Infrastructure. An operation with (a) large, specific, sunk (irreversible) investment costs, (b) associated with economies of scale (and often of scope), and (c) offering services massively demanded (Spiller, 2011). Many other investment projects have one or more of these characteristics, but a typical infrastructure combines all or most of them. An infrastructure can be part of a *network*, or not. A network industry can be infrastructure-based or not. The two concepts are different, see below.

Incumbent. The historical monopolistic provider of service usually protected by a franchise or a similar legal arrangement in the first place, then exposed to entry by competitors.

Inefficiency of perfect competition. According to Economides (1996) 'In the presence of network externalities, it is evident that perfect competition is inefficient. The marginal social benefit of network expansion is larger than the benefit that accrues to a particular firm under perfect competition. Thus perfect competition will provide a smaller network than is socially optimal, and for some relatively high marginal costs perfect competition will not provide the good while it is socially optimal to provide it.' The decentralization of the solution would imply perfect *price discrimination* (of the first type) or proxies of it, but possibly with adverse social implications. Government subsidies to network expansions are closely related to this issue.

Natural monopoly arises from sub-additive average cost functions. Sub-addiditivity (strict) implies that a function having a domain A and an ordered co-domain B, that are both closed under addition, have the property: for all x, y included in A: $f(x+y) < f(x) + f(y)$. This is usually a technological feature (i.e. related to quantities of inputs, and not to their prices). *Private* or *public monopoly*, in contrast, are economic arrangements, usually supported by a set of legal provisions, such as concessions, exclusive franchise, etc (but de facto monopoly can also exist, when entry is possible but unprofitable). Hence private or public monopoly may not be originated by natural monopoly, but natural monopoly usually leads to one of the two arrangements of an industry or part of it, Sharkey (1982).

Network. As a formal concept, the term is related to a graph in mathematics (Iyanaga and Kawada, 1997). A graph is a mathematical object defined by the association of two sets: a set of V of vertices or nodes and a set E of edges or lines. These are two-element subsets of V, and a simple, undirected graph contains lines connecting two nodes in no particular order. There are many other definitions in the mathematical literature. In network economics the analogy is that a user/producer is a node, and some infrastructure or support facility (pipelines, wires, wave frequency, collection point, track, etc) is a line. Typically the two sets are large or very large in network industries, and each set may be partitioned into a number of sub-sets.

Network effects and externality. A typical feature of *two-way network* industries is beneficial demand effects, which are often externalities. This implies that while the conventional demand for the service is inversely related to the price, it is also positively related to the quantity. In other words, for each consumer, the demand depends negatively on price and positively on the expected demand of other consumers. This feature is related to the abovementioned complementarity of lines and nodes. In one-way networks, the demand from other consumers may push economies of scale, and decrease costs.

Oligopoly. Empirically, the most frequent feature of network industries is an oligopoly. In turn, oligopoly can be of the '*compatible*' type or not, and the incentives for individual firms to go in either directions are such that 'Intuitively, a firm benefits from a move to compatibility if (i) the marginal externality is strong; (ii) it joins a large coalition; and (iii) does not thereby increase competition to a significant degree by its action' (Economides, 1996). There is a wide game-theoretic literature linking these issues to ownership (mergers and acquisitions, demergers, etc.). This literature often concludes that extreme inequality of market shares arise naturally from network effects (Economides 2008, p. 483).

One-way/two-way networks. In the latter the direction of exchange through the line, between the two adjacent nodes, can go in either direction, as in switched telephony. In one-way networks this is impossible, for example natural gas. Complementarity applies to both types of networks. Economides and White (1994) observe a similar property for any pair or combination of vertically-related (non-network) industries, which can then be considered a one-way network. This extends the concept to supply chains without a physical infrastructure linking them, outside the scope of this book.

Price-cap. This in principle is a high-powered incentive mechanism, when the firm is allowed by the regulator to retain any profits, provided that it complies with a price formula such as *RPI-X*, where *RPI* is a retail price index and *X* is an ex-ante estimation of productivity change. It can be applied to retail prices (for individual service or for a basket of tariffs) or more frequently now to interconnection charges. The firm is incentivized to cut costs, because it cashes the difference between capped prices and costs. In order to allow room for replacement of investment, price-cap formulas in some countries, notably in the UK, are in the form RPI + K, allowing prices to increase more than inflation (Baldwin et al., 2012).

Price discrimination. The standard taxonomy is due to Pigou (1932). Under first degree price discrimination (PD) the monopolist knows the reservation price of each type of consumer, and is able to fully extract the consumer surplus. Under second degree PD, prices vary with quantity, as with non-linear pricing based on two-part tariffs where the variable part decreases with consumption. Consumer surplus decreases as compared with a unique price, but less than with first degree. Under the third degree PD, consumers are charged different prices according to location, time, or individual attributes.

Pricing in networks. The fact that the network is made up of *complementary* goods implies that the service providers can price the service at either side of the network or at both, and this may considerably enlarge the scope for pricing strategies. Moreover, the fact that the investment is sunk may imply that the incumbent investor, in contrast to the entrant, can price the service below full cost, at least in the short run. Optimal pricing rules for network industries can be very different from other industries.

Private/public monopoly. Without perfect price discrimination a profit-maximizing monopoly will in general supply less than the optimal quantity. In principle a welfare-maximizing public monopoly can internalize the network effects. Thus the relative efficiency consideration for the two organizations of the industry are not altered by *network effects*.

Public enterprise. Production units under the control of government, as instruments for implementing a public sector production plan. Del Bo and Florio (2012) suggest that the plan doesn't need to be optimal in welfare terms, but policies need to be undistorted by corruption, to generate the credible parameters for cost-benefit analysis of privatization or nationalization. United Nations (2008) discuss some management issues of contemporary public enterprises.

Public service. In the French (or Latin) tradition, the *service public* is closely related to the role of national government in providing goods to all citizens through public enterprises and public administrations. It is characterized by (a) equal treatment of users, (b) continuity, that is non-interruptibility of the service, (c) responsiveness to changing needs of citizens and adaptation of technology (Finger and Finon, 2011). In contrast, in the Anglo Saxon tradition the public interest is a generic notion, and can be served by government or markets. In the German tradition the social responsibility of private firms is emphasized, while in Scandinavia (but also in Switzerland and elsewhere) public services are typically related to the local dimension, with municipal firms as providers. In practice, in Europe, these traditions are far from being clear-cut.

Quality coordination. Product differentiation can arise as a way to sustain imperfect completion, and variations in quality are a part of this story. In networks with fragmented ownership there may be an issue of quality, as for example the quality of a telephone call will depend on the state of different lines and nodes belonging to different firms. Under unbundling prices can increase because of double marginalization. Equivalently, quality can decrease as each firm gets a margin from reducing costs related to quality of service. This also often applies to bilateral monopoly. Regulation of quality is notoriously difficult because of information asymmetries (Rovizzi and Thompson, 1992).

Ramsey pricing. This concept has been advocated in regulatory economics to suggest that optimal pricing under *rate-of-return regulation* and *economies of scale* should imply setting mark-ups inversely proportional to demand elasticities. According to Baumol (1987, p. 51): '... self-interest may impel a monopolist to adopt Ramsey prices because those prices are sustainable against entry, meaning that at those prices the monopolist will earn the profits that the constraint allows to him, but any rival firm that undertakes to enter the field will be predestined to lose money even if the incumbent undertakes no strategic (retaliatory) response.'

Rate-of-return regulation. If the regulator knows the cost of the service, it can determine a cost- plus pricing rule, such that no excessive profit is allowed. This is a low-powered incentive mechanism (Laffont and Tirole, 1993), often considered not incentive compatible as the regulated firm has limited incentive to efficiency, and/or would misreport cost information. Excessive investment ('gold-plating') can also be associated to this form of regulation (Averch and Johnson, 1962). In spite of these limitations, in the USA there is a long tradition of this approach to price control, and its performance is not as bad as often assumed.

Sector-specific regulation. One realistic way to look at sector-specific regulation is to consider it as a long-term public contract (Spiller,2011). This implies that, differently from private contracts, politics is part of the story, and the institutional setting should be analysed comprehensively, beyond formal rules. This would include a consideration of opportunism on both sides. On the government's side a typical issue is that of commitment and risk of expropriation, discussed for example by Laffont and Tirole (1993). The latter can take the form of nationalization of assets, or of regulatory squeeze of profits, taking advantage of the irreversibility of sunk investment.

Separation. Separation remedies have been introduced by legislation or regulators to avoid anti-competitive activities of the *incumbent*. These may range from negating competitors the access to essential facilities, asking an excessive price, or downgrading quality of service of the competitor using the facility. Accounting, functional, and ownership separation are the three main remedies, in increasing order of effectiveness, Cave (2006). The first is limited to disclosing information about different business in income statement form. Functional separation is a more vague concept, it usually involves having a separate division or company within the incumbent. Ownership separation amounts to force divestiture of certain infrastructure. Baldwin et al. (2012) discuss possible implications for investment.

Universal service obligation. Regulation can protect consumers by requiring the providers of services of general economic interest to ensure certain delivery mechanisms, for example in rural areas, or to vulnerable users. There are four alternatives: (a) USO is attributed to the *incumbent*, often a privatized or partly government owned historical operator, who was already accustomed to public service missions; (b) all entrants should also comply; (c) a menu of options is offered to the players, and they can either offer USO or not and pay a compensation to the others; (d) USO can be allocated by a public tendering process.

Virtual network. Largely outside the scope of this book, any *compatible* activities that use a common technology platform have been defined as a virtual network, typically computers that share an operative system or other features (Richardson et al., 1998). Some industries, particularly in telecommunications, may include both virtual and infrastructure-based networks.

References

Abbott, M. (2010), 'The Long Term Development of New Zealand's Electricity Supply Industry', *New Zealand Economic Papers*, 44(1):75–89

Alesina, A., Ardagna, S., Nicoletti, G., and Schiantarelli, F. (2005), 'Regulation and Investment', *Journal of the European Economic Association*, 3(4):791–825

Alleman, J. and Rappoport, P. (2005), 'Regulatory Failure: Time for a New Policy Paradigm', *Communications & Strategies*, 60(4):105–21

Allen, D.R. and Rao T.R.N. (2000), *Analysis of Customer Satisfaction Data.* Milwaukee: American Society for Quality

Arano, K. and Blair, B. (2008), 'An Ex-Post Welfare Analysis of Natural Gas Regulation in the Industrial Sector', *Energy Economics*, 30(3):789–806

ARCEP (Autorité de régulation des communications électroniques et des postes). (2011), 'Modèle technico-économique des coûts de la terminaison d'appel fixe en France' http://www.arcep.fr/uploads/tx_gspublication/modele-finalise-070111.pdf (accessed February 2013)

Arellano, M. and Bond, S. (1991), 'Some Tests of Specification for Panel Data: Monte Carlo Evidence and an Application to Employment Equations', *Review of Economic Studies*, 58:277–97

Arentsen, M.J. (2011), 'Trends in Gas', in M. Finger and R. Künneke (eds), *International Handbook of Network Industries. The Liberalization of Infrastructure.* Cheltenham, UK and Northampton, USA: Edward Elgar Publishing Ltd

Armijo, L.E. and Faucher, P. (2002), 'We Have a Consensus': Explaining Political Support for Market Reforms in Latin America', *Latin American Politics and Society*, 44(2):1–40

Armstrong, M. and Sappington, D.E.M. (2006), 'Regulation, Competition, and Liberalization', *Journal of Economic Literature*, 44:325–66

Arocena, P. (2003), 'The Reform of the Utilities Sector in Spain', in C. Ugaz, and C. Waddams Price (eds), *Utility Privatization and Regulation: a Fair Deal for Consumers?* Cheltenham, UK and Northampton, USA: Edward Elgar Publishing Ltd

Aron, D.J. and Burnstein, D. (2003), 'Broadband Adoption in the United States: An Empirical Analysis'. Available at: http://papers.ssrn.com/sol3/papers.cfm?abstract_id=386100 (accessed February 2013)

Asche, F. (2002), 'European Market Integration for Gas? Volume Flexibility and Political Risk', *Energy Economics*, 24(3):249–65

Asche, F., Nilsen, O.B., and Tveteras, R. (2008), 'Natural Gas Demand in the European Household Sector', *The Energy Journal*, 29(3):27–46

Atkinson, A.B. (1970), 'On the Measurement of Economic Inequality', *Journal of Economic Theory*, 2 (3):244–63

Atkinson, A.B. and Stiglitz, J. (1980), *Lectures on Public Economics*. New York: McGraw-Hill Book Co

Attenborough, N. (1962), "Economic Effects of Telephony Price Changes in the UK", NERA Topics #8, available at: http://www.nera.com/67_4616.htm (accessed February 2013)

Auriol, E. and Biancini, S. (2012), 'Powering up Developing Countries Through Integration?', *Cesifo Working Papers N 3872*

Averch, M. and Johnson, L. (1962), 'Behavior of the Firm under Regulatory Constraints', *American Economic Review*, 92:1052–69

Azmat, G., Manning, A., and Van Reenen, J. (2007), 'Privatization, Entry Regulation and the Decline of Labour's Share of GDP: A Cross-Country Analysis of the Network Industries'. *CEPR Discussion Paper* 6348 (June)

Bacchiocchi, E., Florio M., and Gambaro, M. (2011), 'Telecom Reforms in the EU: Prices and Consumers' Satisfaction', *Telecommunications Policy,* 35(4):382–96

Bacci, S. (2008), 'Analysis of Longitudinal Health Related Quality of Life Using Latent Regression in the Context of Rasch Modelling', in Huber, C., *et al* (eds), *Mathematical Methods for Survival Analysis, Reliability and Quality of Life*. Hoboken: John Wiley & Sons, Inc., 275–90

Bagadioglu, N., Waddams Price, C., and Weyman-Jones, T. (2007), 'Measuring Potential Gains from Mergers Among Electricity Distribution Companies in Turkey Using a Non-Parametric D Model', *The Energy Journal*, 28 (2):83–110

Baker, P., Blundell, R., and Micklewright, J. (1989), 'Modelling Household Energy Expenditures Using Micro-Data', *The Economic Journal*, 99(397):720–38

Baker, W. (2006), 'Social Tariffs: A Solution to Fuel Poverty?', Centre for Sustainable Energy, available at: www.cse.org.uk/pdf/pub1059.pdf

Baldwin, R., Cave, M., and Lodge, M. (2012), *Understanding Regulation. Theory, Strategy and Practice, 2nd Edition*. Oxford: Oxford University Press

Baron, D.P. and Myerson, R.B. (1982), 'Regulating a Monopolist with Unknown Costs', *Econometrica*, 50(4):911–30

Barone S., Lombardo, A. (2004), 'Service Quality Design through a Smart Use of Conjoint Analysis', *Asian Journal on Quality* 5(1):34–42

Bartolucci, F., Pennoni, F., and Vittadini, G. (2011), 'Assessment of School Performance through a Multilevel Latent Markov Rasch Model', *Journal of Educational and Behavioral Statistics*, 36(4):491–522

Barzel, Y. (1989), *Economic Analysis of Property Rights*. Cambridge: Cambridge University Press

Barzel, Y. (2002), *A Theory of the State: Economic Rights, Legal Rights, and the Scope of the State*. Cambridge: Cambridge University Press

Bauer, J.M. (2003), 'Spectrum Management and the Mobile Services Industry', Quello Center Working Paper 04-03, East Lansing, Michigan

Baumol, W.J. (1982), 'Contestable Markets: An Uprising in the Theory of Industry Structure,' *American Economic Review*, 72(1):1–15

Baumol, W.J. (1987), 'Ramsey Pricing', in *The New Palgrave Dictionary of Economics*, vol. 4. London: MacMillan

Baumol, W.J., Panzar, J.C., and Willig, RD. (1982), *Contestable Markets and the Theory of Industry Structure.* New York: Harcourt Brace Jovanovich

Baumol, W.J, and Sidak, J.G. (1994), 'The Pricing of Inputs Sold to Competitors,' *Yale Journal of Regulation*, 11(1):171–202

Baumol, W.J. and Swanson, D.G. (2003), 'The New Economy and Ubiquitous Competitive Price Discrimination: Identifying Defensible Criteria of Market Power', *70 Antitrust Law Journal*, 3:661–85

Baumol, W.J. and Willig, R.D. (1986), 'Contestability: Developments Since the Book', *Oxford Economic Papers*, 38(0):9–36

Beato, P. and Mas-Colell, A. (1984), 'The Marginal Cost Pricing as a Regulation Mechanism In Mixed Markets', in M. Marchand, P. Pestieau and H. Tulkens (eds), *The Performance of Public Enterprises*, Amsterdam: North-Holland

Beesley, M. and Littlechild, S. (1983), 'Privatization: Principles, Problems, and Priorities', *Lloyds Rank Review*, 149:1–20

Beesley, M. and Littlechild, S. (1989), 'The Regulation of Privatized Monopolies in the United Kingdom', *RAND Journal of Economics*, 20(3):454–72

Beiner, S., Schmid M., and Wanzenried, G. (2011), 'Product Market Competition, Managerial Incentives and Firm Valuation', *European Financial Management*, 17(2):331–66

Bel, G., Fageda, X., and Warner, M.E. (2010), 'Is Private Production of Public Services Cheaper Than Public Production? A Meta-Regression Analysis of Solid Waste and Water Services', *Journal of Policy Analysis and Management*, 29(3):553–77

Beranes, E. and Bourreau M. (2005), 'An Economist's Guide to Local Loop Unbundling', *Communications & Strategies*, 57:13–32

Bertrand, M. and Mullainathan, S. (2001), 'Do People Mean What They Say? Implications for Subjective Survey Data', *American Economic Review* 91(2):67–72

Besley, T. (2006), *Principled Agents? The Political Economy of Good Government.* New York: Oxford University Press

Bianchi, P. and Labory S. (2006), *International Handbook on Industrial Policy.* Cheltenham, UK and Northampton, USA: Edward Elgar Publishing Ltd

Biava, A. (2007), 'L'action de l'union européenne face au défi de la sécurisation de son approvisionnement énergétique?'. *Politique Européenne*, 22:105–23

Biglaiser, G. and Ma, C.-T.A., (1999), 'Investment Incentives of a Regulated Dominant Firm'. Journal of Regulatory Economics, 16(3):215–35

Biggar, D. (2010), 'Fairness in Public-Utility Regulation: a Theory'. *Agenda: A Journal of Policy Analysis and Reform* 17(1):5–29

Blundell, R.W. and Bond, S.R. (1998), 'Initial Conditions and Moment Restrictions in Dynamic Panel Data Models'. *Journal of Econometrics* 87(1):115–43

Blundell, R.W. and Bond, S.R. (2000), 'GMM Estimation with Persistent Panel Data: An Application to Production Functions'. *Econometrics Reviews* 19:321–40

Boardman, B. (2010a), 'Liberalization and Fuel Poverty', in I. Rutledge and P. Wright, (eds) *UK Energy Policy and the End of Market Fundamentalism.* Oxford: Oxford University Press

Boardmam, B. (2010b), *Fixing Fuel Poverty: Challenges and Solutions*. London: Earthscan

Boardman A.E., Greenberg, D.H., Vining, A.R., and Weimer, D.L. (2005), *Cost–Benefit Analysis. Concepts and Practice*, third Edition. Upper Saddle River: Pearson Prentice Hall

Bognetti, G. and Obermann, G. (2008), 'Liberalization and Privatization of Public Utilities: Origins of the Debate, Current Issues and Challenges for the Future'. *Annals of Public and Cooperative Economics*, 79(3–4):461–85

Bollen, K.A. (1989), *Structural Equations with Latent Variables. Wiley Series in Probability and Mathematical Statistics*. New York: Wiley

Bolton, P. (2010), 'Energy Imports and Exports', *House of Commons' Library*, Parliament UK, United Kingdom

Bonnet, C., Dubois, P., Martimort, D., and Straub, S. (2011), 'Empirical Evidence on Satisfaction with Privatization in Latin America: Welfare Effects and Beliefs'. *The World Bank Economic Review*, http://wber.oxfordjournals.org/content/early/2011/07/14/wber.lhr037.full (accessed February 2013)

Boots, M.G., Fieke, A.M., Rijkers, F.A.M., and Hobbs, B.F. (2004), 'Trading in the Downstream European Gas Market: A Successive Oligopoly Approach'. *The Energy Journal*, 25(3):73–102

Bortolotti, B., D'Souza' J., Fantini' M., and Megginson, W.L. (2002), 'Privatization and the Sources of Performance Improvement in the Global Telecommunications Industry'. *Telecommunications Policy*, 26(5–6):243–68

Bös, D. (1986), *Public Enterprise Economics: Theory and Application*. Amsterdam: North-Holland

Bös, D. (1994), *Pricing and Price Regulation. An Economic Theory for Public Enterprises and Public Utilities*. Amsterdam: Elsevier Science Ltd

Bös, D. (1999), 'Incomplete Contracting and Price Regulation'. *Journal of Public Economics*, 73:353–71

Bosmans, W. (2007), 'A contribution to the current EU debate on energy—Focus on the European gas market', EREM, European Energy Institute and contributors, 1–19

Bourreau, M. and Pinar, D. (2004), 'Service-Based vs Infrastructure-Based Competition in Local Markets'. *Information Economics and Policy*, 16:287–306

Bourreau, M. and Dogan, P. (2005), 'Unbundling the Local Loop'. *European Economic Review*, 49(1):173–99

Boylaud, O. and Nicoletti, G. (2001), 'Regulation, Market Structure and Performance in Telecommunications'. *OECD Economic Studies*, 32:99–142

Bradburd, R. (1995), 'Privatization of Natural Monopoly Enterprises: the Regulation Issue'. *Review of Industrial Organization*, 10(3):247–67

Brau R., Doronzo, R., Fiorio C.V., and Florio M. (2010), 'EU Gas Industry Reforms and Consumers' Prices'. *The Energy Journal*, 21(4):163–79

Brent, R.J. (2006), *Applied Cost- Benefit Analysis*. Cheltenham, UK and Northampton, USA: Edward Elgar Publishing Ltd

Brent, R.J. (2007), *Applied Cost-Benefit Analysis*, Second Edition. Cheltenham, UK and Northampton, USA: Edward Elgar Publishing Ltd

Breton, M. and Kmarbah, M. (2012), 'Transportation and Storage under a Dynamic Price Cap Regulation Process'. *Energy Economics*, 34(7):918–29

Brito, D.L. and Rosellon, J. (2010), 'Pricing Natural Gas in Mexico: an Application of the Little-Mirrlees Rule—The Case of Quasi-Rents'. *Southern Economic Journal,* 76(4):1131–6

Brown, S.P.A and Yücel, M.K. (2008), 'What Drives Natural Gas Prices?' *The Energy Journal,* 29(2):45–60

Brunekreeft, G. (2008), 'Ownership Unbundling in Electricity Markets. A Social Cost Benefit Analysis of the German TSOs'. *Discussion Paper EPRG 08-16, University of Cambridge.* http://ideas.repec.org/p/cam/camdae/0833.html (accessed February 2013)

Buehler S. (2005), The Promise and Pitfalls of Restructuring Network Industries. *German Economic Review,* 6(2):205–228

Bundeskartellamt (2011), 'Sector Inquiry into Electricity Generation and Wholesale Markets', http://www.bundeskartellamt.de/wenglisch/download/pdf/2011-05-05_su_strom_executive_summary_en_final-2.pdf (accessed January 20, 2012)

Buzar, S. (2007), *Energy Poverty in Eastern Europe: Hidden Geographies of Deprivation.* Aldershot: Ashgate

Caillaud, B., Guesnerie, R., Rey, P., and Tirole, J. (1988), 'Government Intervention in Production and Incentives Theory: a Review of Recent Contributions'. *RAND Journal of Economics,* 19(1):1–26

Cameron, P.D. (2007), *Competition in Energy Markets: Law and Regulation in the European Union.* New York: Oxford University Press

Cave, M. (2006), 'Six Degrees of Separation Operational Separation as a Remedy in European Telecommunications Regulation'. *Communications & Strategies* 64(4):89–103

Cave, M. and Crocioni, P. (2007), 'Does Europe Need Network Neutrality Rules?' *International Journal of Communication,* 1:669–79

Cave, M. and Vogelsang, I. (2003), 'How Access Pricing and Entry Interact'. *Telecommunication Policy* 27:717–27

Cawley, R. (2003), 'The European Union and World Telecommunications Market', in G. Madden (ed.) *World Telecommunication Markets, The International Handbook of Telecommunication Economics 3.* Cheltenham, UK and Northampton, USA: Edward Elgar Publishing Ltd

CEC. (2007a), 'Evaluation of the Performance of Network Industries Providing Services of General Economic Interest'. *2006 Report, Commission of the European Community*

CEC. (2007b), 'European Electronic Communication Regulation and Markets'. *12th Report* COM(2007)155

CEEP. (2010), *Public Services Supporting the Very Fabric of European Society.* Activity Report

CEER. (2008), Task Force on Quality of Electricity Supply. Available At http://www.energy-regulators.eu/portal/page/portal/eer_home

CEER. (2008), 4th Benchmarking Report on Quality of Electricity Supply. Available at: http://www.autorita.energia.it/allegati/pubblicazioni/c08-eqs-24-04_4th_benchmarking_report_eqs_10-dec-2008_re.pdf

CEER. (2010), 'Guidelines of Good Practice on Estimation of Costs Due to Electricity Interruptions and Voltage Disturbances'. Ref. C10-EQS-41-03, December

CEER. (2011), '5th Benchmarking Report on Electricity Quality Of Supply. Ref. C11-EQS-47-03

Ceriani, L, Doronzo, R., and Florio, M. (2009), 'Privatization, Unbundling, and Liberalization of Network Industries: A Discussion of the Dominant Policy Paradigm in the EU'. Department of Economics, Business and Statistics Working Paper 2009-09, Università Degli Studi Di Milano

Ceriani, L. and Florio, M. (2011), 'Consumer Surplus and the Reform of Network Industries: A Primer'. *Journal of Economics,* 102(2):111–22

Chamberlin, E.H. (1948), 'An Experimental Imperfect Market'. *Journal of Political Economy,* 56:95–108

Chang Y., Waddams Price, C., and Wilson, C.M. (2006), 'Consumer Search and Switching: Much Still to Gain'. Centre for Competition Policy Working Paper, University Of East Anglia

Chang, T. and Wilson, C. (2006), 'Behavioural Revolution: Workshop on Consumer Behaviour and Bounded Rationality', http://www.esrc.ac.uk/my-esrc/grants/res-578-28-0001/outputs/read/bd4b64c5-94ba-4c81-a17c-9fd6d8a10e58 (accessed February 2013)

Chao, H. (2011), 'Response in Wholesale Electricity Markets: the Choice of Customer Baseline'. *Journal of Regulatory Economics,* 39(1):68–88

Checchi, D., Florio, M., and Carrera, J. (2009), 'Privatization Discontent and Utility Reform in Latin America'. *Journal of Development Studies,* 45(3):333–51

Chisari, O, Estache, A., and Romero, C. (1999), 'Winners and Losers from Utility Privatization in Argentina: Lessons from a General Equilibrium Model', The World Bank Policy Research Working Paper, n. 1824

Chong, A. and López De-Silanes, F. (2003), 'The Truth about Privatization in Latin America', Yale ICF Working Paper No. 03–29

Clements, M.E. (2001), 'Local Telephone Quality-of-Service: The Impact of Regulation and Competition', http://etd.ohiolink.edu/view.cgi?acc_num=osu1273857327 (accessed February 2013)

Clifton, J. and Ahedo, M. (eds) (2010), 'Revisiting Multinational Corporations in the Twenty-First Century'. *Revista Internacional de Organizaciones/International Review of Organizations,* 4, Special Issue: 7–13

Clifton, J., Comín, F., and Díaz-Fuentes, F. (2007), *Transforming Public Enterprise in Europe and North America: Transnation;alisation, Networks and Integration.* London/New York: Palgrave

Clifton J. and Diaz-Fuentes D. (2010a), 'Evaluating EU Policies on Public Services: A Citizens' Perspective'. *Annals of Public and Cooperative Economics,* 81(2):281–311

Clifton, J. and Díaz-Fuentes, D. (2010b), 'From National Monopoly to Multinational: Consequences for Work in Public Services in Europe'. *Revista Internacional de Organizaciones/International Review of Organizations,* 4:107–27

Coase, R. (1937), 'The Nature of the Firm'. *Economica* 4(16):386–405

Coen, D. and Thatcher, M. (2008), 'Network Governance and Multi-Level Delegation. European Networks of Regulatory Agencies'. *Journal of Public Policy* 28(1):49–71

Conway, P. and Nicoletti, G. (2006), 'Product Market Regulation in Non-Manufacturing Sectors in OECD Countries: Measurement and Highlights'. *OECD Economics Department Working Paper No. 530*

Copenhagen Economics (2005), 'Market Opening In Network Industries, Final Report'. DG Internal Market, European Commission, Brussels

Correljé, A. (2008), 'Dilemmas in Network Regulation: the Dutch Gas Industry', in R. Künneke, J. Groenewegen, and A.Correljé (eds) *Innovations in Liberalized Network Industries: Between Private Initiatives and Public Interest*. Cheltenham, UK and Northampton, USA: Edward Elgar Publishing Ltd

Cournot, A. (1897), *Researches into the Mathematical Principles of the Theory of Wealth*. New York: The Macmillan Company

Cremer, H., Marchand, M., and Thisse, JF. (1989), 'The Public Firm as an Instrument for Regulating an Oligopolistic Market'. *Oxford Economic Papers*, 41:283–301

Cronin, J.J. Jr. and Taylor, S.A. (1992), 'Measuring Service Quality: A Reexamination and Extension'. *Journal of Marketing*, 56, July:55–68

Currier, K. M. (2007), 'A Practical Approach to Quality-Adjusted Price Cap Regulation'. *Telecommunications Policy*, 31(8–9):493–501

Dahlby, B. (2008), *The Marginal Cost of Public Funds: Theory and Applications*. Cambridge, USA: The MIT Press

Davies, S. and Waddams, C. (2007), 'Does Ownership Unbundling Matter? Evidence from UK Energy Markets'. *Intereconomics*, 42(6):297–301

Davies, S., Waddams Price, C., and Wilson, C. (2009), 'How Far Does Economic Theory Explain Competitive Nonlinear Pricing in Practice?' CCP Working Paper 09–7

De Battisti, F., Nicolini, G., and Salini S. (2010), 'The Rasch Model in Customer Satisfaction Survey Data'. *Qualitative Technology and Quantitative Management*, 7(1):15–34

De Bijl, P.W.J. (2005), 'Structural Separation and Access in Telecommunications Markets', Competition and Regulation in Network Industries, *Intersentia*, 6(2): 95–115,

De Bijl, P.W.J. and Peitz, M. (2005), 'Local Loop Unbundling in Europe: Experience, Prospects and Policy Challenges'. *Communciations & Strategies*, 57(1):33–57

De Bock, R. and Gijon, J.G. (2011), 'Will Natural Gas Prices Decouple From Oil Prices Across the Pond?' IMF Working Papers 11/143

DEAS, CIRIEC, CSIL, PPMI. (2010), *The Inter-Relationship Between the Structural Funds and the Provision of Services of General (Economic) Interest and the Potential for Cross-Border Service Delivery*. Study Requested by The European Parliament's Committee on Regional Development. Available at: http://www.europarl.europa.eu/meetdocs/2009_2014/documents/regi/dv/pe438616_xm_/pe438616_xm_en.pdf (accessed February 2013)

Deaton , A.S. and Muellbauer, J. (1980), 'An Almost Ideal Demand System', *American Economic Review*, 70(3):312–326

Defeuilley, C. (2009), 'Retail Competition in Electricity Markets'. *Energy Policy*, 37(2):377–86

DEFRA. (2010), Household Living in Fuel Poverty http://archive.defra.gov.uk/sustainable/government/progress/regional/summaries/63.htm (accessed February 2013)

De Fraja, G. (1993), 'Productive Efficiency in Public and Private Firms.' *Journal of Public Economics*, 50(1):15–30

De Fraja, G. (2009), 'Mixed Oligopoly: Old and New'. Discussion Papers in Economics 09/20, Department of Economics, University of Leicester

De Fraja, G. and Delbono, F. (1989), 'Alternative Strategies of a Public Firm in Oligopoly'. *Oxford Economic Papers* 41:302–11

Del Bo, C. and Florio, M. (2012), 'Public Enterprises, Policy Adoption and Planning: Three Welfare Propositions'. *Journal of Economic Policy Reform*, 15(4):263–79

Del Bo, C. and Florio, M. (2012), 'Electricity Investment: An Evaluation of the New British Energy Policy', Paper Presented at The 7th Conference on Energy Economics and Technology Infrastructure for the Energy Transformation, Dresden, April 27th, 2012

De Leo, G., Fiorese, G., and Guarisio, G. (2009), *Energia e Salute Della Terra*. Milano: Fondazione Achille e Giulia Boroli

De Luca, G., Florio, M., Grove, N., and Lorenzini, M. (2012), 'Typologies of Infrastructure Finance in Europe from the Antiquity to the 18th Century', Paper Presented at XVI World Economic History Congress, Stellenbosch, 9–13 July 2012, Session N. 109, and Published at http://www.wehc2012.org/programme.php (accessed February 2013)

Delvaux, B., Hunt, M., and Talus, K. (eds) (2010), *EU Energy Law and Policy Issues*. Brussels: Euroconfidential

Department of Energy and Climate Change. (2011), 'Digest of United Kingdom Energy Statistics (DUKES)' http://www.decc.gov.uk/en/content/cms/statistics/publications/dukes/dukes.aspx (accessed February 2013)

Department of Energy And Climate Change (2011), 'Planning our Electric Future: a White Paper for Secure, Affordable and Low-Carbon Electricity', http://www.decc.gov.uk/en/content/cms/legislation/white_papers/emr_wp_2011/emr_wp_2011.aspx (accessed February 2013)

De Vellis, R.F. (1991), *Scale Development Theory and Applications*. Newbury Park: SAGE Publications

Dewenter, R. and Haucap, J. (2007), *Access Pricing: Theory and Practice*. Amsterdam: Elsevier

Doganoglu, T. and Pereira, P. (2006), 'The Welfare Impact of Competition in Fixed Telephony', Portuguese Competion Authority Working Paper 21. Available At: http://www.concorrencia.pt/ven/estudos_e_publicacoes/working_papers/pages/wp-n-21-the-welfare-impact-of-competition-in-fixed-telephony.aspx?lst=1&pagenr=8 (accessed February 2013)

Doh, J.P. and Teegen H. (2004), 'Balancing Private and State Ownership in Emerging Markets' Telecommunications Infrastructure: Country, Industry, and Firm Influences.' *Journal of International Business Studies* 35(3):233–250

Domah, P.D. and Pollitt, M.G., (2001), 'The Restructuring and Privatisation of the Regional Electricity Companies in England and Wales: a Social Cost Benefit Analysis'. *Fiscal Studies* 22(1):107–46

Doyle, C. (2000), 'Local Loop Unbundling and Regulatory Risk'. *Journal of Network Industries* 1(1):33–54

Drèze, J. and Stern, N. (1990), 'Policy Reform, Shadow Prices, and Market Prices'. *Journal of Public Economics* 42(1):1–45

EBRD. (2003), Transition Report 2003. Integration and Regional Cooperation, London, available at ww.ebrd.com/downloads/research/transition/TR03.pdf

EC. (1980), 'Transparency of Financial Relations between Member States and Public Undertakings', 80/723/EEC, 25 June 1980. Available at: http://eur-lex.europa.eu/lexuriserv/lexuriserv.do?uri=celex:31980l0723:en:html

EC. (1987),'Towards A Dynamic European Economy, Green Paper on the Development of the Common Market for Telecommunications Services and Equipment'. COM (87): 290

EC. (1987), The Single European Act

EC. (1996), 'Implementation of Full Competition in Telecommunications Markets'. 1996/19/EC

EC. (1998), 'On the Application of Open Network Provision (ONP) to Voice Telephony and on Universal Service for Telecommunications in Competitive Environment'. Directive 98/10/EC

EC. (1999), '5th Report on the Implementation of the Telecommunications Regulatory Package.' Available at: http://ec.europa.eu/information_society/policy/ecomm/doc/library/annualreports/5report/finalen.pdf (accessed February 2013)

EC. (2002), Competition In The Markets For Electronic Communications Networks And Services Directive 2002/27

EC. (2002), 'On Universal Service and Users' Rights Relating To Electronic Communications Networks And Services (Universal Service Directive)' 2002/22/EC

EC. (2006), 'The European Electronic Communications Regulations and Markets', 12th Report

EC. (2007), 'Evaluation of The Performance of Network Industries Providing Services of General Economic Interest 2006 Report, Annexes', SEC (2007):1024

EC. (2008), 'Commission Recommendation on the Active Inclusion of People Excluded from the Labour Market', 2008/867/EC

EC. (2009), *Agreement on EU Telecom Reforms Paves Way for Stronger Consumer Rights, an Open Internet, a Single European Telecoms Market and High-Speed Internet Connections for all Citizens,* Brussels

EC. (2009), Directive 2009/136 Amending Directive 2002/22/EC On Universal Service And Users' Rights Relating To Electronic Communications Networks And Services, Directive 2002/58/EC Concerning The Processing Of Personal Data And The Protection Of Privacy In The Electronic Communications Sector And Regulation (EC) No 2006/2004 On Cooperation Between National Authorities Responsible For The Enforcement Of Consumer Protection Laws

EC. (2009), 'Progress Report on the Single European Electronic Communications Market 2008'. Available at: http://ec.europa.eu/information_society/policy/ecomm/library/communications_reports/annualreports/14th/index_en.htm (accessed February 2013)

EC. (2010), *Investing in Europe's Future. Fifth Report on Economic, Social, and Territorial Cohesion*, Luxembourg: Publication Office of the European, Commission. Available at: http://ec.europa.eu/regional_policy/sources/docoffic/official/reports/cohesion5/index_it.cfm (accessed February 2013)

EC. (2010), *Investing in Europe's Future.* Fifth Report on Economic, Social and Territorial Cohesion. Luxembourg: European Union

EC. (2010), 'A Strategy for Smart, Sustainable and Inclusive Growth ', COM (2010)2020

EC. (2011), Reform of the EU State Aid Rules on Services of General Economic Interest Communication from the Commission to the European Parliament, Com (2011)146. Brussels, March 2011. Available at: http://europa.eu/legislation_summaries/competition/state_aid/cc0015_en.htm (accessed February 2013)

Economides, N. (1996), 'The Economics of Networks'. *International Journal Of Industrial Organization*, 14(6):673–99

Economides, N. (2008), 'Public Policy in Network Industries' in Buccirossi P. (ed), *Handbook of Antitrust Economics*, Cambridge: The MIT Press

Economides, N., Seim, K., and Viard, B.V. (2008), 'Quantifying The Benefits of Entry Into Local Phone Service'. *RAND Journal of Economics*, 39(3):699–730

Economides, N. and White, L. (1994), 'Network and Compatibility: Implications for Antitrust'. *European Economic Review*, 38:651–62

ECRB. (2008), Report on the Quality of Electricity Service Standards and Incentives In Quality Regulation. http://www.energy-community.org/pls/portal/docs/364177.pdf

Edwards, G. and Waverman L. (2006), 'The Effects of Public Ownership and Regulatory Independence on Regulatory Outcomes'. *Journal of Regulatory Economics* 29(1):23–67

EIA -Energy Information Administration. (2010), Annual Energy Outlook 2011, DOE/EIA-0383

Ellig, J. (2001), *Dynamic Competition and Public Policy. Technology, Innovation and Anti-Trust Issues.* Cambridge: Cambridge University Press

EP. (2002), 'On Access to, and interconnection of, Electronic Communications Networks and Associated Facilities (Access Directive)'. 2002/19/EC of The European Parliament and of The Council

Erdogdu, E. (2011), 'The Impact of Power Market Reforms on Electricity Price-Cost Margins and Cross-Subsidy Levels: A Cross Country Panel Data Analysis' *Energy Policy*, 39(3):1080–92

Estache, A., Foster, W., and Wodom, Q. (2002), 'Accounting for Poverty in Infrastructure Reform: Learning from Latin America's Experience,' Washington D.C., USA: World Bank Publications

Estache, A. and Goicoechea, A. (2005), 'A 'Research' Database on Infrastructure Economic Performance'. World Bank Policy Research Working Paper 3643

Estache, A., Goicoechea, A., and Manacorda, M. (2006), 'Telecommunications Performance, Reforms, and Governance'. *The World Bank Policy Research Working Paper 3822*

Estache, A., Goicoechea, A., and Trujillo, L. (2006), 'Utilities Reform and Corruption in Developing Countries'. *The World Bank Policy Research Working Paper* 4081

EU. (1957), Treaty on the Functioning of the European Union, Rome

EU. (1996), 'Implementation of Full Competition in Telecommunications Markets', Directive 1996/19/EC, European Parliament and Council, 13 March

EU. (1996b), 'Common Rules for the Internal Market in Electricity', Directive 1996/92/EC, European Parliament and Council, 19 December

EU. (1998), 'Common Rules for the Internal Market in Natural Gas', Directive 1998/30/EC, European Parliament and Council, 22 June

EU. (2003a), 'Common Rules for the Internal Market in Electricity', Directive 2003/54/EC, European Parliament and Council, 26 June

EU. (2003b), 'Common Rules for the Internal Market in Natural Gas', Directive 2003/55/EC, European Parliament and Council, 26 June

EU. (2004), 'Measures to Safeguard Security of Natural Gas Supply' Council Directive 2004/67/EC

EU. (2007), 'An Energy Policy For Europe', COM (2007): 1 Final, SEC (2007): 12, European Parliament and Council, 1 January

EU. (2009a), 'Internal Markets for Gas', COM (2009): Directive 2009/73/EC, European Parliament and Council, 13 July

EU. (2009b), 'Internal Market In Electricity', COM (2009): Directive 2009/72/EC, European Parliament and Council, 13 July

EU. (2009), On for an Internal EU Gas and Electricity Market, Third Legislative Package. Available at: http://ec.europa.eu/energy/gas_electricity/legislation/legislation_en.htm (accessed February 2013)

EU. (2009), Treaty of Lisbon

EU. (2010), Regulations Concerning Measures to Safeguard Security of Gas Supply and Repealing Council Directive 2004/67/EC. 994/2010

Eurobarometer. (2000), Measuring Information Society. EB 53

Eurobarometer. (2000, 2002, 2004), Standard EB, Spring and Autumn Waves

Eurobarometer. (2002), Report No. 53, 'The People of Europe and Services of General Interest'

Eurobarometer. (2004), Report No. 58, 'Consumers' Opinions About Services of General Interest'

Eurobarometer. (2004), The People of Europe and Services of General Interest, Report 53

Eurobarometer. (2006), Report No. 62.1: The Constitutional Treaty, Economic Challenges, Vocational Training, Information Technology at Work, Environmental Issues, and Services of General Interest

Eurobarometer. (2007), Expectations of European Citizens Regarding the Social Reality in 20 Years' Time. FB 227

Eurobarometer. (2008), Neighbours of The European Union, Services of General Interest, Employment and Social Policy, Energy Technologies, and Family Planning. Eurobarometer Survey Series, Commission of the European Communities

Eurelectric. (2003), *Environmental Guidelines*

Eurelectric. (2011), *Power Statistics and Trends*

Edison Electric Institute Data: http://www.eei.org, (data accessed December 16, 2011)

European Court of Justice. (2010), Judgment in Case C-280/08 P Deutsche Telekom AG V Commission. Available at: http://europa.eu/rapid/pressreleasesaction.do?reference=cje/10/104&type=html

European Energy Regulators. (2011), 'National Reporting', http://www.energy-regulators.eu/portal/page/portal/eer_home/eer_publications/national_reports/national%20reporting%202011 (accessed January 20, 2012)

European Environment Agency. (2011), Energy Intensity in the Service Sector (ENER 024) http://www.eea.europa.eu/data-and-maps/indicators/energy-intensity-in-the-service-sector/assessment-1 (accessed February 2013)

Evans, D. and Kula, E. (2011), 'Social Discount Rates and Welfare Weights for Public Investment Decisions under Budgetary Restrictions: The Case of Cyprus', *Fiscal Studies*. 32(1):73–107

Exxon Mobil. (2011), Summary Annual Report. Irving, TX: Exxon Mobil

Fankhauser, S. and Tepic S. (2007), 'Can Poor Consumers Pay for Energy and Water? An Affordability Analysis for Transition Countries'. *Energy Policy*. 35(2):1038–49

Farrer, T.H., (1902), *The State In Its Relation To Trade*. London: Macmillan

FDS. (2010), *Consumer Research on Disconnection Methods*. London: FDS

Federal Ministry of Economy and Technology. (1999) http://www.bmwi.de/english/navigation/technology/telecommunications-and-post.html (accessed February 2013)

Ferrari, A. and Giulietti, M. (eds). (2005), 'La Concorrenza nel Settore Elettrico: l'esperienza internazionale e l'analisi del Mercato Italiano'. *Economia Pubbl*ica, 1–2

Ferrari, P.A., Annoni, P., and Manzi, G. (2010a), 'Evaluation and Comparison of European Countries: Public opinion of Services'. *Quality & Quantity, International Journal of Methodology* 44(6):1191–205

Ferrari, P.A. and Manzi, G. (2010b), 'Nonlinear Principal Components Analysis as a Tool for the Evaluation of Customer Satisfaction'. *Qualitative Technology and Quantitative Management* 7(2):117–32

Ferrari, P.A. and Salini, S. (2011), 'Complementary Use Of Rasch Models And Nonlinear Principal Components Analysis In The Assessment Of The Opinion Of Europeans About Utilities'. *Journal of Classification*. 28(1):53–69

Fichert, F., Haucap, J., and Kai, R. (2007), *Competition Policy in Network Industries*. Berlin: LIT Verlag

Finger, M. and Finon, D. (2011), 'From 'Service Publique' to Universal Service: The Case of European Union', in Finger, M. and Künneke, R. (eds), *International Handbook of Network Industries: The Liberalization of Infrastructure*. Cheltenham, UK and Northampton, USA: Edward Elgar Publishing Ltd, 54–69

Finger, M. and Künneke, R. (eds). (2011), *International Handbook of Network Industries: The Liberalization of Infrastructure*. Cheltenham, UK and Northampton, USA: Edward Elgar Publishing Ltd

Fink, C., Mattoo, A., and Rathindran, R. (2003), 'An Assessment of Telecommunications Reform in Developing Countries'. *Information Economics and Policy*. 15(4):443–466

Fiorio, C.V. and Florio, M. (2010), 'A Fair Price For Energy? Ownership Versus Market Opening in the EU15,' Cesifo Working Paper Series 3124, Cesifo Group Munich

Fiorio, C.V. and Florio, M. (2011), 'Would you Say that the Price you Pay for Electricity is Fair? Consumers' Satisfaction and Utility Reforms in the EU15'. *Energy Economics*. 33(2):178–87

Fiorio, C.V., Florio, M., and Doronzo, R. (2008), 'The Electricity Industry Reform Paradigm in the European Union: Testing the Impact on Consumers', in P. Arestis and M. Sawyer (eds) *Critical Essays on the Privatisation Experience*, International Papers in Political Economy Series. New York: Palgrave Macmillan, 121–59

Fiorio, C.V., Florio, M., and Perrucca, G. (2011), 'Consumers' Satisfaction and Regulation of Local Public Transport: Evidence from European Cities,' Department of Economics, Business and Statistics Working Paper Series 2011–26, Università degli Studi di Milano

Fiorio, C.V., Florio, M. Salini S., and Ferrari, P.A. (2007), 'European Consumers Attitudes On Services Of General Interest: Accessibility, Price And Quality', in M. Marrelli, F. Padovano and I. Rizzo (eds) *Servizi Pubblici: nuove tendenze nella regolamentazione, nella produzione e nel finanziamento*. Milano: Franco Angeli

Florio, M. (2003), 'Does Privatisation Matter? The Long-Term Performance of British Telecom over 40 Years'. *Fiscal Studies*, 24(2):197–234

Florio, M. (2004), *The Great Divestiture. Evaluating the Welfare Impact of British Privatizations 1979–1997*. Cambridge: The MIT Press

Florio, M. (2006), 'Telecom Italia: the Parallel Returns (1997–2006)', in J.L. Briquet and A. Mastropaolo (eds.), *Italian Politics—The Center-Left's Poisoned Victory*. Oxford: Berghahn Books

Florio, M. (ed.) (2007), *Cost-Benefit Analysis and Incentives in Evaluation. The Structural Funds of European Union*. Cheltenham, UK and Northampton, USA: Edward Elgar Publishing Ltd

Florio, M. (ed.) (2011), *Public Investment, Growth and Fiscal Constraints: Challenges for the EU New Member States*. Cheltenham, UK and Northampton, USA: Edward Elgar Publishing Ltd

Florio, M. (2014 forthcoming), 'Applied Welfare Economics'. Oxford, UK: Routledge

Florio, M. and Puglisi, R. (2006), 'Public Ownership, Privatisation and Regulation: Social Welfare Counterfactuals for British Telecom,' UNIMI—Research Papers in Economics, Business, and Statistics Unimi-1017, Università degli Studi Di Milano

Foss, M. (2004), Interstate Natural Gas—Quality Specifications and Interchangeability, Center for Energy Economics, Available at: http://www.beg.utexas.edu/energyecon/lng/documents/CEE_Interstate_Natural_Gas_Quality_Specifications_and_Interchangeability.pdf (accessed February 2013)

Fraquelli, G., Piacenza, M., and Vannoni, D. (2004), 'Scope and Scale Economies in Multi-Utilities: Evidence from Gas, Water and Electricity Combinations', *Applied Economics*, 36(18):2045–58

Frey, B.S. and Stutzer, A. (2002), 'What Economists Can Learn From Happiness Research', *Journal of Economic Literature*, 40:401–35

Friederiszick, H., Grajek, M., and Röller L. (2008), 'Analyzing the Relationship Between Regulation and Investment in the Telecom Sector', *ESMT White Paper*, No. WP-108-01

Frontier Economics (2007), 'Going Their Separate Ways', Bulletin on Energy and Regulation, September 2007. Available at http://www.frontier-economics.com/_library/publications/Frontier%20bulletin%20-%20going%20separate%20ways.pdf (accessed February 2013)

Fumagalli, E., Schiavo, L., and Delestre, F. (2007), *Service Quality Regulation in Electricity Distribution and Supply*. Power Systems Series XIII, 153

Galal, A., Jones, L., Tandon, P., and Vogelsand, I. (1994), *Welfare Consequences of Selling Public Enterprises*. Oxford: Oxford University Press

Garcia, S., Moreaux, M., and Reynaud, A. (2007), 'Measuring Economies of Vertical Integration in Network Industries: an Application to the Water Sector', *International Journal of Industrial Organization*, 25:791–820

Garcia-Murillo, M. (2005), 'International Broadband Deployment: The Impact of Unbundling', *Communications & Strategies*, 57:83

Gas Act (1986), available at: http://www.legislation.gov.uk/ukpga/1986/44 (accessed February 2013)

Gasmi, F., Noumba, U. P., and Virto, L. (2009), 'Political Accountability and Regulatory Performance in Infrastructure Industries: an Empirical Analysis,' *World Bank Economic Review*, 23(3):509–31

Gassner, K., Popov, A., and Pushak N. (2009), 'Does Private Sector Participation Improve Performance in Electricity and Water Distribution?' *The World Bank Trends and Policy Options Series*, 6

Gilbert, N. (2002), *Transformation of the Welfare State: The Silent Surrender of Public Responsibility*. Oxford: Oxford University Press

Glachant, J. and Lévêque, F. (2005), 'Electricity Internal Market in the European Union—What to Do Next'. *Massachusetts Institute of Technology, Center for Energy and Environmental Policy Research Working Papers* 0515

Glachant, J. and Lévêque, F. (2009), 'How to Rationalize the Debate about 'EU Energy Third Package'? Revisiting Criteria to Compare Electricity Transmission Organizations'. European University Institute RSCAS Working Papers 2009/15. Available at: http://cadmus.eui.eu/bitstream/handle/1814/11028/eui_rscas_2009_15.pdf?sequence=1 (accessed February 2013)

GME (2001), *Annual Report*. Roma: GME. Available at: http://www.mercatoelettrico.org/En/MenuBiblioteca/documenti/20120711RelazioneAnnuale2011_en.pdf (accessed February 2013)

Goerten, J. and Clement, E. (2006), 'European Electricity Market Indicators of the Liberalisation Process 2004-2005', Statistics in Focus, Environment and Energy, 6/2006

Gómez-Ibáñez, J.A. (2006), *Regulating Infrastructure: Monopoly, Contracts, and Discretion*. Cambridge: Harvard University Press

Gomez-Lobo, A. (1996), 'The Welfare Consequences of Tariff Rebalancing in the Domestic Gas Market', *Fiscal Studies*, 17(4):49–65

Gonenc, R., Maher, M., and Nicoletti, G. (2001), 'The Implementation and the Effects of Regulatory Reform: Past Experience and Current Issues'. OECD Economics Department Working Paper 251

Gottinger, H. (2003), *Economies of Network Industries*. London: Routledge

Graham, C. (2006), 'The Economics of Happiness: New Lenses for Old Policy Puzzles', in S. Durlauf, and L. Blume (eds), *The New Palgrave Dictionary of Economics, 2nd Edition*. Hampshire: Palgrave Macmillan

Graham, C. and Felton A. (2005), 'Does Inequality Matter in Individual Welfare? An Initial Exploration Based on Happiness Surveys from Latin America'. Center on Social and Economic Dynamics Working Paper Number 38. Available at http://www.brookings.edu/~/media/files/rc/reports/2005/01poverty_graham/01poverty_graham.pdf (accessed February 2013)

Graham, C. and Pettinato, S. (2002), 'Frustrated Achievers: Winners, Losers, and Subjective Well Being in New Market Economies', *Journal of Development Studies*, 38 (4):100–40

Graham, C. and Sukhtankar, S. (2004), 'Does Economic Crisis Reduce Support for Markets And Democracy In Latin America? Some Evidence from Surveys of Public Opinion and Well Being', *Journal of Latin America Studies*, 36(2):349–77

Graham, P. (2006), 'The Heat is on: the Future of Energy in Australia', *CSIRO*, http://csiro0702.interactiveinvestor.com.au (accessed February 2013)

Grassi, S. and Puglisi, R. (2008), 'Regulation and Consumer Satisfaction from Public Services: an Individual Fixed Effects Approach', Department of Economics, Business and Statistics Working Paper 2008–21, Available at: http://ideas.repec.org/p/mil/wpdepa/2008-21.html (accessed February 2013)

Green, R. (2007), 'Electricity and Markets', in D. Helm (ed.), *The New Energy Paradigm*, Oxford: Oxford University Press

Grigoroudis, E. and Siskos, Y. (2010), *Customer Satisfaction Evaluation: Methods for Measuring and Implementing Service Quality*. New York: Springer

Gross, R. and Heptonstall, P. (2010), 'Liberalised Energy Markets: an Obstacle to Renewables?', in I. Rutledge and P. Wright (eds), *UK Energy Policy and the End of Market Fundamentalism*. Oxford: Oxford University Press

Growitsch, C., Jamasb, T., and Pollitt, M. (2009), 'Quality of Service, Efficiency and Scale in Network Industries: an Analysis of European Electricity Distribution', *Applied Economics*, 41(20):2555–70

Grzybowski, L. (2005), 'Regulation of Mobile Telephony across the European Union: An Empirical Analysis'. *Journal of Regulatory Economics*, 28:47–67

Grzybowski, L. (2008), 'The Impact of Regulation on the Retail Price in Fixed Line Telephony Across European Union', *Telecommunication Policy*, 32:131–44

Gual, J. and Trillas, F. (2006), 'Telecommunications Policies. Measurement and Determinants', *Review of Network Economics*, 5(2):249–72

Gutierrez, L.H., (2003), 'The Effect of Endogenous Regulation on Telecommunications Expansion and Efficiency in Latin America', *Journal of Regulatory Economics*, 23(3):257–286

Hall, D. and Lobina, E. (2007), 'International Actors and Multinational Water Company Strategies in Europe, 1990–2003', *Utilities Policy*, 15(2):64–77

Hall, P. (1993), 'Policy Paradigms, Social Learning and the State: the Case of Economic Policy in Britain', *Comparative Politics*, 25(3):275–96

Haney, B. and Pollitt M. (2013), 'New Models of Public Ownership in Energy', *International Review of Applied Economics*, forthcoming

Harper, J. (1997), *Monopoly and Competition in British Telecommincations*. London: Pinter

Hattori, T. and Tsutsui, M. (2004), 'Economic Impact of Regulatory Reforms in the Electricity Supply Industry: a Panel Data Analysis for OECD Countries', *Energy Policy,* 32(6):823–32

Hausman, J.A. and Sidak, J. G. (2005), 'Did Mandatory Undunbling Achieve its Purpose? Empirical Evidence from five Countries', *Journal of Competition Law and Economics,* 1(1):173–245

Hefetz, A. and Warner, M.E., (2008), 'Managing Markets for Public Service: the Role of Mixed Public/Private Delivery of City Services', *Public Administration Review,* 68(1):150–61

Hefetz, A. and Warner M.E. (2011), "Contracting or Public Delivery? The Importance of Service, Market, and Management Characteristics" *Journal of Public Administration Research Theory,* 22(2):289–317

Helm, D. (2003), *Energy, the State, and the Market: British Energy Policy since 1979.* Oxford: Oxford University Press

Helm, D. (2007), *The New Energy Paradigm.* Oxford: Oxford University Press

Henri, C. and Matheu, M. (2001), *Regulation of Network Utilities: The European Experience.* Oxford: Oxford University Press

Hentschel, J., Lanjouw, J.O., Lanjouw, P., and Poggi, J. (2000), 'Combining Census and Survey Data to Study Spatial Dimensions of Poverty', *The World Bank Economic Review,* 14(1):147–65

Hentschel, J. and Lanjouw, P. (2000), 'Household Welfare Measurement and the Pricing of Basic Services', *Journal of International Development,* 12(1):13–27

HM Treasury (2010), Energy Market Assessment http://www.decc.gov.uk/assets/decc/1_20100324143202_e_@@_budget2010energymarket.pdf (accessed January 20, 2012)

Höffler, F. (2005), *Cost and Benefits from Infrastructure Competition: Estimating Welfare Effects from Broadband Access Competition.* Preprints of the Max Planck Institute for Research on Collective Goods, No. 2005/1. Bonn: Max Planck Institute for Research on Collective Goods

Hogan, M.J. (1987), *The Marshall Plan: America, Britain, and the Reconstruction of Western Europe, 1947–1952.* Cambridge: Cambridge University Press

Hoj, J., Kato, T., and Pilat, D. (1995), 'Deregulation and Privatization in the Service Sector', *OECD Economic Studies,* 25:37–74

Holden, L., Løland, A., Lindqvist, O. (2011), 'Valuation of Long-Term Flexible Gas Contracts', *The Journal of Derivatives,* 18(3):75–85

Holmberg, S., Rothstein, B., and Nasiritousi, N. (2009), 'Quality of Government: What you Get', *Annual Review of Political Science,* 12:135–61

Holz, F., Von Hirschhausen, C., and Kemfert, C. (2008), 'A Strategic Model of European Gas Supply (GASMOD)', *Energy Economics,* 30(3):766–88

Holznagel, B, Picot, A., Deckers, S., Grove, N., and Schramm, M. (2010), *Strategies for Rural Broadband: An Economic and Legal Feasibility Analysis.* Wiesbaden: Gabler Verlag

Honore, A. and Stern, J. (2007), *The Pricing of Internationally Traded Gas.* Oxford: The Oxford University Press

House of Commons-Energy and Climate Change Committee. (2011), 'Electricity Market Reform', Fourth Report of Session 2010–12, Volume 1

Hunt, N. and Talus, K. (2009), *EU Energy Directory*. Brussels: Euroconfidential

Iannario, M. and Piccolo, D. (2010), 'A New Statistical Model for the Analysis of Customer Satisfaction', *Quality Technology & Quantitative Management*, 7:149–68

IPA Energy. (2003), 'Can the Poor Pay for Power? The Affordability of Electricity in Southeast Europe'. Available at: http://ww.ebrd.com/country/sector/energyef/about/powersee.pdf

Ipsos-Mori. (2008), *Switching Rates for Vulnerable Customers: Summary Report*. Available at: http://www.ofgem.gov.uk/sustainability/socaction/publications/documents1/mori%20report%20switching%20rates.pdf (accessed February 2013)

ITU. (1994), Yearbook of Statistics—Telecommunication Services. Available at: http://www.itu.int/en/publications/itu-d/pages/publications.aspx?parent=d-ind-yb-2005&media=electronic (accessed February 2013)

ITU. (2008), Terms and Definitions Related to Quality of Service and Network Performance Including Dependability. ITU Recommendations. Available at: http://www.itu.int/rec/t-rec-e.800-200809-i (accessed February 2013)

Iyanaga, S. and Kawada, Y. (1977), *Encyclopedic Dictionary of Mathematics*. Cambridge, MA: The MIT Press, p.1005, 1980

Jaag, C. and Trinkner, U. (2011), 'A General Framework for Regulation and Liberalisation of Network Infrastructures', in M. Finger, and R. Künneke (eds), *International Handbook of Network Industries: The Liberalization of Infrastructure*. Cheltenham, UK and Northampton, USA: Edward Elgar Publishing Ltd

Jamasb, T. and Pollitt, M. (2005), 'Electricity Market Reform in the European Union: Review of Progress Toward Liberalization and Integration', *The Energy Journal*, 26(Special Issue):11–41

Jamasb, T. and Pollitt, M. (2011), (eds), *The Future of Electricity Demand: Customers, Citizens and Loads*. Cambridge: Cambridge University Press

Jamasb, T., Pollitt, M., and Triebs, T. (2008), 'Productivity and Efficiency of US Gas Transmission Companies: a European Regulatory Perspective', *Energy Policy*, 36(9):3398–412

Jarvis, D., Ramesh, M., Wu, X., and Aralal, E. (2011) (eds) *Infrastructure Regulation: What Works, Why and How Do We Know? Lessons from Asia And Beyond*. London: World Scientific Pub Co Inc

Johnsen, T. and Olsen, O. (2011), 'Regulated and Unregulated Nordic Retail Prices', *Energy Policy*, 39(6):3337–45

Johnson, M. D. and Gustafsson, A. (2000), *Improving Customer Satisfaction, Loyalty, and Profit: An Integrated Measurement and Management System*. San Francisco: Jossey-Bass

Jones, L.P., Tandon, P., and Vogelsang, I. (1990), *Selling Public Enterprises: A Cost-Benefit Methodology*. Cambridge: The MIT Press

Jordana, J. and Sancho Royo, D. (2004), 'Institutional Constellations and Regulatory Policy', in E. Levy-Faur and J. Jordana (eds), *The Politics of Regulation*. Glasgow: Edward Elgar Publishing Ltd, 296–321

Jorde, T. and Sidak, J.G. (2000), 'Innovation, Investment, and Unbundling', *Yale Journal on Regulation*, 17(1):1–37

Joskow, P.L. (2000), 'Deregulation and Regulatory Reform in the U.S. Electric Power Sector', in S. Peltzman and C. Winston (eds) *Deregulation of Network Industries: The Next Steps*. Washington, DC: Brookings Institution Press

Joskow, P.L. (2001), 'California's Electricity Crisis', *NBER Working Paper No. W8442*

Joskow, P.L. (2005), 'Regulation and Deregulation after 25 years: Lessons for Research', *Review of Industrial Organization*, 26:169–93

Joskow, P.L. (2006), 'Markets for Power in the United States: An Interim Assessment', *Energy Journal*, 27(1):1–36

Judt, T. (2005), *Postwar: A History of Europe Since 1945*. New York: Allen Lane, Penguin Press

Judt, T. (2010), *Ill Fares The Land*. New York: Allen Lane, Penguin Press

Julius, D. and Mashayekhi, A. (1990), *The Economics of Natural Gas: Pricing, Planning and Policy*. Oxford: The Oxford Institute for Energy

Kahneman, D., Knetsch, J.L., and Thaler, R.H. (1986), 'Fairness and the assumptions of Economics', *Journal of Business, University of Chicago Press*, 59(4):285–300

Kahneman, D. and Tversky, A. (1979), 'Prospect Theory: An Analysis of Decisions under Risk', *Econometrica*, 47(2):263–91

Kaufmann, D., Kraay, A., and Mastruzzi, M. (2005), 'Governance Matters IV: Governance Indicators for 1996–2004'. The World Bank Policy Research Working Paper Series 3630

Keay, M. (2010), 'Can the Market Delivers Security and Environmental Protection in Electricity Generation?', in I. Rutledge, and P. Wright (eds), *UK Energy Policy and the End of Market Fundamentalism*. Oxford: Oxford University Press

KEMA Consulting Gmbh. (2010), *Study on Regulation of Tariffs and Quality of the Gas Distribution Service in the Energy Community: Final Report*. Available at: http://www.energy-community.org/pls/portal/docs/736177.pdf

Kemp, A.G. and Stephen, L. (2007), 'UK Oil and Gas Depletion Policy with Growing Import Dependence', in D. Helm (ed.) *The New Energy Paradigm*. Oxford: Oxford University Press

Kennedy, D. (2005), 'Lifelines in Europe and Eastern Asia'. Washington D.C.: The World Bank, mimeo

Kennedy, D. and Besant-Jones, J. (2004), *World Bank Framework for Development of Energy Trade in South Eastern Europe*. Energy and Mining Sector Board Discussion Paper 12. Washington, DC: The World Bank

Klemperer, P. (1987), 'The Competitiveness of Markets with Switching Costs,' *RAND Journal of Economics*, 18(1):138–50

Klemperer, P. (1995), 'Competition when Consumers have Switching Costs: An Overview with Applications to Industrial Organization, Macroeconomics, and International Trade'. *Review of Economic Studies*, 62(4):515–39

Kiesling, L. (2009), *Deregulation, Innovation and Market Liberalization: Electricity Regulation in a Continually Evolving Environment*. Oxford: Routledge

Koksi, H. and Kretschmer, T. (2004), 'Entry, Standards and Competition: Firm Strategies and the Diffusion of Mobile Telephony', *Review of Industrial Organization*, 26(1):89–113

Kotakorpi, K. (2006), 'Access Price Regulation, Investment and Entry in Telecommunications', *International Journal of Industrial Organization*, 24(5):1013–20

Kuhn, T.S. (1962), *The Structure of Scientific Revolutions*. Chicago: University of Chicago Press

Künneke, R., Groenewegen, J., and Auger, JF. (2009), *The Governance of Network Industries*. Cheltenham, UK and Northampton, USA: Edward Elgar Publishing Ltd

Kwoka, J. (2002), 'Twenty-Five Years of Deregulation: Lessons for Electric Power,' Loyola University Chicago Law Journal 33

Kwoka, J. (2005), 'The Comparative Advantage of Public Ownership: Evidence from U.S. Electric Utilities,' *Canadian Journal of Economics*, 38(2):622–40

Kwoka, J. (2006), 'The Role of Competition in Natural Monopoly: Costs, Public Ownership, and Regulation,' *Review of Industrial Organization*, 29(1–2):127–47

Kwoka, J. (2008), 'Restructuring the US Electric Power Sector: A Review Of Recent Studies.' *Review Of Industrial Organization*, 32(3–4):165–96

Lafontaine, F. and Slade, M. (2007), 'Vertical Integration and Firm Boundaries', *Journal of Economic Literature*, 45(3):629–85

Laffont, J.J. (2005), *Regulation and Development*. Cambridge: Cambridge University Press

Laffont, J.J. and Martimort, D. (2002), *The Theory of Incentives: The Principal-Agent Model*. Princeton: Princeton University Press

Laffont, J.J. and Tirole, J. (1986), 'Using Cost Observation to Regulate Firms,' *Journal of Political Economy*, 94(3):614–41

Laffont, J.J. and Tirole, J. (1993), *A Theory of Incentives in Procurement and Regulation*. Cambridge: The MIT Press

Laffont, J.J. and Tirole, J. (2000), *Competition in Telecommunications*. Cambridge: The MIT Press

Lampietti, J.A., Banerjee, S.G., and Branczik, A., (2007), *People and Power: Electricity Sector Reforms and the Poor in Europe and Central Asia*. Washington: The World Bank

Lavrijssen, S. and Hancher, L. (2009), 'Networks on Track: From European Regulatory Networks to European Regulatory "Network Agencies"', *Legal Issues of Economic Integration*, 36(1):23–55

Layard, R. (2005), *Happiness: Lessons from a New Science*. New York: Penguin Press

Ledyard, J.O. (2008), 'Incentive Compatibility', in S.N. Durlauf and L. E. Blume (eds), *The New Palgrave Dictionary of Economics, 2nd Edition*. New York: Palgrave Macmillan

Li, W. and Xu, L.C. (2004), 'The Impact of Privatization and Competition in the Telecommunications Sector around the World', *Journal of Law & Economics*, 47(2):395–430

Lindskog, H. (2004), 'The Telecommunications Market in Sweden From Monopoly to Competition', Applied Business Research Conference, San Juan, Puerto Rico

Littlechild, S. (2000), 'A Review of UK Electricity Regulation'. *Cambridge Working Papers in Economics* 26

Lott, J.R. (1990), 'Predation by Public Enterprises', *Journal of Public Economics*, 43(2):237–51

Lynch, F. (1984), 'Resolving the Paradox of the Monnet Plan: National and International Planning in French Reconstruction', *The Economic History Review*, 37(2):229–43

Macavoy, P. (2007), *The Unsustainable Costs of Partial Deregulation.* New Haven: Yale University Press

Mandy (2001), Killing the Goose That May Have Laid the Golden Egg: Only the Data Know Whether Sabotage Pays, Journal of Regulatory Economics, 17(2):157–72

Martimort, D. and Straub, S. (2009), 'Privatization and Changes in Corruption Patterns: the Roots of Public Discontent', *Journal of Development Economics,* 90(1):69–84

Martin, R., Roma M., and Vansteenkiste I., (2005), 'Regulatory Reforms in Selected EU Network Industries', *ECB Occasional Paper Series* 28, available at http://www.ecb.int/pub/pdf/scpops/ecbocp28.pdf (accessed February 2013)

Martin, R. and Vansteenkiste I. (2001), 'EU Telecommunications and Electricity Markets-Heading Towards Price Convergence?' *Intereconomics: Review of European Economic Policy,* 36(3):131–40

Martin, S. (1993), 'Endogenous Firm Efficiency in a Cournot Principal Agent Model', *Journal of Economic Theory,* 59(2):445–50

Massarutto, A. (2007), 'Water Pricing and Full Cost Recovery of Water Services: Economic Incentive or Instrument of Public Finance?'*Water Policy,* 9(6):591–613

Massarutto, A., Linares, E., and Paccagnan, V. (2007), 'Liberalisation and Private Involvement in the WWS: The European Experience', in M. Finger, J. Allouche, and P. Louis-Manso (eds), *Water and Liberalisation: European Water Scenarios,* London: IWA 196–216

Maxwell, S. (2008), *The Price is Wrong: Understanding What Makes a Price Seem Fair and the True Cost of Unfair Pricing.* Hoboken, NJ: Wiley

Mazer, A. (2007), *Electric Power Planning for Regulated and Deregulated Markets.* Hoboken: John Wiley & Sons Inc

McAfee, P. (2008), 'Price Discrimination' in *Issues in Competition Law and Policy,* 465–83 Chicago: American Bar Association Section of Antitrust Law

McDonald, D.A. and Ruiters, G. (2012), *Alternatives to Privatization: Public Options for Essential Services in the Global South.* Oxford: Routledge

McKenzie, D. and Mookherjee, D. (2003), 'The Distributive Impact of Privatization in Latin America: Evidence from Four Countries', *Journal of LACEA Economia,* 3(2):161–233

Megginson, W.L. (2005), *The Financial Economics of Privatization.* New York: Oxford University Press

Megginson, W.L and Netter J.F. (2001), 'From State to Market: a Survey of Empirical Studies on Privatization', *Journal of Economic Literature,* 39(2):1–88

Melling, A. (2010), *Natural Gas Pricing and its Future: Europe as the Battleground, Carnegie Report,* available at http://carnegieendowment.org/2010/10/12/natural-gas-pricing-and-its-future-europe-as-battleground/1iat (accessed February 2013)

Melody, W.H. (2011), 'Liberalization in the Telecom Sector', in M. Finger and R. Künneke (eds), *International Handbook of Network Industries: The Liberalization of Infrastructure.* Cheltenham, UK and Northampton, USA: Edward Elgar Publishing Ltd, 103–22

Melody, W.H. and Lemstra, W. (2011), 'Liberalization in Radio Spectrum Management', in M. Finger and R. Künneke (eds), *International Handbook of Network Industries:*

The Liberalization of Infrastructure. Cheltenham, UK and Northampton, USA: Edward Elgar Publishing Ltd, 123–43

Ménard, C. and Peeroo, A. (2011), 'Liberalization in the Water Sector: Three Leading Models', in M. Finger and R. Künneke (eds), *International Handbook of Network Industries: The Liberalization of Infrastructure*. Cheltenham, UK and Northampton, USA: Edward Elgar Publishing Ltd, 310–27

Milgrom, P. and Roberts, J. (1992), *Economics, Organization and Management*. Upper Saddle River: Prentice Hall

Milne, R. (2006), 'ICT Quality of Service Regulation: Practices and Proposals'—ITU http://www.itu.int/itu-d/treg/events/seminars/2006/qos . . . /qos_bkgpaper.pdf

Millward, R. (2005), *Private and Public Enterprise in Europe: Energy, Telecommunications and Transport c.1830–1990*. Cambridge: Cambridge University Press

Millward, R. (2011), 'Public Enterprise in the Modern Western World: An Historical Analysis', *Annals of Public and Cooperative Economics*, 82(4):375–98

Miniaci, R., Scarpa, C., and Valbonesi, P. (2008), 'Distributional Effects of Price Reforms in the Italian Utility Markets', *Fiscal Studies*, 29(1):135–63

Miravete, E. (2003), 'Choosing the Wrong Calling Plan? Ignorance and Learning', *American Economic Review*, 93(1):297–310

Mulder, M. and Zwart, G. (2006), 'NATGAS: A Model of the European Natural Gas Market'. CPB Netherlands Bureau for Economic Policy Analysis Series, CPB Memorandum 144. Available at: http://www.cpb.nl/en/publication/natgas-model-european-natural-gas-market

Murphy, K. (1998), 'A Brief Introduction to Graphical Models and Bayesian Networks'. Available at: http://www.cs.ubc.ca/~murphyk/bayes/bnintro.html (accessed February 2013)

Nagayama, H. (2009), 'Electric Power Sector Reform Liberalization Models and Electric Power Prices in Developing Countries: An Empirical Analysis Using International Panel Data', *Energy Economics*, 31:463–72

Nardi, P. (2010), 'The Interplay between the European Commission, Member State and Utilities in the Debate on Transmission Network Unbundling'. Available at SSRN: http://ssrn.com/abstract=1622731 (accessed February 2013)

Natternman, P.M. and Murphy, D.D. (1998), 'The Finnish Telecommunications Market—Big Bang or Little Whimper?' *Telecommunications Policy*, 22(9):757–73

Nemoto, J. and Goto, M. (2004), 'Technological Externalities and Economies of Vertical Integration in the Electric Utility Industry', *International Journal of Industrial Organization*, 22(1):67–8

Neuhoff, K. (2007), 'Large-Scale Deployment of Renewables for Electricity Generation', in D. Helm (ed.), *The New Energy Paradigm*. Oxford: Oxford University Press

Newbery, D. (2000), *Privatization, Restructuring and Regulation of Network Utilities*. Cambridge: The MIT Press

Newbery, D. (2007), 'Electricity Liberalization in Britain: The Quest for a Satisfactory Wholesale Market Design', *The Energy Journal*, 26(Special I):43–70

Newbery, D. and Pollitt, M.G. (1997), 'Restructuring and Privatisation of the CEGB—Was it Worth it?' *Journal of Industrial Economics*, 45(3):269–304

Nicoletti, G. (2001), 'Regulation in Services: OECD Patterns and Economic Implications', *OECD Economics Department Working Paper* 287

Nickel, S.J. (1981), 'Biases in Dynamic Models with Fixed-Effects', *Econometrica,* 49:1417–26

Nicolini, G. and Salini, S. (2006),'Customer Satisfaction in the Airline Industry: The Case of British Airways'. *Quality and Reliability Engineering International,* 22(5):1–9

NTA (National IT and Telecom Agency), (2007), 'Denmark. Teleprofil 1999.' Available at: http://en.itst.dk/policy-strategy/publications/denmark-teleprofil-99 (accessed February 2013)

Obermann, G., Hall, D., and Sak, B. (2005), 'Services of General Interest in the Internal Market'. Study for the European Parliament's Committee on Internal Market and Consumer Protection, DG Internal Policies of The Union, Working Paper Series Economic and Scientific Policy (PE 358.357). Brussels

Observ'ER (2010), 'Worldwide Electricity Production from Renewable Energy Sources-Twelfth Inventory - Edition 2010', http://www.energies-renouvelables.org/observ-er/html/inventaire/eng/conclusion.asp, (accessed December 16, 2011)

ODTR (2001), The Irish Communications Market Quarterly Review: December 2001 Dublin: Office of the Director of Telecommunications Regulation

OECD. (2000), *OECD Information and Technology Outlook,* Paris: OECD

OECD. (2001), *Understanding the Digital Divide,* Paris: OECD

OECD. (2003), *Communications Outlook,* Paris: OECD, Chapter 7: Quality of Service, pp.181–206

OECD. (2003), *Social Issues in the Provision and Pricing of Water Services.* Paris: OECD

OECD (2005), *Corporate Governance of State-Owned Enterprises: A Survey of OECD Countries,* Paris: OECD

OECD. (2007), *OECD Communications Outlook 2007,* Paris: OECD

OECD. (2008), *OECD Information and Technology Outlook,* Paris: OECD

OFCOM. (2006), The Communications Market. 2006 (August)—Overview. Available at: http://stakeholders.ofcom.org.uk/market-data-research/market-data/communications-market-reports/cm06/overview06/ (accessed February 2013)

Offer, A. (2003), *Why has the Public Sector Grown so Large in Market Societies? The Political Economy of Prudence in the UK, C.1870–2000.* Oxford: Oxford University Press

Offer, A. (2006), *The Challenge of Affluence: Self-Control and Well-Being in the United States and Britain Since 1950.* New York: Oxford University Press

OFGEM. (2008), Gas Distribution Quality of Service Report 2007–2008. Available at: http://www.ofgem.gov.uk/networks/gasdistr/qos/documents1/2007%20to%2008%20qos%20report.pdf (accessed February 2013)

OFGEM. (2009), Review of Protection for Vulnerable Customers from Disconnection. Available at: http://www.ofgem.gov.uk/pages/moreinformation.aspx?docid=139&refer=sustainability/socaction/publications (accessed February 2013)

Palcic, D. and Reeves, E. (2010), 'Organisational Status Change and Performance: The Case of Ireland's National Telecommunications Operator', *Telecommunications Policy,* 34(5–6):299–308

Palcic, D. and Reeves, E. (2011), 'Privatization, Employee Share Ownership and Governance: The Case of Eircom', *Annals of Public and Cooperative Economics*, 82(4):437–54

Panizza, U. and Yanez, M. (2005), 'Why are Latin Americans so Unhappy about Reforms?' *Journal of Applied Economics*, 8(1):1–29

Parasuraman, A., Zeithaml, V., and Berry, L. (1985), 'A Conceptual Model of Service. Quality and its Implications for Future Research', *Journal of Marketing*, 49:41–50

Pareto, W. [1892](2008), 'Dell'utile che procurano al paese le ferrovie'. *Giornale degli Economisti*, February 1892: 154–162. Republished as Pareto (2008) 'On the Benefit which the Railways Provide for the Country', *Giornale degli Economisti*, 67(3):311–19

Parker, D. (1999), 'Regulation of Privatized Public Utilities in the UK: Performance and Governance', *International Journal of Public Sector Management*, 12(3):213–36

Parker, D. (2001), *Privatization and Corporate Performance*. Cheltenham, UK and Northampton, USA: Edward Elgar Publishing Ltd

Parker, D. (2009), 'Editorial: Ppp/Pfi - Solution or Problem?' *Economic Affairs*. 29(1):2–6

Parsons Brinckerhoff. (2010), ' Powering the Nation: a Review of the Costs of Generating Electricity', http://www.pbworld.com/regional/uk_europe_specialty/ (accessed February 2013)

Parker, D. and Willner, J. (2007), 'The Performance of Public and Private Enterprise Under Conditions of Active and Passive Ownership and Competition and Monopoly, *Journal of Economics*, 90(3):221–53

Pearce, D., Atkinson, G., and Mourato, S. (2006), *Cost-Benefit Analysis and the Environment: Recent Developments*. Paris: OECD

Peltzman, S., Levine, M.E., and Noll, R.G. (1989), 'The Economic Theory of Regulation after a Decade of Deregulation', *Brookings Papers on Economic Activity. Microeconomics*, 1989:1–59

Percebois, J. (2008), 'Electricity Liberalization in the European Union: Balancing Benefits and Risks', *The Energy Journal*, 29(1):1–19

Petrecolla, D. (2003), 'Elementos a tener en cuenta en el diseno de un tarifa social', Conferiencias FIEL

Phlips, L. (1983), *The Economics of Price Discrimination*. Cambridge: Cambridge University Press

Picot, A. and Wernick, C. (2007), 'The Role of Government in Broadband Access', *Telecommunications Policy*, 31:660–74

Pigou, A.C. (1932), *The Economics Of Welfare, 4th Edition*, London: Macmillan and Co

Pindyck, R. (1999), 'The Long-Run Evolution of Energy Prices', *The Energy Journal*, 20(2):1–27

Platchkov, L. and Pollitt, M. (2011), 'The Economics of Energy (and Electricity) Demand', in T. Jamasb, and M. Pollitt (eds) *The Future of Electricity Demand: Customers, Citizens and Loads*. Cambridge: Cambridge University Press

Poggi, A. (2007), 'Does Persistence of Social Exclusion Exist in Spain?' *Journal of Economic Inequality*, 5(1):53–72

Poggi, A. and Florio, M. (2010), 'Energy Deprivation Dynamics and Regulatory Reforms in Europe: Evidence from Household Panel Data', *Energy Policy*, 38(1):253–64

Politecnico di Milano—Dipartimento di Energia (2010), 'Costi di produzione di energia elettrica da fonti rinnovabili', Rapporto Commissionato da AEEG, www.autorita.energia.it/allegati/docs/11/103-11arg_rtalla.pdf (accessed February 14, 2012)

Pollitt, M.G. (2002), 'The Declining Role of the State in Infrastructure Investments in the UK', in Berg, S.V, Pollitt, M.G, and Tsuji, M. (eds) *Private Initiatives In Infrastructure: Priorities, Incentives And Performance*. Cheltenham, UK and Northampton, USA: Edward Elgar Publishing Ltd, 67–100

Pollitt, M.G. (2008), 'The Arguments for and Against Ownership Unbundling of Energy Transmission Networks', *Energy Policy*, 36(2):704–13

Pollitt, M.G. (2009), 'Electricity Liberalization in the European Union: a Progress Report' *EPRF Working Paper 0929*, Cambridge Working Paper in Economics 0953, University of Cambridge

Pollock, A. (2005), *NHS Plc: The Privatisation of Our Health Care*. London: Verso Books

Preston, P., Cawley, A., and Metykova, M. (2007), 'Broadband and Rural Areas in the EU: From Technology to Applications and Use', *Telecommunications Policy*, 31(6):389–400

Ravallion, M. (1998), 'Does Aggregation Hide the Harmful Effects of Inequality on Growth?' *Economics Letters*, 61(1):73–7

Reiffen, D. and Ward, M.R. (2002), 'Recent Empirical Evidence on Discrimination by Regulated Firms.' *Review of Network Economics*, 1(1): 2

Reisinger, M. and Schnitzer, M. (2008), 'A Model of Vertical Oligopolistic Competition'. Munich Discussion Paper 2008–08, University of Munich, Department of Economics. Available at: http://ideas.repec.org/p/trf/wpaper/228.html (accessed February 2013)

Riaz, A. (1997), 'The Role of Telecommunications in Economic Growth: Proposal for an Alternative Framework of Analysis'. Media Culture Society, October 1997, 19(4):557–583

Rey, P. and Tirole, J. (2007), 'A Primer on Foreclosure', in M. Armstrong and R. Porter (eds) *Handbook of Industrial Organization Vol. 3*. Amsterdam: North-Holland, 2145–220

Rey, P. and Vergé, T. (2008), 'Economics of Vertical Restraints', in P. Buccirossi (ed.) *Handbook of Anti-Trust Economics*. Cambridge: The MIT Press, 353–90

Richardson, T., Stafford-Fraser, Q., Wood, K.R., and Hopper, A. (1998), 'Virtual Network Computing', *IEEE Internet Computing* 2(1):33–8

Robinson, C. (2007), *Utility Regulation in Competitive Markets: Problems and Progress*. Cheltenham, UK and Northampton, USA: Edward Elgar Publishing Ltd

Roland, G. (2008), *Privatization: Success and Failures*. New York: Columbia University Press

Röller, L.H. and Waverman, L. (2001), 'Telecommunications Infrastructure and Economic Development: a Simultaneous Approach', *American Economic Review*, 91(4):909–23

Rosanvallon, P. (1981), *La crise de l'état-providence*. Paris: Le Seuil

Rosanvallon, P. (2011), *Democratic Legitimacy: Impartiality, Reflexivity, Proximity*. Princeton: Princeton University Press

Ross, A. (1999), 'Does Ownership or Competition Matter? The Effects of Telecommunications Reform on Network Expansion and Efficiency.' *Journal of Regulatory Economics* 15:65–92

Ross, A. (2003), 'The Impact of the Regulatory Process and Price Cap Regulation in Latin American Telecommunications Markets.' *Review of Network Economics* 2(3):270–86

Rothstein, B. (1998), *The Social Democratic State: The Swedish Model and the Bureaucratic Problem of Social Reforms*. Pittsburgh: The University of Pittsburgh Press

Rothstein, B. and Teorell, J. (2008), 'What is Quality of Government? A Theory of Impartial Government Institutions'. *Governance: An International Journal of Policy, Administration and Institutions*, 21(2):165–90

Rothstein, B. (2011), *The Quality of Government Corruption, Social Trust, and Inequality in International Perspective*. Chicago: University Of Chicago Press

RTE Electrical Energy Statistics for France (2010), http://www.rte-france.com/uploads/Mediatheque_docs/vie_systeme/annuelles/Statistiques_energie_electrique/an/statistiques_annuelles_2010_an.pdf (accessed February 2013)

Rott, P. (2007), 'Consumers and Services of General Interest: is EC Consumer Law the Future?' *Journal of Consumer Policy*, 30(1):49–60

Rovizzi, L. and Thompson, D. (1992), 'The Regulation of Product Quality in the Public Utilities and the Citizen's Charter', *Fiscal Studies*, 13(3):74–95

Rutledge, I. and Wright, P. (2010), (eds) *UK Energy Policy and the End of Market Fundamentalism*. Oxford: Oxford University Press

Salini, S. and Kenett, R.S. (2009), 'Bayesian Networks of Customer Satisfaction Survey Data', *Journal of Applied Statistics*, 36(11):1177–89

Salini, S. and Kenett, R. S. (2011), 'Modern Analysis of Customer Satisfaction Surveys: Comparison of Models and Integrated Analysis', *Applied Stochastic Models in Business and Industry*, 27:465–75

Salini, S. and Kenett, R.S. (2011), *Modern Analysis of Customer Satisfaction Surveys: With Applications Using R*. Chichester: John Wiley and Sons

Sappington, D.E.M. and Sidak, J. G. (2003), 'Incentives for Anticompetitive Behavior by Public Enterprises', *Review of Industrial Organization*, 22(3):183–206

Sappington, D.E.M. and Stiglitz, J. (1987), 'Privatization, Information, and Incentives'. *Journal of Policy Analysis and Management*, 6(4):567–82

Schmidt, K. (1996), 'The Costs and Benefits of Privatization. An Incomplete Contracts Approach'. *Journal of Law, Economics, and Organization*, 12(1):1–24

Schmitt, S. and Rammerstorfer, M. (2010), 'Ownership Unbundling and Consumer Prices–A Panel Data Analysis', Paper for the Third Annual Conference on Competition and Regulation in Network Industries, Brussels, 19 November 2010

Schmitz, P.W. (2000), 'Partial Privatization and Incomplete Contracts: The Proper Scope of Government Reconsidered', *Finanzarchiv*, 57(4):394–411

Servaes, J. (1995), 'Privatisation and Regulatory Reform of Telecommunications in the Netherlands: The Case of the Royal PTT Netherlands (KPN)', *Jurnal Komunikasi, Malaysian Journal of Communication*, 11:85–95

Sharkey, W.W. (1982), *The Theory of Natural Monopoly*. Cambridge: Cambridge University Press

Shy, O. (2001), *The Economics of Network Industries*. Cambridge: Cambridge University Press

Sidak, G. and Spulber D. (1998), 'Deregulation and Management Competition in Network Industries'. *Yale Journal on Regulation*, 15(117):117–47

Sidak, J.G. (2006), 'Consumer Welfare Approach to Network Neutrality Regulation of the Internet', *Journal of Competition Law and Economics*, 2(3):349–474

Simpson, J. (2009), 'Spot Natural Gas Prices: A Theoretical Party to Party Bargaining Framework', *Asian Social Science*, 5:25–43

Simpson, J. (2011), 'Future Spot Gas Prices in the US and the UK: Are Movements more Influenced by Country Factors or by Global Factors?' *Journal of Energy Markets*, 4:3–16

Sioshansi, F.P. and Pfaffenberger, W. (2006), *Electricity Market Reform: An International Perspective*. Oxford: Elsevier Science

Sirtori, E. and Vignetti, S. (2011), 'Infrastructure Investment Opportunities in the EU New Member States: The Role of Regional Policies', in M. Florio (ed.), *Public Investment, Growth and Fiscal Constraints: Challenges for the EU New Member States*. Cheltenham, UK and Northampton, USA: Edward Elgar Publishing Ltd, 265–87

Spiller, P. (2011), 'Basic Economic Principles of Infrastructure Liberalization: a Transaction Cost Perspective', in M. and R. Künneke (eds), *International Handbook of Network Industries. The Liberalization of Infrastructure,* Cheltenham, UK and Northampton, USA: Edward Elgar Publishing Ltd

Spulber, D. and Yoo, C. (2009), *Networks in Telecommunications*. Cambridge: Cambridge University Press

Steiner, F. (2000), 'Regulation, Industry Structure and Performance in the Electricity Supply Industry'. *OECD, Economics Department Working Papers* 238

Sterlacchini, A. (2012), 'Energy R&D in Private and State-Owned Utilities: An Analysis of the Major World Electric Companies', *Energy Policy*, 41:494–06

Stern, N. (2006), 'Stern Review on the Economics of Climate Change'. London: HM Treasury

Stern, J. (2009), 'Continental Europe Long-Term Gas Contracts: is a Transition away from Oil-Product Linked Contracts Inevitable and Imminent?' *Oxford Institute for Energy Studies*, NG 34

Stigler, G. (1971), 'The Theory of Economic Regulation', *Bell Journal of Economics and Management Science*, 3:3–18

Sutherland, E. (2007), 'The Regulation of the Quality of Service in Mobile Networks', *Info*, 9(6):17–34

Sweeting, A. (2007), 'Market Power in the England and Wales Wholesale Electricity Market 1995–2000', *The Economic Journal*, 117:654–85

Sylos Labini, P. (1962), *Oligopoly and Technical Progress*. Cambridge: Harvard University Press

Talus, K. (2010), 'Term Upstream Natural Gas Contracts and EC Competition Law—Efficiencies Under Article 81(3) and Objective Justifications Under Article 82', in B. Delvaux, M. Hunt, and K. Talus (eds), *EU Energy Law and Policy Issue, 2nd Edition*, Brussels: Euroconfidential

Teichman, J. (1995), *Privatization and Political Change in Mexico.* Pittsburgh: University of Pittsburgh Press

Tejeiro, J.B., and Cañete, O.R. (1999), 'The Privatization and Regulation of Telecommunications in Spain', *Annals of Public and Cooperative Economics*, 70(2):331–50

Teligen, Strategy Analytics Ltd. (2008), 'Report on Telecoms Price Developments from 1998 to 2008'. Produced for The European Commission Directorate General for Information Society. Available at: http://ec.europa.eu/information_society/policy/ecomm/doc/library/ext_studies/price_developments_1998_2008/tariff_trends_report_1998_2008.pdf (accessed February 2013)

ten Heuvelhof, E., Stout, H., De Jong, M., and Kars, M. (2009), *Strategic Behaviour In Network Industries*. Cheltenham, UK and Northampton, USA: Edward Elgar Publishing Ltd

Thaler, R.H. (1985), 'Mental Accounting and Consumer Choice', *Marketing Science* 4: 1999–214

Thaler, R.H. (1994), *Quasi Rational Economics*. New York: Russell Sage Foundation

The Swedish Energy Markets Inspectorate (2011), 'The Swedish Electricity and Natural Gas Markets 2010', http://www.ei.se/energy-markets-inspectorate/library/, (accessed January 20, 2012)

Thomas, S. (2003), 'The Seven Brothers', *Energy Policy,* 31(2003):393–403

Thomas, S. (2004), 'The Rise, Fall and Rise Again of British Gas', mimeo, Greenwich: University of Greenwich, Public Service International Research Unit-PSIRU

Thomas, S. (2006), 'The British Model in Britain: Failing Slowly', *Energy Policy,* 34(5):583–600

Thomas, S. (2010), The Seven Brothers: Then There Were Three. Unpublished paper

Thorngren, B. (2006), Net Neutrality: Not as Neutral as it Sounds. *Ericsson Business Review* 1

Tobin, J. and Houthakker, H.S. (1950), 'The Effects of Rationing on Demand Elasticities'. *The Review of Economic Studies,* 18(3):140–53

Torriti, J., Leach, M., and Devine-Wright, P. (2011), 'Demand-Side Participation: Price Constraints, Technical Limits and Behavioural Risks', in T. Jamasb, and M. Pollitt, (eds), *The Future of Electricity Demand: Customers, Citizens and Loads*. Cambridge: Cambridge University Press

Ugaz, C. and Waddams Price, C. (2003), *Utility Privatisation and Regulation: A Fair Deal for Consumers?* Cheltenham, UK and Northampton, USA: Edward Elgar Publishing Ltd

United Nations (2008*), Public Enterprises: Unresolved Challenges and New Opportunities* Available at: http://www.unpan.org/Portals/0/60yrhistory/documents/Publications/Public%20Enterprises%20-%20Unresolved%20Challenges.2005.pdf (accessed February 2013)

Urakami, H. (2007), 'Economies of Vertical Integration for the Japanese Water Industry', *Jahrbuch Für Regionale Wissenschaft,* 27:129–41

Van Damme, E. (2004), 'Pragmatic Privatisation: The Netherlands 1982–2002', *TILEC Discussion Paper No. 2004–007,* http://arno.uvt.nl/show.cgi?fid=12111 (accessed February 2013)

Van Dijk Consulting. (2010), 'Broadband Internet Access Cost (BIAC): Final Report'. European Commission Information Society and Media Directorate-General, available at http://ec.europa.eu/information_society/eeurope/i2010/docs/eda/biac_2009.pdf (accessed February 2013)

Van Witteloostuijna, A., Brakmanb, S., and Van Marrewijk, C. (2007), 'Welfare Distribution Effect of a Price Reduction in the Dutch Gas Transport Market: A Scenario Analysis of Regulatory Policy, Market Form and Rent Allocation', *Energy Policy*, 35:6299–308

Varian, Hal R. (1987), *Intermediate Microeconomics: A Modern Approach*. New York: W.W. Norton & Company, Inc.

Varian, H.A. (1985), 'Price Discrimination and Social Welfare', *The American Economic Review*, 75(4):870–5

Varian, H.A. (1989), 'Price Discrimination', in R. Schmalensee and R. Willig (eds) *Handbook of Industrial Organization*, Edition 1, Volume 1(1), Amsterdam: Elsevier

Vavra, T.G. (1997), *Improving Your Measurement of Customer Satisfaction: A Guide to Creating, Conducting, Analyzing, and Reporting Customer Satisfaction Measurement Programs*. Milwaukee: American Society for Quality

Verhelpen, A. (2010), 'The Assessment of Vertical Long-Term Gas Supply Contracts and the Tension Between the Upstream and Downstream Markets', in B. Delvaux, M. Hunt, and K. Talus (eds), *EU Energy Law and Policy Issue, 2nd Edition*, Brussels: Euroconfidential

Viani, B.E. (2007), 'Consequences of Vertical Separation and Monopoly: Evidence from the Telecom Privatizations'. *AEI-Brookings Joint Center for Regulatory Studies Working Papers* 06–20

Vickers, J. (1995), 'Competition and Regulation in Vertically Related Markets', *Review of Economic Studies*, 62:1–17

Vickers, J. and Yarrow, G. (1988), 'Regulation of Privatized Firms in Britain', *European Economic Review*, 32(2/3):465–72

Vickers, J. and Yarrow, G. (1993), *Privatization: An Economic Analysis*. Cambridge: The MIT Press

Viscusi, W.K., Harrington, J.E., and Vernon, J.M. (2005), *Economics of Regulation and Antitrust*, 4th Edition. Cambridge: The MIT Press

Vogelsang, I. (1990), 'Optional Two-Part Tariffs Constrained by Price Caps', *Economics Letters*, 33(3):287–92

Waddams, C. (2009), 'Competition Remedies in Consumer Markets', *Loyola Consumer Law Review*, 21(4):439–95

Waddam Price, C. (2005), 'The Effect of Liberalizing UK Retail Energy Markets on Consumers', *Oxford Review of Economic Policy*, 21(1):128:144

Waddams Price, C. (2011), 'Energy, Fuel Poverty and Demand (Maintaining Affordability with Sustainability and Security of Supply)' in T. Jamasb, and M. Pollitt, M. (eds) *The Future of Electricity Demand: Customers, Citizens and Loads*. Cambridge: Cambridge University Press

Waddams Price, C. and Hancock, R. (1998), 'Distributional Effects of Liberalising UK Residential Utility Markets', *Fiscal Studies*, 19(3)295–319

Waddams Price, C. and Pham, K. (2009), 'The Impact of Electricity Market Reform on Consumers', *Utilities Policy*, 17(1):43–8

Wallsten, S. (2001), 'An Empirical Analysis of Competition, Privatization, and Regulation in Africa and Latin America', *Journal of Industrial Economics*, 49(1):1–19

Wallsten, S. (2003), 'Of Carts and Horses: Regulation and Privatization in Telecommunications Reforms', *Journal of Policy Reform*, 6(4):217–31

Wallsten, S. (2004), 'Privatizing Monopolies in Developing Countries: the Real Effects of Exclusivity Periods in Telecommunications', *Journal of Regulatory Economics* 26(3):303–20

Warner, M.E. (2011), 'Competition or Cooperation in Urban Service Delivery?' *Annals of Public and Cooperative Economics*, 82(4):421–35

Whitehead, J.C. et al. (2008), 'Combining Revealed and Stated Preference Data to Estimate the Nonmarket Value of Ecological Services: An Assessment of the State of the Service', *Journal of Economic Surveys*, 22(5):872–908

WHO (2004), 'Energy Sustainable Development and Health', in Chapter 3 of *Access to Electricity and Heating*, Geneva: WHO Background Papers

Williamson, J. (1989), 'What Washington Means by Policy Reform', in J. Williamson (ed.) *Latin American Readjustment: How Much Has Happened*. Washington: Peterson Institute for International Economics

Williamson, J. (2000), 'What Should the Bank Think about the Washington Consensus?' *The World Bank Research Observer*, 15(2):251–64

Willig, R. (1976), 'Consumer's Surplus without Apology', *The American Economic Review*, 66(4):589–97

Willner, J. (2008), 'Liberalisation, Competition and Ownership in the Presence of Vertical Relations', *Empirica*, 35:449–64

Willner, J. [1996](2000), 'Privatisation of Natural Monopolies: a Comment', *Review Of Industrial Organization*, 11(6): 869–82; Reprinted in D. Parker (ed.) (2000), *Privatisation and Corporate Performance: The International Library of Critical Writings in Economics*. Cheltenham, UK and Northampton, USA: Edward Elgar Publishing Ltd, 14–27

Willner, J. (2003), 'Privatization: a Sceptical Analysis', in D. Parker and D. Saal (eds), *International Handbook on Privatization*. Cheltenham, UK and Northampton, USA: Edward Elgar Publishing Ltd, 60–86

Willner, J. (2011), 'Reforming a Network Industry: Consequences for Cost Efficiency and Welfare'. Department of Economics Business and Statistics Working Paper, 2011–33. Available at: http://ideas.repec.org/p/mil/wpdepa/2011-33.html (accessed February 2013)

Wilson, C.M. and Waddams Price, C. (2010), 'Do Consumers Switch to the Best Supplier?' *Oxford Economic Papers* 62(4):647–68

Wodon, Q. (2000), 'Low Income Energy Assistance and Disconnection in France', *Applied Economics Letters*, 7:775–79

Wolak, F. (1996), 'The Welfare Impacts of Competitive Telecommunications Supply: A Household-Level Analysis'. *Brookings Papers on Economic Activity: Microeconomics*, 1996: 269–350

Woo, C. (2001), 'What Went Wrong in California's Electricity Market?' *Energy Policy*, 26(8):747–58

World Bank, The. (2000), *Maintaining Utility Service for the Poor*. Available at: http://siteresources.worldbank.org/ecaext/resources/publications/maintaining-utility-services/14804.pdf

World Bank, The. (2002), *Sourcebook for Poverty Reduction Strategies: Core Techniques and Crosscutting Issues*, Washington DC: The World Bank

World Nuclear Association. (2011), 'Comparison of Life Cycle Greenhouse Emissions of Various Electricity Generation Sources'. Available at: http://world-nuclear.org/uploadedfiles/org/reference/pdf/comparison_of_lifecycle.pdf (accessed February 2013)

Woroch, G.A. (2002), 'Local Network Competition'. *Handbook of Telecommunications Economics,* 1:641–716

Yi-Chong, X. (2005), 'Models, Templates and Currents: The World Bank and Electricity Reform', *Review of International Political Economy*, 12(4):647–73

Zafirova, Z. (2010), 'Natural Gas Short-Term Trading Markets: Legal Ins and Policy Outs', in B. Dekvaux, M. Hunt, K. Talus (eds), *EU Energy Law And Policy Issue, 2nd Edition,* Brussels: Euroconfidential

Zanella, A. (2001), 'Valutazione e modelli interpretativi di customer satisfaction: una presentazione d'insieme' Istituto Di Statistica, Università Cattolica Del Sacro Cuore, Milano, Serie E.P. N. 105

Zhang, Y., Kirkpatrick, C., and Parker, D. (2002), 'Electricity Sector Reform in Developing Countries: An Econometric Assessment of the Effects of Privatization, Competition and Regulation'. University of Manchester, Institute for Development Policy and Management (IDPM), Working Papers 30593. Available at: http://Ideas.Repec.Org/A/Kap/Regeco/V33y2008i2p159-178.Html

Zwart, G.T.J. (2009), 'European Natural Gas Markets: Resource Constraints and Market Power', *The Energy Journal,* 30:151–66

Index

Bold entries refer to figures and tables.

Index